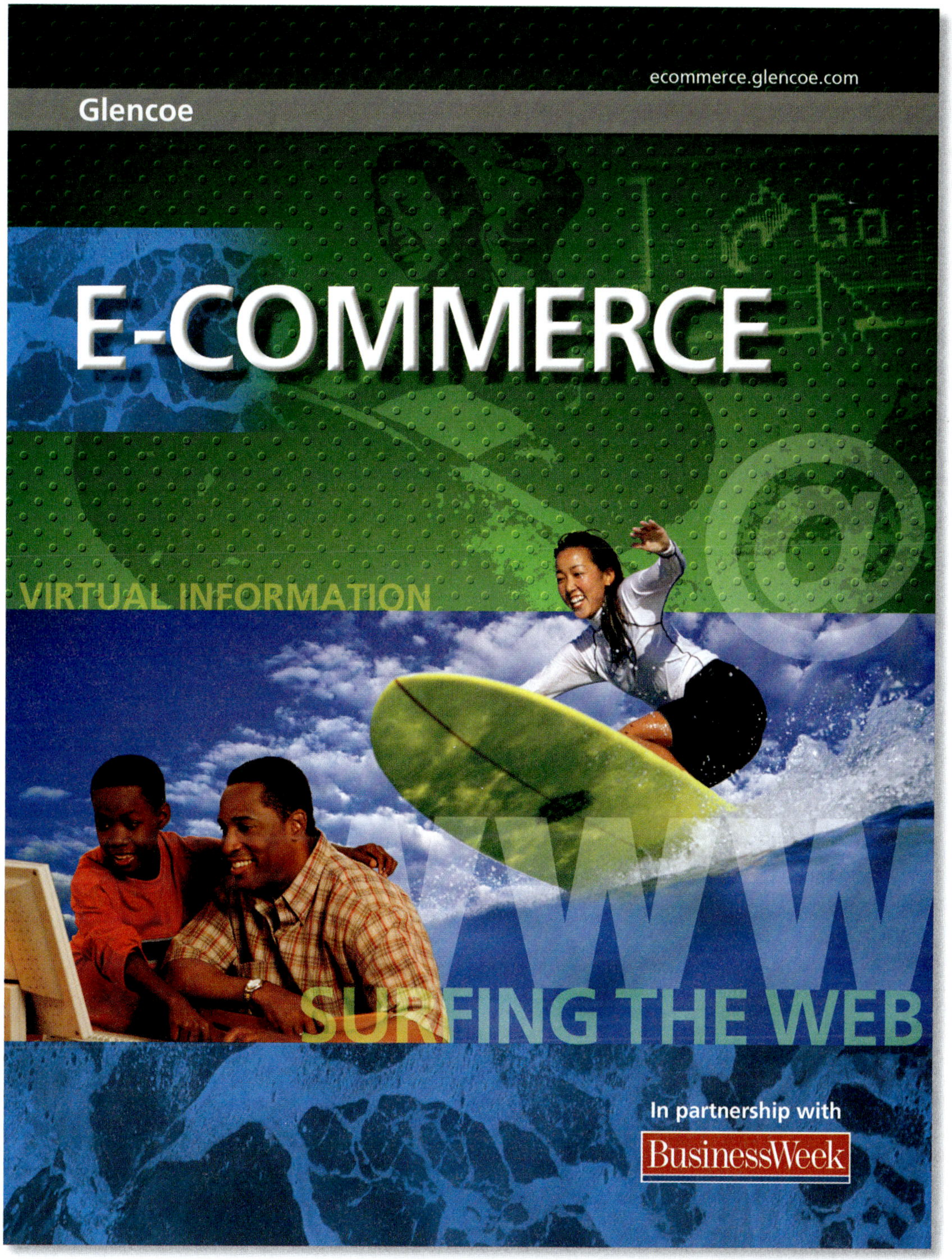
ecommerce.glencoe.com
Glencoe
E-COMMERCE
VIRTUAL INFORMATION
WWW
SURFING THE WEB
In partnership with
BusinessWeek

James E. Miles
Chip Dolcé

D1762770

McGraw Hill Glencoe

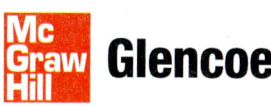

New York, New York   Columbus, Ohio   Chicago, Illinois   Peoria, Illinois   Woodland Hills, California

Copyright © 2006 by The McGraw-Hill Companies, Inc. All rights reserved. Except as permitted under the United States Copyright Act, no part of this publication may be reproduced or distributed in any form or by any means, or stored in a database or retrieval system, without prior written permission of the publisher.

Printed in the United States of America.

Send all inquiries to:
Glencoe/McGraw-Hill
21600 Oxnard Street, Suite 500
Woodland Hills, CA 91367

ISBN 0-07-861333-7 (Student Edition)
ISBN 0-07-866542-6 (Teacher Annotated Edition)

1 2 3 4 5 6 7 8 9   079   09 08 07 06 05 04

About the Authors

JAMES E. MILES, a nationally recognized secondary marketing educator, has been teaching marketing and business education for 37 years at Pittsford Sutherland High School, Nazareth College of Rochester, and Monroe Community College, both in New York. He currently teaches e-commerce at both the secondary and post-secondary levels. James Miles was honored in 2000 with the National Business Education Secondary Teacher of the Year award. He has been a DECA and FBLA advisor, served as president of several professional organizations, and is coach for the championship Pittsford LifeSmarts Team. As a Master Teacher with the International Center for Leadership in Education, James Miles has facilitated workshops for business and marketing educators in 14 states, published journal articles, presented at local, state, and national conferences, and authored several textbooks. He has a bachelor's degree in business and distributive education, a master's degree from the University of Rochester, and numerous additional credits in administration and supervision.

CHIP DOLCÉ is a 7th grade principal with the Gates Chili Central School District in Rochester, New York. His work has focused on instructional technology and literacy. He has served as both a District Instructional Technologist as well as a Teacher Trainer. He is a national presenter for instructional technology, conducting workshops and seminars across the country on topics such as e-commerce, handheld technology, and using technology to support a balanced literacy program. He has also worked as an Internet Applications Manager for an Inc. 500 Web-development firm. Chip Dolcé holds a bachelor's degree in Secondary Social Studies education from St. John Fisher College and master's degrees from the State University of New York at Brockport and St. John Fisher College.

E-Commerce Reviewers

ROLAND CARPENTER
Cedar Crest High School
Lebanon, PA

ROGER CRIDER
Franklin Regional Senior High School
Murrysville, PA

GREG D'AMBROSIO
Monroe-Woodbury High School
Central Valley, NY

GORDIE GREGG
Butler Technology and Career Development Schools
Fairfield Township, OH

CELIA HOBBINS
Mifflinburg Area High School
Mifflinburg, PA

DONALD HURWITZ
Deep Creek High School
Chesapeake, VA

ROBERT HUTCHISON
Middleton High School
Middleton, WI

BONNIE MALCOLM
Lincoln North Star High School
Lincoln, NE

CATHY MANNING
Westhill High School
Stamford, CT

CANDACE MAXIAN
Middletown High School
Middletown, DE

GINNY MERCHANT
Weston High School
Weston, CT

LARRY MIDDLETON
Montgomery County High School
Christiansburg, VA

DEB MOORE
Sunrise Mountain High School
Peoria, AZ

DEBORAH NUDELMAN
H. Frank Carey High School
Franklin Square, NY

SANDY OLSON
Central Community College
Grand Island, NE

MARC PAYEUR
Hillsboro-Deering High School
Hillsboro, NH

KAREN PHIPPS
Ansonia High School
Ansonia, CT

KIM RADFORD
Blacksburg High School
Blacksburg, VA

HELEN STAHL
Lafayette High School
Williamsburg, VA

GARY TENHULZEN
Arvada High School
Arvada, CO

Brief Table of Contents

Unit 1 THE HISTORY, NATURE, AND IMPACT OF E-COMMERCE
- Chapter 1   The Internet and E-Commerce
- Chapter 2   The Nature of E-Commerce
- Chapter 3   Retailing on the Internet
- Chapter 4   Global E-Commerce

Unit 2 YOU AND E-COMMERCE
- Chapter 5   Building a Career in E-Commerce
- Chapter 6   Ethical, Legal, and Social Responsibilities in E-Commerce

Unit 3 BUSINESS STRUCTURES AND THE BUSINESS PLAN IN E-COMMERCE
- Chapter 7   Business Structures and Economics in E-Commerce
- Chapter 8   Revenue Models and the Business Plan in E-Commerce

Unit 4 WEB SITE DEVELOPMENT
- Chapter 9   Creating a Web Site
- Chapter 10   Building a Web Site
- Chapter 11   Web Site Management

Unit 5 MARKETING IN THE DIGITAL WORLD
- Chapter 12   Fundamentals of Internet Marketing
- Chapter 13   Distribution in E-Commerce
- Chapter 14   Customer Service and Web Site Personalization
- Chapter 15   Advertising for E-Commerce

Table of Contents

To the Student xii

UNIT 1   THE HISTORY, NATURE, AND IMPACT OF E-COMMERCE .. 2

BusinessWeek Byte WebSmart 3

Chapter 1   The Internet and E-Commerce 4

SECTION 1-1 Internet Basics 6
　　　　　　 Section 1-1 Review 11
SECTION 1-2 Connecting to, Searching, and Using the Internet 12
　　　　　　 Section 1-2 Review 16
SECTION 1-3 E-Commerce in Action 17
　　　　　　 Section 1-3 Review 20

Assessment 21
　　Worksheet 1-1: Using Search Engines 21
　　Worksheet 1-2: Learning About E-Commerce 22

Chapter 1 Review and Activities 23

Chapter 2   The Nature of E-Commerce 26

SECTION 2-1 Characterizing E-Commerce in Business 28
　　　　　　 Section 2-1 Review 33
SECTION 2-2 Conducting Business on the Web 34
　　　　　　 Section 2-2 Review 38

Assessment 39
　　Worksheet 2-1: Which Brand Is for You? 39
　　Worksheet 2-2: Doing Business on the Internet 40

Chapter 2 Review and Activities 41

Chapter 3   Retailing on the Internet 44

SECTION 3-1 Retailing—Then and Now 46
　　　　　　 Section 3-1 Review 51
SECTION 3-2 The E-Tail Experience 52
　　　　　　 Section 3-2 Review 58

Table of Contents

Assessment 59
    Worksheet 3-1: Internet Influence 59
    Worksheet 3-2: Merchandise Motivation 60

Chapter 3 Review and Activities 61

Chapter 4 — Global E-Commerce 64

SECTION 4-1 Going Global 66
    Section 4-1 Review 72

SECTION 4-2 The Impact of E-Commerce on International Trade 73
    Section 4-2 Review 78

Assessment 79
    Worksheet 4-1: Cultural Diversity 79
    Worksheet 4-2: Can the Internet Improve World Trade? ... 80

Chapter 4 Review and Activities 81

SELF-ASSESSMENT: What Is Your Potential? 84

BusinessWeek It's Your Project What Is True? 88

UNIT 2 — YOU AND E-COMMERCE 90

BusinessWeek Byte Privacy Matters: Perils and Promise of Online Schmoozing 91

Chapter 5 — Building a Career in E-Commerce 92

SECTION 5-1 Basics of the E-Commerce Workplace 94
    Section 5-1 Review 99

SECTION 5-2 Searching for an E-Commerce Job 100
    Section 5-2 Review 106

Assessment 107
    Worksheet 5-1: Describing Your Future Lifestyle 107
    Worksheet 5-2: Learning About an E-Commerce Job 108

Chapter 5 Review and Activities 109

Table of Contents

Chapter 6   Ethical, Legal, and Social Responsibilities in E-Commerce 112

SECTION 6-1   Internet Law and Ethics 114
     Section 6-1 Review 118

SECTION 6-2   Privacy Issues and the Internet 119
     Section 6-2 Review 121

SECTION 6-3   Internet Security 122
     Section 6-3 Review 128

Assessment 129
     Worksheet 6-1: Cybercrime 129
     Worksheet 6-2: Fact or Fiction? 130

Chapter 6 Review and Activities 131

BusinessWeek It's Your Project   Who Has the Right to Know? 134

UNIT 3   BUSINESS STRUCTURES AND THE BUSINESS PLAN IN E-COMMERCE 136

BusinessWeek Byte   Architects: Ivan Seidenberg, Verizon 137

Chapter 7   Business Structures and Economics in E-Commerce 138

SECTION 7-1   Business Structures 140
     Section 7-1 Review 145

SECTION 7-2   Components of the New Economy 146
     Section 7-2 Review 150

Assessment 151
     Worksheet 7-1: Do You Have What It Takes? 151
     Worksheet 7-2: What Do Statistics Really Reveal? 152

Chapter 7 Review and Activities 153

Chapter 8   Revenue Models and the Business Plan in E-Commerce. . 156

SECTION 8-1   E-Commerce Revenue Models 158
     Section 8-1 Review 163

Table of Contents

| SECTION 8-2 | E-Commerce Business Plan | 164 |

Section 8-3 Review 172

Assessment 173
    Worksheet 8-1: Creating a Road Map for Your Business 173
    Worksheet 8-2: Your E-Business Plan 174

Chapter 8 Review and Activities 175

SELF-ASSESSMENT: Decide Your Future 178

BusinessWeek It's Your Project What's Your Goal? 180

UNIT 4   WEB SITE DEVELOPMENT 182

BusinessWeek Byte   Rob Glaser: Real Close 183

Chapter 9   Creating a Web Site 184

SECTION 9-1   Conceiving a Web Site 186
    Section 9-1 Review 191

SECTION 9-2   Planning a Web Site 192
    Section 9-2 Review 196

Assessment 197
    Worksheet 9-1: Planning Is Key 197
    Worksheet 9-2: Customer Care 198

Chapter 9 Review and Activities 199

Chapter 10   Building a Web Site 202

SECTION 10-1   Fundamentals of Web Design 204
    Section 10-1 Review 213

SECTION 10-2   Creating an Attractive Site 214
    Section 10-2 Review 218

Assessment 219
    Worksheet 10-1: Design It! 219
    Worksheet 10-2: Decode It! 220

Chapter 10 Review and Activities 221

Table of Contents

Chapter 11   Web Site Management 224
- **SECTION 11-1** Positioning a Web Site 226
  - Section 11-1 Review 229
- **SECTION 11-2** Back-End Management Tools 230
  - Section 11-2 Review 234
- **Assessment** 235
  - Worksheet 11-1: Type It! 235
  - Worksheet 11-2: Behind the Scenes 236
- **Chapter 11 Review and Activities** 237

BusinessWeek It's Your Project   What's Good About It? 240

UNIT 5   MARKETING IN THE DIGITAL WORLD 242

BusinessWeek Byte   These Sites Are a Shopper's Dream 243

Chapter 12   Fundamentals of Internet Marketing 244
- **SECTION 12-1** Marketing Basics 246
  - Section 12-1 Review 253
- **SECTION 12-2** Market Research 254
  - Section 12-2 Review 258
- **Assessment** 259
  - Worksheet 12-1: Shop Till You Drop—No More! 259
  - Worksheet 12-2: School Spirit 260
- **Chapter 12 Review and Activities** 261

Chapter 13   Distribution in E-Commerce 264
- **SECTION 13-1** Channels of Distribution 266
  - Section 13-1 Review 270
- **SECTION 13-2** Physical Distribution 271
  - Section 13-2 Review 276
- **Assessment** 277

Table of Contents

    Worksheet 13-1: Can You Get There from Here?........................ 277
    Worksheet 13-2: The Customer Wants Delivery Now!.................. 278
Chapter 13 Review and Activities.. **279**

Chapter 14   Customer Service and Web Site Personalization....... 282

SECTION 14-1 Providing a Customer Interface................................ 284
               Section 14-1 Review.. 290
SECTION 14-2 Customizing of a Web Site.................................... 291
               Section 14-2 Review.. 296
Assessment... 297
    Worksheet 14-1: Secret Agents?.. 297
    Worksheet 14-2: Customer Service at School............................ 298
Chapter 14 Review and Activities.. **299**

Chapter 15   Advertising for E-Commerce............................ 302

SECTION 15-1 The Basics of Building an Online Brand......................... 304
               Section 15-1 Review.. 309
SECTION 15-2 Advertising Your Web Site.................................... 310
               Section 15-2 Review.. 316
Assessment... 317
    Worksheet 15-1: Careers in Digital Advertising.......................... 317
    Worksheet 15-2: Brand: New... 318
Chapter 15 Review and Activities.. **319**
SELF-ASSESSMENT: A Day in the Life of the Internet......................... 322
BusinessWeek It's Your Project   What's Your World?..................... 324

Glossary 326
Index 346

To the Student

Welcome to *E-Commerce*

Welcome to *E-Commerce*. Get ready to learn about one of the fastest growing types of business in the world. E-commerce is a subject you probably know something about as so many of you already use the Internet for purchases. This book will let you into the other side of the e-commerce world—setting up your own Web site and selling your own products.

Understanding the Unit

The units introduce you to the new industry we now call e-commerce: its history, impact, career potential, and structures as well as designing Web sites and Internet marketing. Each unit opens with a preview and an excerpt from *BusinessWeek* and closes with a provocative project. The 15 chapters in *E-Commerce* are organized into five units:

UNIT 1: The History, Nature, and Impact of E-Commerce

UNIT 2: You and E-Commerce

UNIT 3: Business Structures and the Business Plan in E-Commerce

UNIT 4: Web Site Development

UNIT 5: Marketing in the Digital World

xii To the Student

To the Student

Previewing the Unit

Unit Opener Photo

The unit opener photo illustrates a concept that is relevant to the upcoming unit. Ask yourself, "How does the photo relate to the content of the unit?"

BusinessWeek Byte

In each unit the *BusinessWeek* Byte offers a stimulating inside look at a major Internet issue or personality. This is followed by "Think About It," a question related to the article and designed to get you to think about what you read.

In This Unit...

The titles of the unit chapters are listed on the left-hand side of the unit opener spread. Think about what you can learn in each chapter. Directly below the chapter titles is an *E-Commerce Online* feature which tells you where you can get more information on the subjects covered in the unit.

BusinessWeek It's Your Project

The unit opener ends with a preview of a challenging project that will appear at the end of each unit.

To the Student

Closing the Unit

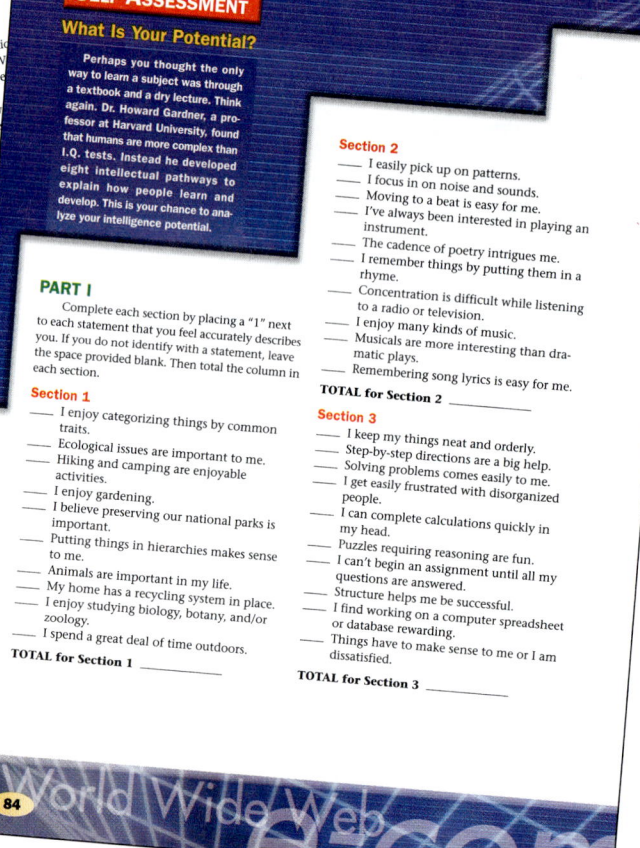

BusinessWeek It's Your Project

Each unit closes with a hands-on activity called It's Your Project. The project focuses on a broad social issue as it affects you personally.

Self-Assessment

Three of the five units also close with a Self-Assessment. This is an opportunity to consider your own intellectual, career, and social development.

To the Student

Understanding the Chapter

Each unit of *E-Commerce* includes two to four chapters. Each chapter focuses on one specific area of e-commerce such as retailing on the Internet, global e-commerce, or building your own Web site.

Previewing the Chapter

The chapter opener resources are designed to capture your interest and set a purpose for reading.

Prereading Strategies
This offers a thinking exercise to help you begin to consider the upcoming material in each chapter.

Chapter Opener Photo
The chapter opener photo focuses on the chapter topic. You might ask yourself, "How does this photo relate to the chapter title?"

Chapter Preview
This paragraph provides a broad overview to introduce the chapter.

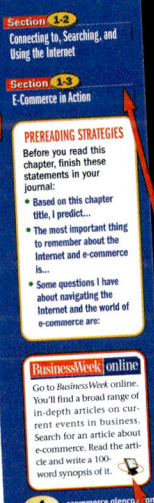

BusinessWeek Online
This feature provides an exercise that will complement the chapter you are starting.

Using the Sections
Each chapter of *E-Commerce* is divided into two or, occasionally, three sections. By using the activities and resources in each section, you can maximize learning.

Drill Down
This feature offers a timeline of inventions and discoveries relevant to the Internet uses and methods you are about to investigate. Get the Big Picture, at the end of each Drill Down, gives you food for thought related to the Drill Down.

To the Student **xv**

To the Student

Understanding the Features

Special features in each chapter are designed to stimulate your interest and promote your understanding of chapter content. Features incorporate activities, such as critical-thinking questions, to help you integrate what you have learned.

Drill Down

Drill Down, in each chapter opener, presents a timeline of discoveries and inventions relevant to the aspect of e-commerce discussed in the chapter. This feature helps you put digital technology into a broader perspective. The Get the Big Picture critical-thinking question gives you an opportunity to integrate the timeline and understand its application to the chapter.

CONNECTION: LANGUAGE ARTS

The Connection feature demonstrates the relationship between academic class work and the real world of e-commerce. Each feature will give you a practical problem to solve that connects e-commerce to the subject of Math, Language Arts, Social Studies, or Science.

Web Site Success

This feature tells an interesting anecdote about a creative Internet use or a personality in the digital world. The feature includes a critical-thinking question to help you build a bridge between the feature and the material in the text.

AS YOU READ...

You Will Learn lists the knowledge you can expect to gain.

Why It's Important explains how the chapter concepts relate to your world.

Key Terms lists significant terms presented in the section.

Photographs and Figures

Photographs, illustrations, charts, and graphs reinforce content. Each has a caption to guide you, and each caption ends with a thought-provoking question.

Quick Talk at the end of each section gives you an opportunity to brainstorm ideas with a partner.

To the Student

Worksheets

At the end of each chapter's text, before the review section, special worksheets provide review and skill-building activities related to the chapter content. Your teacher may choose to assign one or both of these worksheets to assess how well you understand the chapter. Whether or not the worksheets are used as evaluation tools, they will provide you with crucial knowledge of real-world applications. Two one-page worksheets give you the opportunity to complete an activity or exercise and apply the chapter content in a variety of interesting formats. These worksheets are designed to give you hands-on experience using the skills you will need to actually create and manage an e-commerce business of your own.

Name _____ Date _____

Worksheet 1-1

Using Search Engines

Search engines are the telephone books for the twenty-first century. In order to search for products, services and information online, you'll need to be familiar with a variety of search tools. Using the Internet, research six search engines: Google, Yahoo!, MSN, Netscape, and two of your choice.

1. Create a chart, using a word-processing or spreadsheet program, to visually explain the following:
   a. Similarities and differences between the six search engines.
   b. The common search engine commands and how they work in each, such as +, -, " ", or wildcards.
   c. Ease or difficulty of use.

2. Which search engine do you like the best? Which do you like the least? Explain why.

3. Different search engines yield different results. Search for the terms "flower delivery" and "airline tickets." What are the top two results in each engine? Add your findings to the chart you created.

4. Imagine you're an Internet researcher for a new online company and your boss needs to know how you conducted your Internet search on a particular topic. Select a topic you want to research. Document your methods of research, and then prepare a one-page summary of your techniques to find information from credible sources.

Chapter 1 The Internet and E-Commerce **21**

To the Student **xvii**

To the Student

Understanding Assessments

At the end of each chapter, *Chapter Review and Activities* presents a chapter summary with key terms, recall and critical-thinking questions, and a variety of activities that target academic and workplace skills.

Chapter Summary

The Chapter Summary is a bulleted list of the main points of each section, related to the chapter objectives. This provides you with a quick reminder of each section's material.

Review of Key Terms

This review requires you to match the key terms in the chapter with their correct definitions to determine whether you understand their meanings.

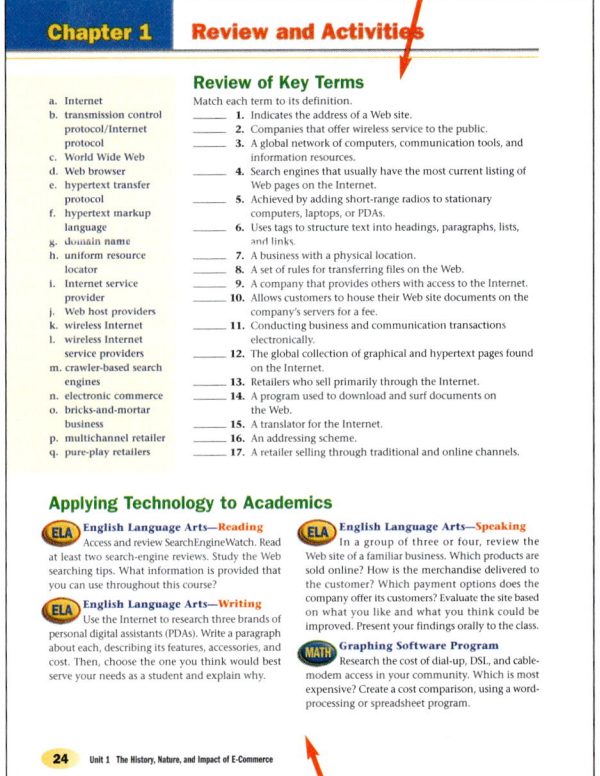

Applying Technology to Academics

This title could just as easily read "Applying Academics to Technology" because its purpose is to show you the interrelationship between them and to demonstrate once again the value of academics in the workplace. The section is divided into four parts:

- English Language Arts—Reading
- English Language Arts—Writing
- English Language Arts—Speaking
- Math

To the Student

Critical Thinking
This section asks you to think about points raised in the chapter and logically extend your thinking to answer challenging questions.

Competitive Event
This section asks you to prepare a presentation responding to a set of circumstances. These circumstances mirror life in the real-world workplace, so your presentation needs to be grounded in reality. This provides experience in dealing with issues you may very well encounter during your career. Each Competitive Event tells you the criteria on which your performance will be evaluated.

BusinessWeek Online
This activity gives you an opportunity to discover the latest trends in e-commerce and explore how e-commerce is changing the world of business. Each chapter's activity focuses on a different aspect of e-commerce around the world. You'll use *BusinessWeek* online as a starting point in your research to base your projects on the most up-to-date information.

E-Commerce Web Site
The *E-Commerce* Web site at **ecommerce.glencoe.com** draws on the vast resources of the Internet to enhance your understanding of the course materials as well as to expand your exploration of career topics.

The Student Center includes the following:
- Chapter objectives
- Interactive practice tests for each chapter
- E-flashcard games to assist you in learning and using key terms
- Web links and resources to use when exploring career options
- Career tools such as self-assessments, résumé writing tips, job-search tips, and educational resources
- Disability support links
- *BusinessWeek* activities for illustrating and researching various aspects of e-commerce

To the Student

Reading Strategies

How can you get the most from your reading? Effective readers are active readers—readers who become actively involved with the text. Think of this textbook as a tool to help you learn more about the world, especially the specialized, technological world of the Internet and e-commerce. It is non-fiction writing describing real workplace ideas, people, and events. Use the reading strategies in the *Power Read* box at the beginning of each chapter along with the strategies in the margins of each section that pose questions for you to consider.

Make educated guesses about each section's content by combining clues in the text with what you already know. Predicting helps you anticipate questions and stay alert to new information.

Ask yourself:
- What does this section heading mean?
- What seems to be this section's topic?
- How does this section tie into what I've read so far and what I already know?
- Why is this information important in understanding the subject?

Draw parallels between what you are reading and the circumstances of your own life.

Ask yourself:
- What do I already know about the topic?
- How do my own experiences compare to the information I'm reading?
- Can I apply this information in my life? How?
- Why is this information important in understanding the subject?

Ask yourself questions as you go along to help you clarify the meaning of what you read.

Ask yourself:
- Do I understand what I've read so far? If not, reread it. Focus on specifics.
- What is this section about? Paraphrase the central ideas.
- What does this mean? If you don't know, look it up.
- Why is this information important in understanding the subject?

React to what you are reading. Form opinions and make judgments about the section *while you are reading*—not just after you've finished.

Ask yourself:
- Does this information make sense to me?
- What can I learn from this section?
- How can I use this information to start planning for my own career?
- Why is this information important in understanding the subject?

To the Student

More Reading Strategies
Use this menu for more reading strategies to get the most from your reading.

BEFORE YOU READ...

SET A PURPOSE
- Why are you reading the textbook?
- How does the subject relate to your life?
- How might you be able to use what you learn in your own life?

PREVIEW
- Read the chapter title to preview the topic.
- Read the subtitles to see what you will learn about the topic.
- Skim the photos, charts, graphs, or maps. How do they support the topic?
- Look for key terms that are boldfaced. How are they defined?

DRAW FROM YOUR BACKGROUND
- What have you read or heard concerning new information on the topic?
- How is the new information different from what you already know?
- How will the information that you already know help you understand the new information?

AS YOU READ...

PREDICT
- Predict events or outcomes by using clues and information that you already know.
- Change your predictions as you read and gather new information.

CONNECT
- Think about people, places, and events in your own life. Are there any similarities with those in your textbook?
- Can you relate the textbook information to other areas of your life?

QUESTION
- What is the main idea?
- How do the photos, charts, graphs, and maps support the main idea?

VISUALIZE
- Pay careful attention to details and descriptions.
- Create graphic organizers to show relationships that you find in the information.

NOTICE COMPARE AND CONTRAST SENTENCES
- Look for clue words and phrases that signal comparison, such as *similarly, just as, both, in common, also,* and *too*.
- Look for clue words and phrases that signal contrast, such as *on the other hand, in contrast to, however, different, instead of, rather than, but,* and *unlike*.

NOTICE CAUSE-AND-EFFECT SENTENCES
- Look for clue words and phrases, such as *because, as a result, therefore, that is why, since, so, for this reason,* and *consequently*.

NOTICE CHRONOLOGICAL SENTENCES
- Look for clue words and phrases, such as *after, before, first, next, last, during, finally, earlier, later, since,* and *then*.

AFTER YOU READ...

SUMMARIZE
- Describe the main idea and how the details support it.
- Use your own words to explain what you have read.

ASSESS
- What was the main idea?
- Did the text clearly support the main idea?

- Did you learn anything new from the material?
- Can you use this new information in other school subjects or at home?
- What other sources could you use to find more information about the topic?

UNIT 1

THE HISTORY, NATURE, AND IMPACT OF E-COMMERCE

In This Unit...

Chapter 1
The Internet and E-Commerce

Chapter 2
The Nature of E-Commerce

Chapter 3
Retailing on the Internet

Chapter 4
Global E-Commerce

E-Commerce Online
To learn more about the history, nature, and impact of e-commerce, visit the *E-Commerce* Web site at ecommerce.glencoe.com.

BusinessWeek Byte

WebSmart

At age 38, Michael S. Dell is the master of electronic business. From his headquarters in Round Rock, Texas, he runs a $50 million-a-day online distribution channel—equivalent to five Amazon.coms. More than anyone, Dell has shown how the Web, intelligently used, drives efficiency.

Now he's carrying his network magic to the next level. At a time when many tech companies are paring costs, Dell is piling on new technology. He's tagging parts with radio-powered ID chips. At plants from China to Texas, online orders are transformed quickly into radio signals. These instruct Dell's automatic parts-picking machines to round up the components for each PC. They also transmit assembly blueprints to workers and tack the shipping of the finished product. Dell managers can monitor the entire process online.

Dell also is using his e-commerce machine to sell a vast range of new offerings, from PDAs and printers to plasma-screen TVs. It's working as he planned, with sales soaring 16 percent in the most recent quarter.

Source: Excerpted with permission from "WebSmart," *BusinessWeek*, E.Biz 25, September 29, 2003.

To read this *BusinessWeek* article in its entirety, go to **ecommerce.glencoe.com**. Click on the Student Center, find *BusinessWeek* Articles, and go to Unit 1.

THINK ABOUT IT

Is Dell, as an e-business, running any differently than a traditional bricks-and-mortar business?

BusinessWeek It's Your Project

Is there something in life that makes you skeptical—technology, politics, media, and so on? The Unit 1 project, "What Is True?" on page 88, asks you to discover the history, nature, and impact on a topic's truth.

ecommerce.glencoe.com

Chapter 1
The Internet and E-Commerce

Section 1-1
Internet Basics

Section 1-2
Connecting to, Searching, and Using the Internet

Section 1-3
E-Commerce in Action

PREREADING STRATEGIES

Before you read this chapter, finish these statements in your journal:

- Based on this chapter title, I predict...
- The most important thing to remember about the Internet and e-commerce is...
- Some questions I have about navigating the Internet and the world of e-commerce are:

BusinessWeek online

Go to *BusinessWeek* online. You'll find a broad range of in-depth articles on current events in business. Search for an article about e-commerce. Read the article and write a 100-word synopsis of it.

The Online World

The Internet is one of history's most revolutionary advances because it impacts virtually every aspect of daily life: communication, information, education, entertainment, and business.

Drill Down

▶ **1454 Start the Presses**
German inventor Johann Gutenberg develops the printing press, which causes an immediate increase in book circulation and spreads new ideas around the world.

▶ **1800 Capturing Power**
Alessandro Volta, an Italian scientist, creates batteries that can store electrical energy to power machinery.

▶ **1876 Telephone Line**
Alexander Graham Bell files a patent application for the telephone, which ultimately revolutionizes communication.

▶ **1908 Wheels in Motion**
Henry Ford introduces the Model T Ford, which becomes the first mass-produced automobile for public use.

▶ **1969 A New Frontier**
On July 20, 1969, Neil Armstrong relays the first message from the moon when he announces: "Houston, Tranquility Base here. The Eagle has landed."

▶ **1972 Communication Revolution**
Larry Roberts writes the first e-mail management program to list, read, and receive messages.

Get the Big Picture

With a group of your classmates, discuss the above timeline's central theme. How do you think the Internet has influenced communication?

POWER READ

Be an active reader and use these reading strategies:

PREDICT what the section will be about.
CONNECT what you read with your own life.
QUESTION as you read to make sure you understand the content.
RESPOND to what you've read.

5

Section 1-1

Internet Basics

AS YOU READ...

YOU WILL LEARN
- how the Internet began.
- the key components of the Internet.
- the basic ways to use the Internet.

WHY IT'S IMPORTANT

The Internet greatly impacts both our personal and professional lives. An understanding of how the Internet came to be and its vast capabilities can help you to realize the scope of its influence.

KEY TERMS
- Internet
- transmission control protocol/Internet protocol
- World Wide Web
- Web browser
- hypertext transfer protocol
- hypertext markup language
- domain name
- uniform resource locator

THE BASICS OF THE INTERNET AND WORLD WIDE WEB

It's a meeting place, a discount store, a political rally, and a global library. It's the way you communicate with friends, make travel plans, buy birthday gifts, express political views, read today's news, or research a term paper. What is it? Some call it cyberspace; others know it as the **Internet**—a global network of computers, communication tools, and information resources.

What Is the Internet?

The Internet is a system comprising vast telecommunications networks, satellites, and fiber optics that allows you to access information from other computers and their users. Government agencies such as the U.S. National Science Foundation and the National Aeronautics and Space Administration (NASA), along with companies such as AT&T, Sprint, NetCom, and MCI, provide many of the high-speed networks that compose the backbone of the Internet. It carries data such as an e-mail message to your sister in Florida, NASA's recent photos, or files shared between cancer researchers in Sweden and the United States. In its infancy, the Internet's backbone may have resembled a spine with ribs spanning its length, but now it takes the shape of an intricate fishing net enveloping the earth.

 GLOBAL NETWORK Using satellites, fiber optics, and telecommunications networks, the Internet connects computers around the planet. *Where is all the information found on the Internet stored?*

Figure 1.1

Beginnings of ARPANET

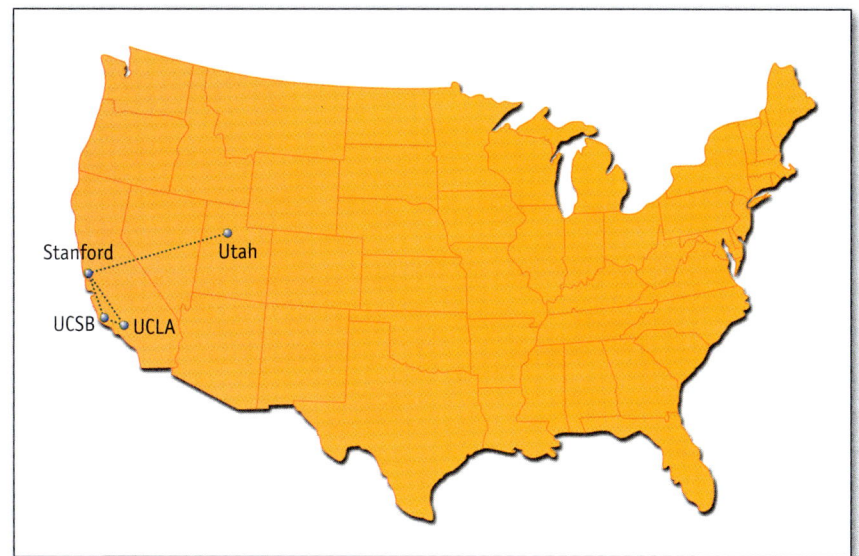

INTERNET ORIGINS ARPANET, the basic foundation for today's Internet, connected computers at Stanford University, the University of California at Los Angeles, the University of California at Santa Barbara, and the University of Utah. *What has powered the incredible growth of the Internet?*

BIRTH OF THE NET The basic foundation for the Internet you use today began with the Advanced Research Projects Agency (ARPA), a segment of the U.S. Department of Defense. The agency created a network called ARPANET, which connected four major computers at Stanford University, the University of California at Los Angeles, the University of California at Santa Barbara, and the University of Utah. ARPANET allowed scientists, academics, military bases, and computer professionals to share information and communicate with one another. Launched in 1969, the network was designed to facilitate ongoing communication in the event of a nuclear war. **Figure 1.1** illustrates ARPANET in its infancy. The Internet delivered information in a *packet switching network,* which allowed one machine to deliver packets of data communications to numerous machines at one time as opposed to a telephone's *circuitry switching.*

By 1973 ARPANET had expanded to include dozens of universities and approximately 2,000 users and had made its first international connections, to the University College of London and the Royal Establishment in Norway. In the 1980s the Internet grew beyond its roots as a research and scientific facilitator and began to increase its commercial activity. Today the Internet facilitates sharing and discussions among educational and research institutions, businesses, and government organizations around the globe. Over 170 million people in the U.S. now use the Internet. There are 544 million users worldwide, and each year these figures increase.

WHO'S IN CONTROL? No one authority controls the Internet. The collaborative spirit that began with the early ARPANET continues even today. Governing groups comprising various companies, governments, and individuals define how new technologies should be implemented and how the Internet should work. For example, the Internet Engineering Task Force, an international community of network designers, operators, vendors, and researchers, works to ensure

PREDICT

How does the word "powerful" describe the Internet?

CONNECT

How would your life be different if the Internet hadn't been developed?

that the Internet evolves smoothly. In the next section, you'll learn more about the standardized technologies, languages, and tools used on the Internet.

Internet Anatomy

Internet technologies allow you to transfer documents, view graphical files, log on to remote computers, and participate in virtual discussions. These advanced components rely on the basic language of transmission control protocol/Internet protocol (TCP/IP).

TCP/IP Transmission control protocol/Internet protocol is the common underlying language or protocol through which systems communicate on the Internet. Think of it as a translator that enables all applications and devices to speak the same language. When you set up your computer to access the Internet, a copy of the TCP/IP program is installed for you. Every other computer on the Internet also has a copy.

TCP/IP is a two-layer program. The first layer, transmission control protocol, manages the assembling of messages or files into smaller packets that are transmitted over the Internet and received by a TCP layer that reassembles the packets into the original messages or files. This data-transmission method is called packet switching. The second layer, Internet protocol, handles the address part of each packet so that it reaches the right destination.

Remember, TCP/IP is the basic language for Internet communication. Next you'll explore some higher-level Internet components, all of which employ TCP/IP to function.

WORLD WIDE WEB Before the Internet became the sophisticated system used today, it needed an organizational structure. London-born Tim Berners-Lee created the communications system now known as the **World Wide Web** (WWW) while working at CERN, the world's largest particle physics laboratory. When his system went live in 1991, it brought structure and organization to the boundless information available online, and today the World Wide Web represents a vast global collection of graphical and hypertext Internet pages that can be read, viewed, and interacted with via computer.

The Web is the most popular and widely used segment of the Internet. Though some people may use the terms "Internet" and "World Wide Web" interchangeably, they are not the same: The Internet is the global system of computer networks that supports the collection of resources known as the World Wide Web.

To deliver these graphical and hypertext resources to your computer, the Web browser was developed. A **Web browser** is a program, such as Internet Explorer or Netscape Navigator that is used to view, download, surf, or access Web documents. Web browsers contain applications that can read **hypertext transfer protocol** (HTTP or *http://*)—the language that moves hypertext files across the Internet and defines the rules for transferring those files, which may include text, graphic images, sound, video, and other multimedia. Web browsers read pages coded in a standard language such as **hypertext markup language** (HTML), an easy-to-learn language that uses tags to structure text into headings,

Figure 1.2

Sample HTML Document and Displayed Document

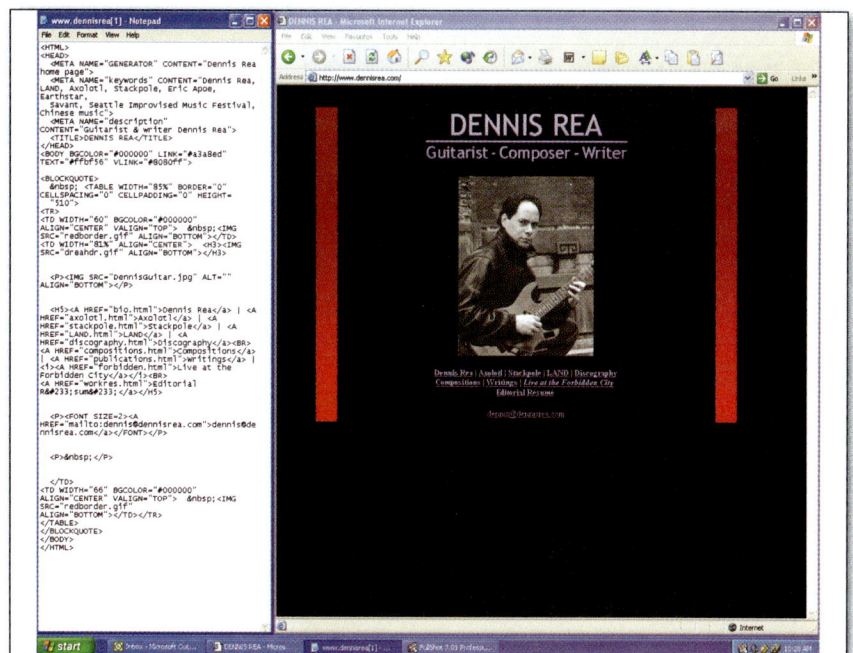

HYPERTEXT ON DISPLAY Web browsers translate information written in HTML to create all the elements of a Web page. *How does a Web browser know what to display?*

paragraphs, lists, and links. HTML is the standard language for the Web browser, and its tags tell a Web browser how to display tables, formatting preferences, images, and other visual elements. Tags consist of a collection of brackets (< >) that, at their most basic level, turn a function on and off. For example, if you wanted to boldface a word, the tags would look like this: .... The "/" turns off the code, or, in this case, the boldface, so it does not keep applying that code to subsequent characters. (See Chapter 10 to learn more about HTML.)

Figure 1.2 illustrates an HTML document alongside the document as displayed in a Web browser. Browsers can also interpret pages that are coded using newer languages such as XHTML (HTML's successor) and XML (a next-wave language that allows you to define your own tagging system).

INTERNET ADDRESSES The Internet contains millions of computers, and each has its own identification. It's important to know where information is coming from and where it's going. Therefore, an Internet Protocol, or IP address, is assigned to each connected computer. This address is represented by a dotted decimal notation, such as 123.22.232.109, and is assigned by a regional Internet registry (RIR). To date, there are four continent-specific RIRs. For example, ARIN serves North America and African countries located south of the equator.

Since it's hard to remember IP numbers, domain names are used to identify Internet addresses. A **domain name** is an addressing scheme employing words and phrases to identify and locate computers on the Internet. For example, the domain name Rollerblade.com is registered by the New Jersey–based Rollerblade company to represent its IP

QUESTION

What is HTML, and why is it important to Web browsers?

Figure 1.3

Commonly Used Domain Name Extensions

NAMING DOMAINS Internet domain names vary depending on the type of business, organization, school or other resource listed. *Provide examples of reasons you might access the different domains listed in Figure 1.3.*

EXTENSION	TYPE OF ORGANIZATION	EXAMPLE
.edu	Educational institutions	utexas.edu University of Texas
.com	Commercial entities	homedepot.com The Home Depot
.org	Nonprofit organizations	redcross.org American Red Cross
.net	Internet administrative entities	earthlink.net EarthLink Internet Service Provider
.gov	Government sites	firstgov.gov U.S. government's official Web site

address. The Domain Name System (DNS) translates this registered domain name into the underlying IP numbers. Companies use domains to store all the documents that will appear on their Web sites, including the home page and related hypertext documents. To find these pages, or addresses, on the Web, you'll need to know the **uniform resource locator** (URL).

A URL indicates the address of a Web site. It consists of two primary parts: the hypertext transfer protocol and the domain (or server). Note that the suffix found at the end of a URL signifies the category of a Web site, such as a commercial site (.com) or a government site (.gov). A Web site might attach its country's suffix to the end of the URL. **Figure 1.3** contains a list of the most commonly used suffixes.

E-MAIL Another widely used component of the Internet is electronic mail, or *e-mail*. E-mail is a system of worldwide electronic communication through which a user can compose a message at one computer and send it to a recipient at another computer. Much like making phone calls, e-mail is a preferred method of modern communication.

With all the benefits of new communication, there are also file-management issues to consider: first and foremost, the size of an attachment to an e-mail message. The last thing you want to do is send an e-mail message with a large attachment that can crash the recipient's computer or slow his or her productivity. A four-kilobyte (4K) attachment takes one second to send and receive. If you send a 40K attachment, it will take 40 seconds to send and receive with a standard modem. In general, you should avoid sending attachments that are larger than 200K.

FTP You can obtain files over the Internet by using the file transfer protocol (FTP). Thousands of FTP sites have been set up to allow the public to download and view files. You can even set up your own FTP sites using a FTP server. While many people choose to send files via an e-mail message, you can use FTP when attachments are too large for an e-mail server.

Section 1-1 Review

RESPOND to what you've read.

1. How are the operations of the Internet monitored? _____

2. What are Web browsers, and why are these programs necessary? _____

3. How can a domain name reveal clues about an organization? _____

4. What is ARPANET? What government department created it, and what was its original purpose? _____

Quick TALK Imagine that a new student with no prior knowledge of the Internet has enrolled in your school. *Brainstorm with your partner to create a short explanation of the Internet and its basic purposes for the student.*

Section 1-2

Connecting to, Searching, and Using the Internet

AS YOU READ...

YOU WILL LEARN
- the various options available to connect to the Internet.
- the factors to consider when choosing an ISP.
- how search engines help users to locate businesses and services.
- how the Internet is used by an assortment of user groups.

WHY IT'S IMPORTANT

Understanding the differences among the various types of Internet service providers can help you to select the most appropriate one for your needs. Finding the right Internet service provider allows you to use a search engine to access data and take full advantage of the Internet's power.

KEY TERMS
- Internet service provider
- Web host providers
- wireless Internet
- wireless Internet service providers
- crawler-based search engines

PREDICT

What is an Internet service provider?

LINKING TO THE INTERNET

Connecting to the Internet is easier and faster than ever. Need to check your e-mail messages from your cell phone? No problem. Want to see how your fantasy sports team is doing? Log on. Interested in scanning the latest baseball scores from your laptop in the park? Done. Need to build your Web site from your desktop using a cable modem? Can do. The information superhighway's access ramps are all around you.

Types of Internet Connections

You want a connection that is fast, reliable, and affordable when you use the Internet. Before you start shopping for an Internet connection, take an inventory of the behavior behind the usage. Perhaps you need the Internet to surf for personal reasons. Or perhaps you need a connection that supports a small or large business. The following are some examples of ways in which you can connect to the Internet.

INTERNET SERVICE PROVIDERS AND WEB HOSTING SERVICES An **Internet service provider** (ISP) is a company that provides other companies or individuals with access to or a presence on the Internet. ISPs such as AT&T WorldNet, Netcom, AOL, and EarthLink have the equipment and the telecommunications infrastructure needed to offer numerous points of Internet access for the various geographic areas they serve. Many ISPs may also provide e-mail accounts and assist in the design, creation, and administration of Web sites. Separate companies that provide Web-site hosting services, such as Apollo Hosting or Infinity Host, are called **Web host providers**. For a monthly fee, Web host providers allow customers to house their Web-site documents on the company's servers.

Today there are an estimated 7,000 ISPs worldwide. More than 60 percent of these are U.S. companies. When choosing an ISP, customers should look for both reliable service and the type of connection they desire. These days connecting to the Internet can involve a lot more than a phone line and a dial-up modem. The following sections outline some of the most popular Internet connectivity options.

DIAL-UP CONNECTIONS When the Internet became accessible to the general public, people often used analog modems to connect to it.

12 Unit 1 The History, Nature, and Impact of E-Commerce

An *analog modem* communicates over standard telephone lines by converting computer (digital) data into sine waves. At the receiving end, the data is then converted back into digital. Most people can use a modem because it runs over a normal telephone line to provide Internet access. While you're online, however, your phone may not be available for calls. Many customers solve this problem by installing a second phone line for their computer or installing an application that sends an alert to their computer desktop for incoming phone calls.

A modem's speed is determined by how fast it transmits data. Data-transmission rates can range from 75 to 56,000 bits per second and beyond. As the type of data transmitted via the Internet has become larger and more complex (including pages with lots of tables, video clips, and photos), phone lines have become a less desirable method for connecting online. Long wait times and highly complex Web pages have made broadband connections more attractive to both individuals and businesses.

BROADBAND SERVICES What's the buzz about broadband? Broadband enables large amounts of electronic data to be transmitted quickly, which is particularly important to businesses in the technology, entertainment, and communications industries. If you wanted to get more water down a pipe faster, you would get a broader pipe. The same formula holds true when it comes to the Internet. The following sections cover the most commonly used broadband options.

- A local phone company or an ISP might provide a *digital subscriber line* (DSL) to operate over your phone line. DSL won't disrupt your telephone service and is typically ten times faster than a standard dial-up modem. It's also referred to as a T line, another reliable connectivity choice that's used by many businesses, colleges, and universities. A T-1 line transmits data at 1.544 megabits per second, and another version of T line, the T-3, operates at 44.734 megabits per second. T-3s are often used by ISPs to provide connectivity to their subscribers or by businesses that set up their own private Internet networks.

- Do you have cable television? If so, consider using *cable modem service* to connect to the Internet. This service requires a special cable modem and possibly a network interface card, which the cable company can install for you. Not only does cable modem service provide a faster Internet connection, but it also improves audio- and video-streaming quality.

- *Integrated services digital network* (ISDN) is the oldest form of broadband and operates over standard telephone lines and fiber-optic circuits. ISDN enables its user to download graphics, sound files, and multimedia about four times faster than a standard dial-up modem. Businesses often opt for ISDNs because they are very reliable and create fewer errors in transmission, but individuals are less likely to select it because the service costs more than $100 per month. One ISDN line can handle up to eight devices, two of which can be operating simultaneously, including a personal computer, a telephone, a fax, and a video clip.

CONNECT

What type of ISP do you presently use? What type of provider would you use if price were no object?

CONNECTION
LANGUAGE ARTS

What Do They Mean?
Every industry has its own jargon, and the world of e-commerce has its own language that business-people need to learn and use. *Predict what the following e-commerce terms mean: adopter, kiosk, and real time.*

QUESTION

What are some factors to consider when choosing an ISP?

- *Satellite broadband service* transmits high-speed data via satellite to a dish antenna at a home or business and uses either a two-way or a one-way satellite connection. One-way satellite connections require both a modem and a telephone connection to an ISP. The newer two-way satellite systems allow users to download and upload directly through their satellite dishes. Satellite connections allow for download speeds that are ten to 20 times faster than dial-up services. It is a more expensive service, but it may be the only fast option if you live outside the service area of cable or DSL providers.

WIRELESS INTERNET Interested in checking the latest weather report before a hike at your state park? Are you miles from your home computer? Try a wireless Internet connection. **Wireless Internet** works much in the same way as a cordless telephone, adding short-range radios to stationary computers, laptops, and handheld *personal digital assistants* (PDAs). With this connection, you can exchange data without wires at rates of up to 11,000 kilobytes per second over distances from several hundred feet indoors to ten miles outdoors.

Wireless Internet service providers (WISPs), such as Boingo and Pronto, offer wireless connection services to the public. Cafés and airport lounges often offer wireless network access free of charge.

Search Engines

Browser search tools have become the telephone book for the Internet generation. It's important that an e-commerce business stands out when the search tool retrieves and displays a listing of similar online businesses. In order to boost a Web page's ranking, *meta tags* are used. (Chapter 10 talks in-depth about meta tags as a device to maximize a Web site in a search engine.) Instead of just a few hundred people in a city seeing your business name, address, and phone number, millions of potential customers can access a link to your Web site. You can use a *search engine*, such as AltaVista or Excite, to retrieve information on the Web. Search engines use key words to access lists of documents containing those key words.

MOBILE OFFICE One of the main benefits of wireless technology is mobility. *In what types of jobs might wireless Internet access be a useful tool?*

Figure 1.4

Search Engine Tips

SYMBOL	FUNCTION	EXAMPLE
+	Finds the pages with all the words you enter	+NPR+car talk
–	Finds the first key word, but not the second key word	Dreamweaver-XM 2004
" "	Find a specific phrase	"Colorado River Rafting"

REFINING YOUR SEARCH These tips can simplify the process of searching for information on the Internet. *Why is being specific so important when using search engines?*

You could say that there is an art to surfing the Web. Remember, the more specific your key words, the more tailored your Web search. Unless you want to scroll through hundreds or even thousands of pages to get your desired information, see **Figure 1.4** for some tips on using search engines.

THE DIFFERENCE BETWEEN SEARCH TOOLS When searching the Internet, there is no single Web site that will enable you to find all of the information that is available online. Fortunately, you can use multiple tools to search for products, services, and information. Web sites such as SearchEngineWatch list hundreds of resources and tips that can help you find information on the Internet. Some search tools have distinguished themselves from the competition by providing both ease of use and enhanced functionality.

An *Internet directory* is a comprehensive listing of Web sites. Yahoo! and Microsoft's MSN are two examples of user-friendly Internet directories. These Web sites employ hundreds of people who continually add and update the thousands of links they provide for their visitors.

Internet directories are different from **crawler-based search engines**, which use automated computer programs to scan Internet databases in search of new or revised Web pages. These engines are updated automatically, using computer programs that search the Internet for new content. Google.com, a crawler-based engine, generally has the most current listings of Web pages, which is especially helpful for users who are in search of the most up-to-date information. A disadvantage for the user, however, is that the results of a Google.com search are not displayed in as user-friendly a format as those of an Internet directory such as Yahoo!.

There are various means you can use to search for information on the Internet. Use the "+" symbol when you want a search engine to find pages that contain all the words you've entered. Use the "–" symbol when you want the search engine to find pages that contain one word, but not another. Use "quotation marks" when you want the search engine to identify pages in which the terms appear exactly in the order you've specified. Use the word "or" when you want it to find any of the terms you've specified. Lastly, consider using wildcards, such as "*," "?," or "%," when you have only part of the term or word and want the search engine to locate pages with the entire term or word.

Unless you're a tech-savvy teen who already knows how to build your own Web site and won't settle for anything less than the highest speed Internet connection, some of what you've read may sound a bit complicated. However, keep in mind that computers, the Internet, and search engines are tools designed to serve you and be used by you. The more you know about how they work, the more you'll be able to make them work for you. As tools they are indispensable to conducting e-commerce, whether you're looking for the best price on your favorite band's new CD or setting up your own Web site to sell CDs.

Section 1-2 Review

RESPOND to what you've read.

1. Why might a user select a cable-modem service over a dial-up connection? _____

2. Where can you find a wireless ISP? How can you access one? _____

3. What might make the more heavily trafficked search engines preferable to users? _____

4. Name two types of Internet search tools, and explain how they are different. _____

Quick TALK With your partner, brainstorm four ways in which the Internet is routinely used. *How was each of these tasks performed prior to the Internet?*

Section 1-3

E-Commerce in Action

FUNDAMENTALS OF E-COMMERCE

So you've decided to buy a new mountain bike? You're ready to research the brands, check out consumer reviews, compare prices, select your accessories, and make your purchase. Years ago, this would have required days of shopping at various stores, but today the only trip you need to take is an excursion to the PC in your family room. What's the trick? Turn on your computer, launch a Web browser, and access hundreds of resources that will enable you to research and buy.

How is all this possible? It is possible through **electronic commerce**, or e-commerce—the conducting of business and communication transactions by electronic means. Simply put, e-commerce refers to purchases from online stores, which are also called virtual or cyber stores. Examples of e-commerce transactions include the following:

- You purchase a CD from Amazon.com.
- A department store buys tables for resale from a custom-furniture manufacturer's Web site.
- Your friend participates in an online auction.

When you think e-commerce, think of an exciting phenomenon that comprises a variety of advertising efforts, marketing strategies, and technological innovations. In the early 1990s e-commerce represented only $60 billion in sales, but today it generates revenues of $900 billion per year and has added a new dimension to the business world—one that you'll explore throughout the remainder of this textbook. Before you jump into the nuts and bolts, take a look at the roots of electronic communication to learn how it all began.

The Evolution of Electronic Communication

Did you know that electronic communication dates back to Samuel F.B. Morse, the man who invented the system of dots and dashes known as Morse Code? Telegraphy was the seed for the next discovery—connecting the continents. Messages needed to be sent across continents, and a cable company took the initiative to begin this journey. In 1866, with miles of cable on board, a ship called the *Great Eastern* left Valentia, Ireland, and headed to Newfoundland to lay down the cable on the ocean floor. Over the course of numerous broken cables thousands of feet below sea level and years of investment, cross-continent electronic communication became a reality. In 1866 two telegraph lines connected messages across the Atlantic Ocean, and communicators continued to use the electronic telegraph system well

---

AS YOU READ...

YOU WILL LEARN
- how e-commerce evolved from electronic communication.
- the characteristics that define e-commerce.
- the basic purpose of e-commerce.
- how e-commerce can increase a business's revenues.

WHY IT'S IMPORTANT

E-commerce has revolutionized the business world, making it essential for companies in virtually every industry to have an online presence. Understanding the mechanics of the electronic marketplace will help you to appreciate its nature, scope, and importance.

KEY TERMS
- electronic commerce
- bricks-and-mortar business
- multichannel retailer
- pure-play retailer

PREDICT

How can e-commerce increase a business's revenues?

into the twentieth century, when satellite transmission became a sound alternative to cable.

Electronic communication became more prevalent after the 1960s, as each decade built on the discoveries of the last. Generally, you can classify the computer revolution into two stages: the introductory stage and the permeation stage. The creation, polish, and readiness of the computer paved the way for the introductory stage.

By the 1990s computers were a mainstay in workplaces, libraries, schools, and the home. Online information was power, and the World Wide Web became the perfect venue for distributing that information. The advent of the Internet and the World Wide Web also created new dimensions of global interdependence and communication expectations.

When the National Science Foundation lifted its restrictions on the commercial use of the Internet in 1991, the business community paid attention. In the mid-90s companies such as Pizza Hut, Reebok, and FedEx built Web sites allowing their customers to order pizzas, check out the latest in shoes, and track package shipments. These sites provided a new self-service environment in which customers could browse, purchase, and pay for products or services without any human interaction. At that time International Data Corporation (IDC) surveyed 1,000 large U.S. companies, asking managers to characterize their current and planned usage of the Internet. Of those surveyed, 30 percent viewed the Internet as a "major industry development."

Over the decades the computer has slowly crept from a mechanism used by the military and scientists to a popular vehicle for business and personal use. Now the masses use computers to send electronic greeting cards, order stamps, pay bills online, auction goods, conduct research, start businesses, apply for jobs, and so on. Meanwhile,

> **CONNECT**
> When might you consider making an online purchase instead of a traditional transaction?

Web Site Success

AN INTERACTIVE, ONGOING DEALMAKER
Barry Diller

Barry Diller is a household name in the media and Web worlds. He was responsible for creating Fox Broadcasting Company and its motion-picture unit as well as serving as chief executive of Paramount Pictures Corporation. With nearly $4 billion in cash and stock, Diller's company, InterActiveCorp (IAC), engages the world in interactive commerce—via both freestanding and online companies. IAC owns Ticketmaster, Expedia, Hotels.com, Match.com, and CitySearch, just to name a few. He has created an online conglomerate of Internet, television, and telephone companies. And with money in hand, Diller is probably scouting his next move right now.

Thinking Critically
Why do you think media mogul Barry Diller has expanded his interests from bricks-and-mortar retail to Internet retail?

computers have gone from occupying an entire room to fitting inside your front pocket.

Computers have also begun to replace humans in the workplace to a certain degree. In some instances the digital automated voice mechanisms generated from computers have replaced telephone operators, bank tellers, and assembly-line workers. At the same time, new tech-related jobs, such as systems analysts, network engineers, online producers, and so on, are continuing to permeate the workforce.

Characteristics of E-Commerce

In an article published in *Atlantic Monthly*, Peter F. Drucker, a social science professor at Claremont Graduate School, wrote "e-commerce is to the Information Revolution what the railroad was to the Industrial Revolution." Both inventions have indeed expanded our horizons. The construction of steel and steam allowed us to access distant spaces via train. Today e-commerce has revolutionized business in several ways.

E-commerce transcends geographic boundaries, attempts to level the playing field, and allows modern businesses to utilize new avenues of advertising, selling, and distribution. For example, a manufacturer of custom jewelry in Ireland can now reach customers in Miami or Munich using the Web. This company's Web presence and virtual real estate can equal that of a large competitor.

In addition, e-mail advertisements and specials may help to increase sales and generate interest in a company's products at relatively low costs. Online surveys and chat rooms can help a business to understand customer demographics, buying trends, and product preferences. Meanwhile, customers from any geographic location can easily select and purchase products online with a few simple clicks of a mouse.

In an attempt to ride the e-commerce wave, many traditional businesses have created an online presence. For example, The Home Depot began strictly as a **bricks-and-mortar business**—a business with an actual physical location, or storefront. In 2000 the company launched HomeDepot.com to introduce flexibility and convenience to its customers. Now it's considered a **multichannel retailer**—a retailer that sells its products via traditional channels (catalog, bricks-and-mortar, and telephone) as well as via an online channel.

Why are companies in such a rush to get their products online? W.W. Grainger, a leading supplier of facilities-maintenance products, reported that when the company's Web site went live, it produced $10.1 million in revenue, an increase of 339 percent over the previous year's first quarter. With these results, it's no wonder companies are scrambling to stake a claim in the cyber market. While some companies generate sales from both real-world and online storefronts, others operate strictly online. A company such as NetGrocer.com doesn't have a physical storefront, but it competes with local grocery stores. **Pure-play retailers** sell primarily through the Internet.

E-commerce is the fastest-growing form of commerce in the world, prompting major changes in markets, industries, individual businesses, and society. For up-to-date numbers on the e-commerce sector, check out the U.S. Census Bureau findings.

> **QUESTION**
>
> How do you define e-commerce?

Purpose of E-Commerce

While the purpose of e-commerce may vary from business to business, the basic goal is to reach and transact business with customers using electronic means. Individual businesses may define purposes unique to their strategies. For example, a bricks-and-mortar business may define the objective of its e-commerce plan as a means to supplement revenues and improve product recognition. Other companies launch Web sites to increase the efficiency of their internal business activities in addition to reaching new customers. The following items represent business activities that might be improved using electronic means:

- transmitting orders
- transaction processing
- payment processing
- communication with customers on order status

No matter what the defined goal, e-commerce is bringing new opportunities, challenges, and adventures to today's business world.

Section 1-3 Review

RESPOND to what you've read.

1. How can the introductory and permeation stages of the computer revolution be differentiated?

   _____
   _____
   _____

2. Describe how e-commerce has changed business. _____

   _____
   _____
   _____

3. What jobs has e-commerce created? _____

   _____
   _____
   _____

Quick TALK With your partner, list the reasons why a bricks-and-mortar business in your city may not presently participate in e-commerce. *Plan a logical argument to convince the business owner to become a multichannel retailer.*

Name _____ Date _____

Worksheet 1-1

Using Search Engines

Search engines are the telephone books for the twenty-first century. In order to search for products, services and information online, you'll need to be familiar with a variety of search tools. Using the Internet, research six search engines: Google, Yahoo!, MSN, Netscape, and two of your choice.

1. Create a chart, using a word-processing or spreadsheet program, to visually explain the following:

    a. Similarities and differences between the six search engines.

    b. The common search engine commands and how they work in each, such as +, -, " ", or wildcards.

    c. Ease or difficulty of use.

2. Which search engine do you like the best? Which do you like the least? Explain why.

    _____
    _____
    _____
    _____
    _____

3. Different search engines yield different results. Search for the terms "flower delivery" and "airline tickets." What are the top two results in each engine? Add your findings to the chart you created.

    _____
    _____

4. Imagine you're an Internet researcher for a new online company and your boss needs to know how you conducted your Internet search on a particular topic. Select a topic you want to research. Document your methods of research, and then prepare a one-page summary of your techniques to find information from credible sources.

    _____
    _____
    _____
    _____
    _____
    _____
    _____

Name _____ Date _____

Worksheet 1-2

Learning About E-Commerce

The ability to conduct business and communication transactions electronically has revolutionized the world. It's important to understand how e-commerce evolved, how it works today, and what it will be like in the future.

1. Create a timeline, including text and graphics, of electronic communication's evolution, charting dates, historical events, and important people.

2. Identify a bricks-and-mortar business in your community that has a Web presence. (Examples might include The Home Depot, Borders, or Walgreens.) Using the Internet, research the company's Web site. List three features that make the site user friendly and areas you think could be improved upon. How does the Web site compare to the bricks-and-mortar store? Where would you rather shop? Explain why.

   _____
   _____
   _____
   _____
   _____
   _____
   _____
   _____
   _____
   _____
   _____
   _____

3. Predict what e-commerce will be like in ten years. Working with a team, create a visual presentation of your predictions. Use computer graphics, make a collage with pictures from a magazine, paint, or draw freehand. Share your illustration with the class.

   _____
   _____
   _____
   _____
   _____
   _____
   _____
   _____
   _____
   _____

Chapter 1 Review and Activities

Chapter Summary

Section 1-1

- The Internet began with the Department of Defense and four major universities connecting computers to allow scientists, the military, and academics to share information and communicate with one another.

- Some of the key components of the Internet include the World Wide Web, e-mail, and FTP.

- You can use the Internet to view, download, surf, or access documents on the Web.

Section 1-2

- There are many options to choose from when connecting to the Internet. You'll need to use an Internet service provider (ISP) such as AT&T WorldNet, Netcom, AOL, or EarthLink. You can then connect using dial-up or broadband services such as DSL, cable modem, ISDN, or satellite.

- When choosing an ISP, customers should look for one that offers the type of connection they desire and provides a fast and reliable connection, good customer service, and reasonable prices.

- Browser search tools have become the telephone book for the Internet generation. People use search engines to find a favorite business name, address, or phone number.

- No single Web site will enable you to find all the information that exists online. Internet directories and crawler-based search engines represent two ways in which users can access the information they need.

Section 1-3

- The Internet has transformed communication around the world, bringing goods, services, and information to users' fingertips in new and immediate ways.

- E-commerce transcends geographic boundaries, attempts to equalize competition, and advertises and markets to a global marketplace.

- The basic goal of e-commerce is to reach and transact business with customers via electronic means.

- E-commerce can increase a business' revenues, improve product recognition, and increase the efficiency of its internal business activities.

Chapter 1 Review and Activities

Review of Key Terms

a. Internet
b. transmission control protocol/Internet protocol
c. World Wide Web
d. Web browser
e. hypertext transfer protocol
f. hypertext markup language
g. domain name
h. uniform resource locator
i. Internet service provider
j. Web host providers
k. wireless Internet
l. wireless Internet service providers
m. crawler-based search engines
n. electronic commerce
o. bricks-and-mortar business
p. multichannel retailer
q. pure-play retailers

Match each term to its definition.

_____ 1. Indicates the address of a Web site.
_____ 2. Companies that offer wireless service to the public.
_____ 3. A global network of computers, communication tools, and information resources.
_____ 4. Search engines that usually have the most current listing of Web pages on the Internet.
_____ 5. Achieved by adding short-range radios to stationary computers, laptops, or PDAs.
_____ 6. Uses tags to structure text into headings, paragraphs, lists, and links.
_____ 7. A business with a physical location.
_____ 8. A set of rules for transferring files on the Web.
_____ 9. A company that provides others with access to the Internet.
_____ 10. Allows customers to house their Web site documents on the company's servers for a fee.
_____ 11. Conducting business and communication transactions electronically.
_____ 12. The global collection of graphical and hypertext pages found on the Internet.
_____ 13. Retailers who sell primarily through the Internet.
_____ 14. A program used to download and surf documents on the Web.
_____ 15. A translator for the Internet.
_____ 16. An addressing scheme.
_____ 17. A retailer selling through traditional and online channels.

Applying Technology to Academics

English Language Arts—Reading
Access and review SearchEngineWatch. Read at least two search-engine reviews. Study the Web searching tips. What information is provided that you can use throughout this course?

English Language Arts—Writing
Use the Internet to research three brands of personal digital assistants (PDAs). Write a paragraph about each, describing its features, accessories, and cost. Then, choose the one you think would best serve your needs as a student and explain why.

English Language Arts—Speaking
In a group of three or four, review the Web site of a familiar business. Which products are sold online? How is the merchandise delivered to the customer? Which payment options does the company offer its customers? Evaluate the site based on what you like and what you think could be improved. Present your findings orally to the class.

Graphing Software Program
Research the cost of dial-up, DSL, and cable-modem access in your community. Which is most expensive? Create a cost comparison, using a word-processing or spreadsheet program.

Chapter 1 Review and Activities

Critical Thinking

1. No one authority controls the Internet. As the use of the Internet continues to grow, should some organization, agency, or person take control? Why or why not?
2. Explain how the electronic marketplace has revolutionized the way in which business is conducted nationally and around the world.

COMPETITIVE EVENT

Declining sales and increasing competition concern your boss, the owner of a local sporting-goods store. She has asked you to help her think of ways to solve these issues. You research the competition and realize it has a strong Web presence. A few days from now, you're scheduled to meet with her and discuss implementing an online function in the store's business. Prepare a brief presentation explaining the basic concepts of e-commerce to her.

EVALUATION

She will evaluate your idea based on how well you meet the following criteria:
1. Explain the nature of e-commerce.
2. Discuss trends in e-commerce.
3. Discuss issues in e-commerce.
4. Explain the economic impact of e-commerce.
5. Make an oral presentation.

INTERNET ACTIVITY

To engage in online activities, visit the *E-Commerce* Web site at **ecommerce.glencoe.com**.

BusinessWeek online

TOPIC: Future Trends in Satellite Broadband

The Internet has evolved to where computer users can get connected without ever being connected. Satellite broadband service is among the latest developments in wireless high-speed Internet options, allowing any user with a clear view of the sky to get online.

ACTIVITY Go to the Student Center at *ecommerce.glencoe.com*. Click on *BusinessWeek* Activities and open Chapter 1. There you'll learn more about satellite broadband and its potential as an e-commerce tool.

Chapter 2
The Nature of E-Commerce

Section 2-1
Characterizing E-Commerce in Business

Section 2-2
Conducting Business on the Web

PREREADING STRATEGIES

Before you read this chapter, finish these statements in your journal:

- Based on this chapter title, I predict…
- The most important thing to remember about the purpose of e-commerce is…
- Some questions I have about how businesses operate in the world of e-commerce are:

BusinessWeek online

Go to *BusinessWeek* online. Find an article about e-commerce. Write down the words in the article you don't understand, look up their meanings, and write definitions for them.

Limitless Possibilities

While business transactions have existed since the dawn of civilization, e-commerce remains a relatively unexplored territory of limitless possibilities. Navigating your way through the world of e-commerce is fraught with success and failure.

Drill Down

▶ **1492 The World Goes Round**
German mapmaker Martin Behaim creates the first globe, going against the then-popular belief that the earth was flat.

▶ **1714 Mercury Rising**
Gabriel Fahrenheit develops the mercury thermometer, which impacts virtually every scientific field of study.

▶ **1755 Look It Up**
On April 15, Samuel Johnson publishes the first English language dictionary after nine years of research and writing.

▶ **1804 Locomotion**
English mining engineer Richard Trevithick invents the steam-powered locomotion engine, which launches the Railroad Era.

▶ **1939 The Computer Age**
Konrad Zuse develops the first freely programmable electromechanical computer.

▶ **1976 The E-mail Era**
Queen Elizabeth II becomes the first head of state to send an e-mail message.

Get the Big Picture

Using Internet maps, such as MapQuest, Yahoo! Maps, and MSN Maps and Directories, try to locate someone using your home address as the starting point. Do you get the same route with the three different map versions?

POWER READ

Be an active reader and use these reading strategies:

PREDICT what the section will be about.
CONNECT what you read with your own life.
QUESTION as you read to make sure you understand the content.
RESPOND to what you've read.

27

Section 2-1

Characterizing E-Commerce in Business

AS YOU READ...

YOU WILL LEARN
- how e-commerce has changed the way people do business.
- advantages of e-commerce over traditional ways of doing business.
- business problems unique to e-commerce.

WHY IT'S IMPORTANT

E-commerce has impacted the economy in countless ways. Successful e-commerce entrepreneurs are aware of the specific advantages and disadvantages of conducting business online and plan accordingly.

KEY TERMS
- e-business
- brand loyalty
- mass customization
- value chain
- elastic demand

PREDICT

Why might a bricks-and-mortar business want to add an e-commerce component?

E-Commerce Environment

Business in the old economy focused almost exclusively on manufacturing physical goods. Today, in the new, or Electronic, economy, knowledge production is the primary business driver. The microchip—that small silicon semiconductor that processes electric functions—has revolutionized the way business is conducted, from online bill payments to cyber auctions to Internet dating.

According to the United States Census Bureau, the Electronic economy has vast economic implications. In order to keep up, today's companies must function within a rapidly changing global environment of interest rates, laws, regulations, business protocol, social concerns, and consumer preferences.

A company's *supporting infrastructure* lays out the foundation for its e-commerce activities. Infrastructure is the basic organizational foundation for structure of a system, and the infrastructure supporting e-commerce includes the physical tools needed to get the job done, such as computers, satellites, wire, and cable, as well as software, support services, and people.

E-Commerce versus E-Business

Is e-commerce the same thing as e-business? No. Although people may use the words interchangeably, they are not using them correctly. In Chapter 1 you learned that e-commerce is business conducted via electronic methods. *E-commerce also refers to electronic transactions involving the transferring of ownership of goods or services.* The buyer and seller have to agree on an exchange, whether monetary or non-monetary. For example, you can download the basic version of Adobe Acrobat Reader for free, and it is considered an e-commerce transaction even though no money crosses the wire.

THE DEFINITION OF E-BUSINESS Electronic business, or **e-business**, is any process a business conducts over a computer network. Government, not-for-profit organizations, and corporations all conduct e-business. As the Census Bureau points out, e-business comprises electronic processes focused on production, customers, management, or internal procedures. Given the distinctions between e-commerce and e-business, take the quiz in **Figure 2.1**, and match each business process to its corresponding term.

Figure 2.1

What's the Difference Between E-Commerce and E-Business?

MATCHING UP E-commerce and e-business are such new concepts that the Census Bureau had to define them to measure their effects on the economy. The Census Bureau offered the following examples to explain what constitutes e-business and e-commerce. *Can you match the activities below to the correct concept?*

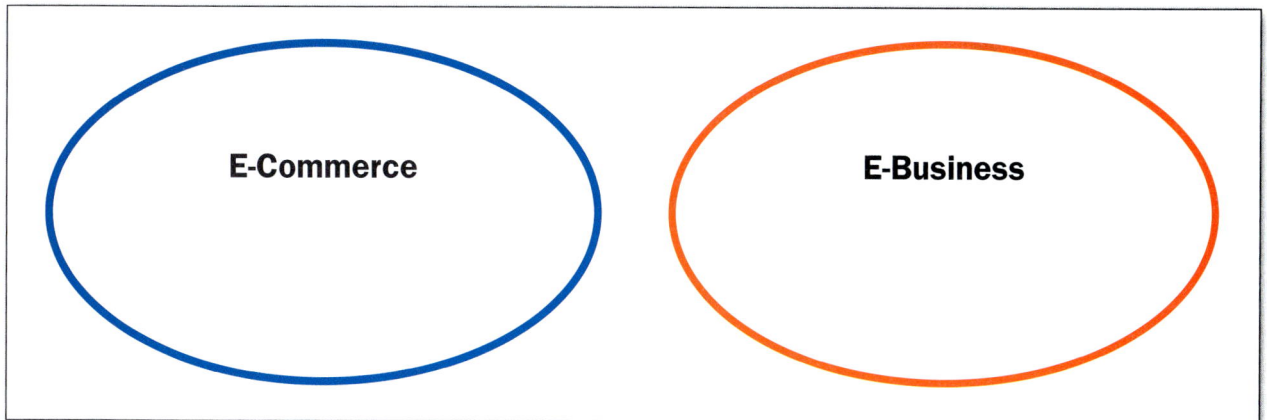

- **A.** A business buys office supplies online or through an electronic auction.
- **B.** Production-focused processes include procurement, ordering, automated stock replenishment, payment processing, and other electronic links with suppliers, as well as production control and processes more directly related to the production process.
- **C.** An individual withdraws funds from an automated teller machine (ATM).
- **D.** Customer-focused processes include marketing, electronic selling, processing of customer orders and payments, and customer management and support.
- **E.** Internal or management-focused processes include automated employee services, training, information sharing, video conferencing, and recruiting.
- **F.** An individual purchases a book over the Internet.
- **G.** A government employee reserves a hotel room over the Internet.

Source: U.S. Census Bureau

Benefits and Characteristics of E-Commerce

The evolution of Web technologies continues to modify culture and business. *BusinessWeek* reported that what e-commerce has done for books, music, and travel continues with newer online industries like jewelry sales, online bill payment, real estate, and software. E-commerce has become a virtual open marketplace for the sleepless. Limitless buyers at all hours of the day look for just the right good or service to satisfy a particular want or need.

Figure 2.2

Why People Buy Online

WHY BUY? Cyberstores must offer incentives to keep shoppers motivated to buy online. The following survey charts the various reasons why customers make purchases online. *After reviewing the table, summarize the major reasons.*

REASONS CITED	PERCENTAGE OF SURVEYED SHOPPERS
Save time by not going to store	70%
Can shop when stores are closed	69%
Avoid the holiday crowds	68%
Might be able to find better prices	59%
Can find products online more easily	52%
Find products not available in stores	50%
Easier to compare prices	47%
Have gifts sent directly to recipient	36%
Can avoid wrapping gifts	13%
Can earn loyalty points	13%
Purchase from wish list	10%

Source: Jupiter Research/IPSOS

Credit: ecommerce-guide.com®, "Giving Good Customer Service During the 2003 Holiday Season." November 13, 2003.

LIMITLESS BUYERS Fifteen years ago, the number of customers a small, Miami-based custom-jewelry manufacturer could attract was probably limited by its geographic location. In the era of e-commerce, however, the number of potential buyers is limitless. Customers in Munich, New York City, and London can now purchase holiday gifts for their friends from the jeweler's Web site. The company's Web presence provides it with virtual real estate that can potentially rival that of its largest competitor, equalizing the playing field and opening new markets.

OPEN 24/7 When those same customers in Munich, New York City, and London decide it's time to purchase their holiday gifts, they don't have to worry about when the store opens its doors for business. There's no need to call ahead to find out the store's hours or the types of jewelry the company sells. The electronic marketplace is open 24/7, and in this respect, it has made the retail marketplace more customer-focused. **Figure 2.2** charts the reasons people choose to shop online, and it shows what the leading motivator is: convenience. While all-hours e-commerce offers a distinct advantage to customers, it can pose various challenges to businesses, which must now be able to operate effectively at any time.

ADVERTISE, MARKET, AND ANALYZE E-commerce offers businesses a multitude of effective and inexpensive ways to reach new and existing customers. E-mail advertisements and coupons may help to increase sales and generate interest in new products at relatively low costs. Online surveys and chat rooms can help a business owner to understand customer demographics, buying trends, and product preferences. With these tools and data, companies can customize their products and prices to target specific audiences.

Equally important, the technologies that support Web sites enable businesses to track customers' purchasing habits. A business can determine, for example, whether a low price always leads to a sale or whether brand loyalty is affected by increased selection. **Brand loyalty** refers to a customer's preference for a particular product. If a customer will not accept a substitute, even when there are many other comparable products available, he or she is demonstrating brand loyalty. A classic example is the soda rivalry: Are you a Coca-Cola or Pepsi drinker?

In addition, Web-site technologies, such as HyperTracker, allow a business to track statistics like purchase dates and times, popular price ranges, best-selling products, and failed transactions. This knowledge helps business owners to shape their Web presence, product design, and inventory to meet the demands of their customers.

MANAGING INVENTORY The management of inventory can be tricky. Suppose you're the manager of a souvenir store located at Yankee Stadium. How many New York Yankees t-shirts should you order? How many ball caps should you have on hand for the World Series? Where will you store 20,000 t-shirts? How much will you have to pay to store these items until they are sold?

The online retailer may be able to bypass many of these difficult decisions. As customers purchase t-shirts or caps online from a Yankees souvenir Web site, shipment may be made directly from the manufacturer, eliminating the cost of inventory storage.

As the manager of the store located in Yankee Stadium, will any of your merchandise disappear due to shoplifting? If so, how many products and sales will be lost? Businesses spend thousands of dollars installing surveillance cameras, using holographic tags, and hiring security guards

CONNECT
Name a specific instance in which you or a family member demonstrated brand loyalty.

Web Site Success

THE BATTLE ON THE BEAT
News 24/7

The morning's newspapers recap yesterday's news, while new-media journalism provides almost instantaneous accounts. MSNBC and CNN call themselves 24/7 operations—they offer up-to-date news 24 hours a day, 7 days a week. Timeliness and teamwork enable news stories to hit readers fast. A writer and a producer cull through wire information and video feeds. The associate producer and the multimedia designer create the audio and visual elements needed to tell the story. By the time the story goes live, several people in the newsroom will have contributed to some part of it.

Thinking Critically
How does competition influence the world of online journalism?

to protect their merchandise. The virtual store, where products are displayed via photographs, bypasses this problem.

MEETING CUSTOMER NEEDS In the old economy, goods such as automobiles and textiles relied on mass-production techniques. The made-to-stock business model ruled the economy, meaning that companies held large inventories of mass-produced products. With e-commerce, however, mass customization is the ticket.

What does mass customization mean? Break it down: Mass means "mainstream," or "majority," and customize means to cater to individuals' specialized choices. Thus, **mass customization** is the production of goods that offer specialized choices to mainstream buyers. For example, imagine you stop by Jamba Juice for your favorite smoothie—the Orange Dream. You like all the ingredients, but you want to add a splash of strawberry juice. Your order enters the system, and the person in the back makes your personalized, "made-to-order" drink. Typically, these beverages aren't premade and available from a machine, like soda or coffee. Instead, the clerk makes the drink just for you. This same business model works like a charm on the Web. Sites like BlueNile.com and Mondera.com allow customers to customize jewelry purchases by selecting diamonds, settings, and prices that meet their specific needs.

From a business perspective, mass customization influences all aspects of the value chain. A **value chain** is the sequence of design, production, and marketing efforts a business conducts to deliver its products at the right price and time. (Section 2-2 covers value chain in more detail.) Because the electronic marketplace demands speed and is highly competitive, the production and delivery cycle is more demanding than ever. When a company streamlines its value chain, productivity and profitability generally increase. In a sense, mass customization is the business model that never sleeps.

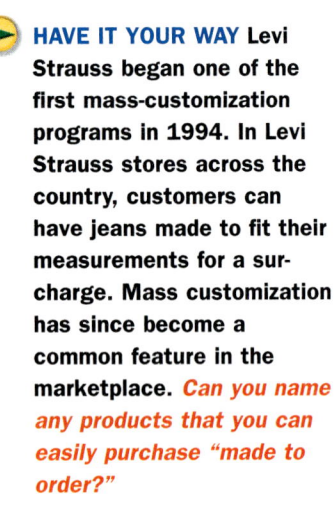

HAVE IT YOUR WAY Levi Strauss began one of the first mass-customization programs in 1994. In Levi Strauss stores across the country, customers can have jeans made to fit their measurements for a surcharge. Mass customization has since become a common feature in the marketplace. *Can you name any products that you can easily purchase "made to order?"*

STARTING OUT Although it might take some time and effort to purchase a domain name, you can register and start building an e-commerce site today. In addition, you can select a business model, take photographs of your products, and establish and ship goods. The fact is, a virtual storefront is much faster, easier, and cheaper to set up than a bricks-and-mortar store. Online businesses can also be grown quickly, whether through alliances with other e-businesses or by offering additional goods and services. Instead of expanding retail space to showcase new merchandise, an e-commerce company can simply add more Web pages to its online catalog.

PRICING COMPETITIVELY Price is often the factor that makes or breaks a sale. The online environment makes it easier than ever to comparison shop. Consumers can use search tools to locate the same item being sold at different prices. Epinions.com, ConsumerReports.org, and ZDNet.com offer tools to help consumers locate the products they want at the best prices. This environment fosters **elastic demand**, in which pricing changes create a change in the amount of goods or services consumers are willing to buy at a certain price.

Just as consumers benefit from this highly competitive marketplace, businesses routinely profit by using online intelligence to price their products to meet consumer expectations. For example, if you planned to launch a new brand of inline skates, you could research competing skates, prices, and features online in order to price and market your product appropriately to your target audience. You might monitor a competitor's Web site for product releases or marketing initiatives to ensure timeliness and fair pricing.

> **QUESTION**
> How can e-commerce weaken or add to a local economy?

Section 2-1 Review

RESPOND to what you've read.

1. What does it mean for a company to "manage" its inventory? How is an online company's inventory easier to manage than that of a bricks-and-mortar business? _____
_____
_____

2. How can a company streamline its value chain? _____
_____
_____

3. Why is it difficult to charge a higher price for an item than your competitors in the e-commerce marketplace? _____
_____

> **Quick TALK** Debate the pros and cons of mass customization and made-to-order items with your partner. *Decide which perspective you'll take, then prepare a brief argument for your side prior to the discussion.*

Chapter 2 The Nature of E-Commerce

Section 2-2

Conducting Business on the Web

AS YOU READ...

YOU WILL LEARN
- to identify e-commerce models and how they operate.
- how one large retail business has streamlined its value chain.
- some of the ways in which online businesses strive to capture sales.

WHY IT'S IMPORTANT

There are a variety of business models available for online businesses to follow. You can maximize your chances for e-commerce success by familiarizing yourself with the models and using them to your advantage.

KEY TERMS
- business model
- business-to-business
- business-to-consumer
- consumer-to-consumer
- consumer-to-business
- business-to-government
- government-to-consumer

PREDICT

How might an e-commerce venture collaborate with another business?

THE PROCESS OF BUYING AND SELLING ONLINE

You're shopping for a new laptop. After researching several different models on the manufacturers' Web sites, you find a perfect, lightweight laptop with lightning-quick processing speed. Instead of placing the order online, you decide to call the company's toll-free number and place the order. Do you consider this an e-commerce transaction? Since no goods or services transferred ownership over the computer network, this isn't an example of an e-commerce transaction. Although this situation appears to contain all the components you'd need for an e-commerce deal—the hardware, the buyer, and the seller—it isn't e-commerce; the process entails more. The rest of this chapter explores the different types of business models common in e-commerce.

E-Commerce Business Models

Imagine the transactions that are occurring online everyday. Your friend buys a sweater from AmericanEagle.com. You file your income-tax return and submit your tax payment electronically to the IRS. A computer manufacturer buys electronic components from its supplier in China. Your friend places a bid for a collectible baseball card on eBay. If there's a way to transact business online, you can bet someone is doing it right now.

B2B. B2C. B2what? These acronyms represent various ways in which companies conduct online transactions. Each acronym stands for a type of business model in which two parties transact some form of business. A **business model** is a system of policies, operations, resources, and technologies used to generate revenue. The following are the most commonly used business models on the Web:

- business-to-business (B2B)
- business-to-consumer (B2C)
- consumer-to-consumer (C2C)
- consumer-to-business (C2B)
- business-to-government (B2G)
- government-to-consumer (G2C)
- intermediary hub

If you plan to launch a business on the Web, you'll want to pay close attention and select the best model for your e-commerce venture.

BUSINESS-TO-BUSINESS The **business-to-business** (B2B) model applies when a business transacts information, goods, or services with another business. The oldest and fastest-growing form of e-commerce, B2B allows e-business to collaborate, partner, and share research and complex data with other businesses. For example, a furniture manufacturer in New Jersey may need a payroll service to help process its weekly payroll, a janitorial service to clean its facilities, and various suppliers to provide raw materials for its finished goods. This manufacturer might contract Paychex as its payroll service, National Maintenance Services as its janitorial service, and a variety of pine and oak suppliers in this B2B model.

In B2B circles, Wal-Mart is widely known for its highly effective integrated value chain of internal and external processes that deliver products to customers. You will recall from Section 2-1, a value chain is a system of integrated processes, as illustrated in **Figure 2.3**. Consider the variety of activities that Wal-Mart must complete to make a sale. The company must find suppliers for products, negotiate prices for its purchases, receive and warehouse products, distribute goods to stores, price products, stock its store shelves, handle returns, and keep inventory stocked. The purchasing, invoicing, accounting, inventory, shipping, and transportation systems at Wal-Mart have been coordinated with the systems at its stores and warehouses, manufacturers, and suppliers to increase efficiency, streamline processes, and reduce costs for all parties involved. By developing and refining this model, Wal-Mart has become the top retail business in the world.

To understand how this model works, assume that you buy a baseball glove at Wal-Mart. The cash register reads the bar code on the price tag. This action immediately notifies its central warehouse that the Wal-Mart retail store needs a new baseball glove to replace the one you just purchased. At the same time, the Wal-Mart central warehouse notifies the manufacturer and the raw suppliers that the store needs a new baseball glove. So, in essence, the transaction at the Wal-Mart cash register notifies everyone in the supply chain about the baseball glove. This demonstrates the power of using electronic data exchange to run a

Figure 2.3

Links in the Energy Chain

STEP BY STEP Each step in the value chain for electricity represents a different company or a different business unit of a company that plays a role in providing energy for business and residential customers. At each step—production, trading, transmission, and system operation—these companies or units establish links with each other to deliver their product efficiently. *As a consumer, what does each link in the value chain of a product or service represent to you?*

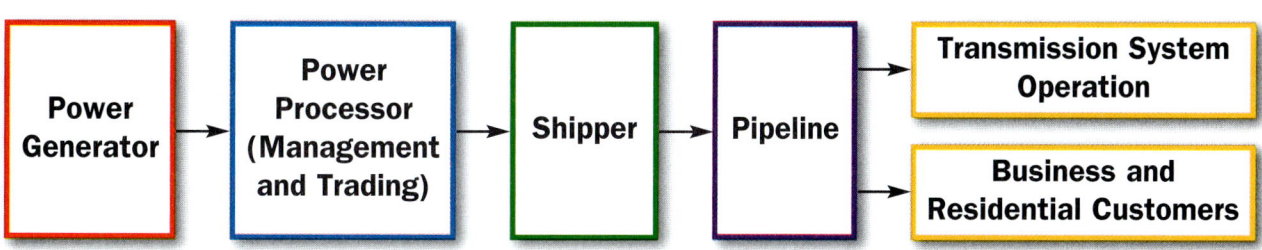

CONNECT

Name some examples of ordinary B2C and C2C purchases you make in your daily life.

successful business. Using the value-chain model, Wal-Mart is able to replenish items on its store shelves in less than three days. This highly effective, integrated system allows for quick turnaround, lower inventory costs, and discounted prices.

BUSINESS-TO-CONSUMER If you have ever used the Web to purchase a concert ticket or a fleece jacket, you've engaged in an e-commerce transaction. A **business-to-consumer** (B2C) model applies to any business or organization that uses the Internet to sell its products or services to consumers. Amazon.com (shown in **Figure 2.4**) and Dell.com are two successful B2C companies. B2C also includes businesses that provide consumer services, such as online banking (NetBank.com), travel services (Expedia.com), health information (WebMD.com), and real estate (REMAX.com). A study by Forrester Research predicts that B2C e-commerce in the United States will grow from $38.8 billion to $184.5 billion in just four years.

Turning browsers into buyers, building customer loyalty, and fulfilling customer orders in a timely and satisfactory way are all concerns for B2C sites. Studies have shown that the longer a customer stays on a Web site, the greater the chance that he or she will buy. Successful online businesses keep customers busy once they arrive at their sites by offering a range of products, services, and information. An e-commerce site needs to offer easy navigation and a secure way to buy products, as well as a reason for customers to come back to make future purchases. Did the order arrive quickly? Was the right product delivered? Customer satisfaction and business loyalty play key roles in a company's success.

CONSUMER-TO-CONSUMER Looking for some used camping equipment? Have you picked up an appliance or a pair of headphones at a garage sale? You can transact these items online as well. The **consumer-to-consumer** (C2C) model allows consumers to interact with one another online to transact goods and services. Online auctions

Figure 2.4

A Successful B2C Model

FROM AMAZON TO YOU
Amazon.com is one of the most successful online B2C retailers. *Which features can you identify on the home page of Amazon.com that might draw customers into the site?*

© 2004 Amazon.com, Inc. All rights reserved.

are the most common forum for consumer-to-consumer e-commerce. The relationship comprises the auctioneer and the bidder, and the auctioneer determines the product's starting price. Consumers analyze the product, decide how much they value it, and then make their bids. The consumer who places the highest bid wins the purchase.

The most successful online auction site is eBay. According to a recent *BusinessWeek* article, 30 million people buy and sell more than $20 billion in merchandise a year on eBay—more than the gross domestic product of all but 70 countries in the world. How does a Web site such as eBay make money when it simply acts as a portal for consumers to transact with one another? Auction-style models make money from licensing, transaction, commission, and advertising fees. Since there is no inventory, eBay avoids warehousing costs and pays only for its office space, software applications, and the salaries of its employees.

CONSUMER-TO-BUSINESS In the **consumer-to-business** (C2B) model, the customer initiates the transaction, posting an intent or desire to buy a certain product at a certain price. It is a reverse-auction scenario: The business responds to the consumer's requested product and its price. Priceline.com provides such a service for air travel, hotels, and car rentals. For example, you would like to travel from New York to Los Angeles on a particular October day for no more than $250. Priceline will perform a quick electronic query with all its participating airlines to see if one will agree to your price and book the request. If an airline agrees, Priceline charges your credit card, completes the transaction, and displays your travel times and flights. In this business model, Priceline makes its money from transaction fees and advertising.

OTHER E-COMMERCE MODELS A variety of other models for conducting business transactions on the Web are evolving quickly as different needs arise. One useful model is the **business-to-government** (B2G) model, connecting the private sector to the government marketplace. Businesses wishing to bid on government contracts are able to expedite the process by obtaining information and processing forms with governmental agencies.

Another way in which business is transacted on the Web is via the **government-to-consumer** (G2C) model, which allows consumers to easily access relevant information from government agencies. Instead of standing in line at the post office to make sure your annual income-tax payment is delivered by April 15, you can pay your tax bill online at IRS.gov. **Figure 2.5** on page 38 shows how the New Zealand Treasury department allows households and businesses to submit payments online.

Some businesses don't buy or sell products or purchase services themselves. Instead they act as an intermediary, bringing buyers and sellers together. This business model, known as the *intermediary hub*, can operate vertically or horizontally. *Vertical hubs* match buyers and sellers within a particular industry. *Horizontal hubs* focus on providing the same types of products or services across various industries. A vertical hub such as Covisint, which was formed by DaimlerChrysler, Ford, General Motors, and Renault-Nissan, works with manufacturers, suppliers, and industry trade groups to help automakers find suppliers, negotiate purchase prices, reduce costs, and increase efficiency.

CONNECTION: LANGUAGE ARTS

A Matter of Planning
Trying to set up a site's online transaction capabilities with cheap software might not be the greatest idea. Small businesses are often less apt to invest in the software needed for these transactions because of the up-front investment costs required. *If your site was planning to offer online transactions, what issues should you consider before launching into e-commerce?*

QUESTION
How does an online-auction site generate income?

Figure 2.5

G2C in New Zealand

EXAMPLES OF NEW ZEALAND'S E-GOVERNMENT TRANSACTIONS If you're a resident of New Zealand and need to pay a tax bill or customs duties, your transactions can be conducted via electronic means. *Which advantages and disadvantages can you identify for the citizen?*

Inland Revenue	Collects standard tax payments from households, businesses, and other entities. The level and time at which payments are made is set by statute. Inland Revenue has a high level of control over the payment method. Security of information is important.
Customs	Collects customs duty from importers. The statute determines the size of payments. Payments depend on often irregular shipping patterns. Information must be collected and audited. Customers may be based overseas. Security of information is important.
Courts	Collects fees for court use, as well as fines and penalties imposed by courts. Security of information is important. Matches transaction information with the fees and fines imposed. Ability to pay easily via a range of methods is useful to maximizing collection.
Fisheries	Levies for the cost of fisheries management and collects penalties imposed for overfishing. Quota payments are regular and of uniform size; transaction matching is important. Penalties are likely to be irregular and variable in size.
Police	Collects traffic infringement fees and revenue from the sale of unclaimed property.

Source: New Zealand Treasury

Section 2-2 Review

RESPOND to what you've read.

1. Why do e-commerce businesses strive to keep potential customers on their sites as long as possible?

2. What are some key challenges facing the B2C model?

3. How might the G2C model be helpful to the average consumer?

Quick TALK With a partner, think of ways to keep visitors busy on your gourmet food site. *Which strategy would be the most effective and why?*

Name _____ Date _____

Worksheet 2-1

Which Brand Is for You?

Customers exhibit loyalty to their favorite brands. Whether you drink Coca-Cola or Pepsi or use Google or Yahoo!, the brand often makes a big difference. Are you or your friends loyal to any particular brands?

1. Conduct a survey to determine how and why teens remain loyal to certain brands. Choose two product categories (e.g., soft drinks, radio stations, computers, jeans, or hamburgers) for the survey. Interview at least 15 people. Ask the following questions:

    a. Which brand do you purchase?

    b. How often do you purchase it?

    c. When you are ready to purchase again, will you choose the same brand?

    d. Are you loyal to this brand?

2. Chart your results using a word-processing or spreadsheet program. Combine your results with your fellow students' results. Which products generate the most brand loyalty? Which ones generate the least loyalty?

    _____
    _____
    _____

3. Are you loyal to a particular brand of product? Explain.

    _____
    _____
    _____

4. What motivates you to purchase a product? Explain.

    _____
    _____
    _____

5. Select a product marketed to teens. Come up with ways for the product to obtain brand loyalty.

    _____
    _____
    _____

Name _____ Date _____

Worksheet 2-2

Doing Business on the Internet

Each week millions of transactions take place week on the Internet. There are numerous ways to do business online. Chapter 2 outlines seven common models. If you plan to start a business on the Web, you'll need to have a good understanding of the models.

1. Select four of the seven e-commerce models. Using the Internet, find an example of each type on the Web. _____
   _____
   _____
   _____
   _____
   _____

2. Answer the following questions for each example you find. Using a word-processing or spreadsheet program, create a chart to present your findings.

   - What product or service is provided?
   - Explain whether or not the site is easy to navigate.
   - Is there a secure way to transact business? If so, how?
   - Describe the features designed to hold your interest as a visitor to the site.

3. Imagine a business you would like to launch online. Which model would you use for your business? Explain why. _____
   _____
   _____
   _____
   _____
   _____

4. Describe what your Web site might look like for your business. Which features would you include to attract customers to your site? _____
   _____
   _____
   _____
   _____
   _____

Chapter 2 Review and Activities

Chapter Summary

Section 2-1

- E-commerce's ability to transcend geographical boundaries, equalize competition, and market inexpensively to the global marketplace has changed the way people do business.

- Advantages of e-commerce include an unlimited number of potential buyers, 24/7 availability, a multitude of effective and inexpensive ways to advertise, the ability to track and tally customer purchasing habits, a decrease in inventory issues, and the ability to meet consumer needs almost instantly. The economic impact of e-commerce may vary from one community to another.

- Operating effectively in a 24/7 environment and the accelerated speed required to produce and deliver products to consumers are problems common to e-commerce business owners.

Section 2-2

- There are seven commonly used business models on the Web. If you're going to do business online, it's important to choose the best model for your e-commerce vision.

- Wal-Mart has streamlined its value chain by coordinating its purchasing, invoicing, accounting, inventory, shipping, and transportation systems with the systems at its stores, warehouses, manufacturers, and raw material suppliers.

- Successful online businesses turn browsers into buyers, build customer loyalty, and provide excellent customer service in order to increase sales. Ease of navigation through the site and a secure way to purchase products also help to capture additional business.

Chapter 2 Review and Activities

Review of Key Terms

a. business-to-government
b. consumer-to-business
c. elastic demand
d. business model
e. business-to-business
f. brand loyalty
g. government-to-consumer
h. value chain
i. consumer-to-consumer
j. e-business
k. mass customization
l. business-to-consumer

Match each term to its definition.

_____ 1. A system of policies, operations, and technologies designed to generate revenues.
_____ 2. When a customer will not accept a substitute product, even when there are many similar products from which to choose.
_____ 3. Any process a business conducts over a computer network.
_____ 4. An e-commerce model in which a business uses the Internet to sell products to consumers.
_____ 5. eBay is an example of this type of e-commerce business model.
_____ 6. An e-commerce model in which the customer initiates the transaction.
_____ 7. The production of goods that offer specialized choices to mainstream buyers.
_____ 8. The sequence of design, production, and marketing efforts designed to deliver products at the right price and time.
_____ 9. The oldest and fastest-growing form of e-commerce.
_____ 10. The e-commerce model used when you pay your taxes online.
_____ 11. When changes in price create a change in the amount of goods consumers are willing to buy at that price.
_____ 12. Connecting the private sector to the government marketplace.

Applying Technology to Academics

English Language Arts—Reading
Programmers, graphic artists, Web designers, systems analysts, and computer technicians are just a few of the jobs the Internet offers. Choose one that interests you, and locate an Internet site that provides information about the job. How important do you think reading is for a career in this field? Explain.

English Language Arts—Writing
Access a department store's Web site, and select four different items it sells. Create a list of raw materials used to manufacture those items. Select one of those raw materials and research its origins. Write a paragraph explaining your findings.

English Language Arts—Speaking
In groups of three or four, review an e-commerce site based in your community. Define the purpose of the site. What does it sell online? How does it deliver merchandise? What payment options does it offer? Evaluate the site based on your likes and dislikes. Outline the improvements it could make. Present your findings orally to the class.

Calculation
Imagine you market sporting goods and equipment on the Web. You have just received an order for 15,600 ball caps for the upcoming NASCAR race. Each cap sells for $8.95. Your cost is $3.25 per cap. How much money will you make on this sale?

Chapter 2 Review and Activities

Critical Thinking

1. Can you envision a time when business will only be conducted online? Why or why not?
2. Which e-commerce activities are most interesting to you? Ask three people this same question, and compare the answers. Which e-commerce business model is the most popular among your interviewees? Explain their reasoning.

COMPETITIVE EVENT

You're a management consultant for a consulting firm. A local bricks-and-mortar company has employed you to design an e-commerce component. The local Chamber of Commerce wants to know how e-commerce is changing the local economy, and the firm's president has asked you to prepare a presentation. In this presentation, you will need to cover the types of e-commerce transactions available. Create your presentation in outline form, and run it past the firm's president (i.e., your teacher, in this case).

EVALUATION

You will be evaluated on how well you meet the following performance indicators:

1. Explain the nature of e-commerce.
2. Explain employment opportunities in e-commerce.
3. Describe current business trends.
4. Explain the economic impact of e-commerce.
5. Explain the nature of effective communications.

INTERNET ACTIVITY

To engage in online activities, visit the *E-Commerce* Web site at **ecommerce.glencoe.com**.

BusinessWeek online

TOPIC: Health Care: Doing Business on the Web

It should be no surprise that the Internet has become a vast store of medical information. Consumers can look up details of almost any illness or procedure to become more informed patients. In some cases, patients' personal medical records can be accessed online, enabling doctors to consult one another miles apart.

ACTIVITY Go to the Student Center at **ecommerce.glencoe.com**. Click on *BusinessWeek* Activities and open Chapter 2. There you'll learn more about how medical information is treated online and the advantages and disadvantages of accessing that information online.

Chapter 3
Retailing on the Internet

Section 3-1
Retailing—Then and Now

Section 3-2
The E-Tail Experience

PREREADING STRATEGIES

Before you read this chapter, finish these statements in your journal:

- Based on this chapter title, I predict…
- The most important thing to remember about online retailing is…
- Some questions I have about the online-retailing sector are:

BusinessWeek online

Go to *BusinessWeek* online. Read an article about an online retail business. In your journal, describe a strategy the business uses to attract customers, how you've seen the strategy applied in your own experience, and how effective you think it is.

The Mode of Modern Retailing

E-commerce companies face many of the same challenges and enjoy many of the same rewards as conventional bricks-and-mortar companies. Ultimately, the goals remain the same—to satisfy customers' wants and needs, to keep customers coming back, and to make a profit.

Drill Down

1712 Full Steam Ahead
Thomas Newcomen patents the atmospheric steam engine, which becomes a dominant method of transportation throughout the 18th century.

1859 Striking Oil
The first commercial oil well is drilled in Pennsylvania. Oil ultimately becomes a cornerstone of the world's economy.

1879 Let There Be Light
Thomas Edison invents the first incandescent lightbulb.

1899 Cleaning House
American J.S. Thurman patents the motor-powered vacuum cleaner.

1923 TV Dinners
Clarence Birdseye invents frozen food, which gains increasing popularity after the invention of television.

1991 Weaving the Web
Tim Berners-Lee develops the World Wide Web, launching the Internet era.

Get the Big Picture

Many new developments in technology quickly caught on with consumers by offering them greater convenience. Select one of the developments above, do some research on it, and, in your journal, write about how it saved people time and labor.

POWER READ

Be an active reader and use these reading strategies:

PREDICT what the section will be about.
CONNECT what you read with your own life.
QUESTION as you read to make sure you understand the content.
RESPOND to what you've read.

Section 3-1

Retailing—Then and Now

AS YOU READ...

YOU WILL LEARN
- how retail business has evolved since the 1800s.
- what makes today's retailing environment a customized, self-service experience.
- how the major retailer categories are defined.

WHY IT'S IMPORTANT

To attract customers and win their share of the market, retailers develop new strategies to outdo their competitors. Understanding how the retail business has evolved since the 1800s will help you understand how retailing and e-commerce became what they are today.

KEY TERMS
- retailers
- wholesalers
- e-tailing
- services retailers
- non-store retailers

PREDICT

Why was the development of the shopping mall in the 1970s so revolutionary?

RETAILING BEFORE E-COMMERCE

First and foremost, retailers want your business. **Retailers** are establishments that sell goods and services to the general public. The retailing process is the final step in the distribution of products. Target; Sears, Roebuck, and Co.; and Supercuts are examples of successful retailers. In contrast, **wholesalers** sell products to distributors or retailers and not usually to the end-user or consumer. A typical example of a wholesaler is Affiliated Foods Midwest, which supplies approximately 800 food retailers with a full line of grocery products.

If you've been to a Wal-Mart or Ralphs supermarket recently, there were probably several conveniences you took for granted: the number of departments, the vast selection of products, the availability of clerks, and the variety of payment methods. These are all aspects of the retail experience today. Shopping, however, has not always been so convenient.

The History of Retailing

The first retailers in the U.S. appeared in the 1700s as street vendors and small, family-owned shops. Throughout the 18th and early 19th centuries, most customers dealt directly with craftsmen, farmers, or local manufacturers when bartering or buying items such as horse carriages, food, and clothing.

The rise of the general store in the mid-1800s brought together a selection of staples in one convenient location. General stores often consisted of rural outposts that you had to travel a great distance to reach and which offered limited supplies. If you wanted to obtain special items such as watches, jewelry, or china, you might have to order them through a catalog and have to wait several weeks for delivery.

With the development of industry and the U.S. rail system in the late 19th century, manufacturers could mass-produce goods and ship them to market more easily. They required new means of merchandising their goods, and general stores soon gave way to the first department and chain stores. By the 1920s, mail-order houses like Sears, Roebuck, and Co. and Montgomery Ward began selling their goods directly to the public through retail outlets.

From the 1920s to the 1950s, "Main Street" shopping was based on the small department store, the drugstore, the five-and-dime, and the coffee shop. During the 1950s and 1960s, when large segments of the population began migrating to the suburbs, strip shopping centers or *strip malls*—rows of shops offering various products located along a busy street—were constructed and quickly became the retailers of choice.

Retail chains like McDonald's, Best Buy, and Payless ShoeSource© moved into strip malls to meet suburban demands for their products.

By the 1970s, customers wanted a more convenient shopping experience that would enable them to visit all of their favorite retail stores under one roof. The strip mall soon evolved into the shopping mall, a single location housing a variety of businesses, from restaurants to clothing stores to movie theaters. Companies such as Sears and J.C. Penney established themselves as the "anchor" tenants of these malls—tenants with a proven ability to attract a large number of shoppers.

In the 1980s a new revolution in retailing occurred. Consumers sought an array of products at good prices, and companies such as Kmart, Target, and Wal-Mart expanded their bases to fill this need. Often called *big-box retailers*, these companies built single-story structures of approximately 10,000 square feet or more, with parking lots that could accommodate hundreds of cars. Today, many of these stores have their own warehouse and distribution systems that lower their operating costs and result in lower prices.

Along with the big-box stores came the *category killers*—large stores that specialize in a particular type of product, such as toys, hardware, books, or sporting goods. Some examples of category killers are Toys "R" Us, Barnes & Noble, and Oshman's. Like big-box retailers, category killers have special distribution systems that enable them to keep costs down. They are called "category killers" because, by offering the lowest prices available, they are able to "kill" their competition.

The most modern innovation in retailing is **e-tailing**, or the buying and selling of retail goods on the Internet. E-tailing enables consumers to choose from an almost infinite variety of products and purchase them without leaving their own homes.

Before going further into the e-tail experience, however, it's important to look at the primary types of retail establishments that exist today.

Retailing Today

To compete in the marketplace, retailers must now offer a variety of means to make shopping more convenient and accessible to consumers.

> **CONNECT**
> What are some of the conveniences you take for granted when you shop at a discount or department store?

EVOLUTION OF THE MALL The strip malls of the 1950s and '60s evolved into the large indoor shopping malls of the '70s and '80s, although strip malls are still common. *Where do you do most of your shopping? What do you like or dislike about the outdoor strip mall and the indoor mall?*

Some consumers might prefer the convenience of ordering a pair of jeans online at Gap.com rather than going to a store. Other shoppers might prefer to visit Target where they can try on and feel a pair of jeans before buying them. Yet other shoppers might prefer browsing through a Land's End catalog they received in the mail and ordering a pullover sweater by phone. All these shopping methods make today's retailing environment a customized, self-service experience.

Creating a successful retail experience requires a keen understanding of what consumers want the most. As a consumer, what motivates you to buy—price, selection, personalized service, convenience? Different kinds of retail establishments offer different kinds of experiences. The major categories of retailers, summarized in **Figure 3.1**, are:

- specialty stores
- discount stores
- non-store retailers
- department stores
- services retailers

Figure 3.1
Types of Retailers

DISCOUNT STORES In recent years, discount stores have become increasingly successful. *Can you identify the reasons for their current popularity, economic or otherwise?*

TYPES	EXAMPLES	CHARACTERISTICS
Specialty Stores	- Toys "R" Us - Barnes & Noble - Ace Hardware	- limited number of merchandise categories - narrow variety, but deep assortment - high level of service - typically less than 8,000 square feet - targets a specific group of customers
Department Stores	- J.C. Penney - Macy's - Sears	- broad variety of products, with less depth in assortment - limited service - medium cost
Discount Stores	- Wal-Mart - Target - Big Lots	- mostly self-service, but many have knowledgeable sales associates - can kill a category for other retailers - ability to negotiate for good prices, terms, and delivery from manufacturers
Services Retailers	- banks - hospitals - health clubs	- non-traditional retailers - intangible performances, actions, or services - perishable, cannot be stored or resold, and has demand peaks and lows - inconsistency: no two performances are exactly alike
Non-Store Retailers	- NetGrocer.com - Spiegel catalog - Home Shopping Network	- use methods such as infomercials, broadcasting, direct-response advertising, traditional and electronic catalogs, Web sites, door-to-door solicitation, in-home demonstrations, trade-show booths, and vending-machine distribution

If you're interested in opening a retail store someday, you'll want to review these categories to understand their advantages and disadvantages.

SPECIALTY STORES *Specialty stores,* such as Toys "R" Us, Borders, Ace Hardware, and REI, are stores that specialize in specific kinds of products or product lines and offer a wide assortment within their given categories. Most specialty stores are staffed by sales clerks and managers who have a high level of expertise in their areas. These stores can demand volume discounts from their suppliers because they make large and frequent purchases in order to keep their shelves stocked.

Specialty stores have developed new techniques for consumers to test their products before buying them. Stores such as Best Buy and Circuit City allow customers to listen to CDs or try out electronic equipment. At an REI store, you can climb an indoor wall to try out rigging devices and climbing shoes. Some gourmet food stores provide free samples of food and coffee for customers. Sales strategies such as these make the retail experience more pleasant for consumers while enticing them to buy products they might otherwise not have considered.

DEPARTMENT STORES Sears, Foley's, Macy's, and J.C. Penney fall into the category of *department stores*. These stores offer a variety of products and choices within each product line and a floor plan that provides specialized departments such as men's apparel, housewares, and appliances. Operating a department store requires a large area and a considerable staff. With increased competition from discount stores, many department stores engage in the heavy use of promotions and sales to attract customers.

DISCOUNT STORES *Discount stores* may not offer specialized expertise in the products they sell, but they do offer incredibly low prices. When it comes to discount stores, Wal-Mart has set the standard. Since its founding in 1962, the company has become the world's largest retailer. National distribution and inventory-management systems keep costs down for companies like Wal-Mart and, consequently, the consumer.

While some consumers are thrilled by the low prices and broad product selection, others prefer not to shop at discount stores because of their effect on the local economy. As big-box retailers and category killers, discount stores drive many small, local stores that can't compete with the low prices out of business, putting people out of work.

SERVICES RETAILERS Retailing does not always involve a tangible product such as a bicycle or a set of stereo speakers. **Services retailers** sell services such as haircuts, medical care, or financial planning. These retailers don't depend on maintaining a large inventory of products. Instead, their businesses depend on the quality and consistency of their services. Banks, dental offices, pet groomers, and lawn-maintenance operations are all examples of services retailers. Services retailers on the Internet include e-banking, online photo services, and online stock trading. Services retailers play an important role in our economy by providing specialized skills and expertise most consumers lack and need.

Web Site Success

TRASH OR TREASURE
eBay

The world's largest swap meet, eBay, went public in September 1998 and in 2004 was transacting almost $60 million worth of goods per day. This person-to-person marketplace trades everything from vintage wedding dresses and jewelry to cars and even real estate. Now people might think twice before rummaging through their family's storage space and throwing those dusty items in the garbage bin. This Internet start-up has been successful in creating and growing a worldwide community of shoppers.

Thinking Critically
Do you think the success of eBay has increased or decreased the number of businesses?

QUESTION

What are some types of services retailers on the Internet?

NON-STORE RETAILERS **Non-store retailers** are businesses that use means other than traditional storefronts to sell their products, such as infomercials, catalogs, door-to-door solicitation, trade shows, and vending machines. The Home Shopping Network is an example of a highly successful non-store retailer. E-tailing, which uses electronic means of buying and selling products, is a form of non-store retailing. Non-store retailers are able to lower costs by selling directly to consumers, as shown in **Figure 3.2**, without the cost of maintaining a storefront.

Non-store retailing has existed for a long time, since the days of the Sears, Roebuck catalog, but a number of factors have contributed to its accelerated recent growth.

- Working women are turning to faster, more convenient methods to accomplish their shopping tasks.
- Finding and securing a location for a store is increasingly difficult and expensive.
- The cost of running a physical location can reduce profits.

As you learned in Chapter 1, businesses that use physical stores are called *bricks-and mortar businesses,* businesses that use a variety of methods to sell goods, such as Sears, Roebuck, are called *multichannel retailers,* and those that sell on the Internet only, such as Amazon.com, are called *pure-play retailers*. While the real-world storefronts of Wal-Mart, Best Buy, and Payless Shoes are here to stay, cyberstores such as Amazon.com, HomeDepot.com, and CDNow.com are challenging their place in the field of retailing by providing consumers with greater selection and convenience than ever before.

Figure 3.2

Non-Store Retailers

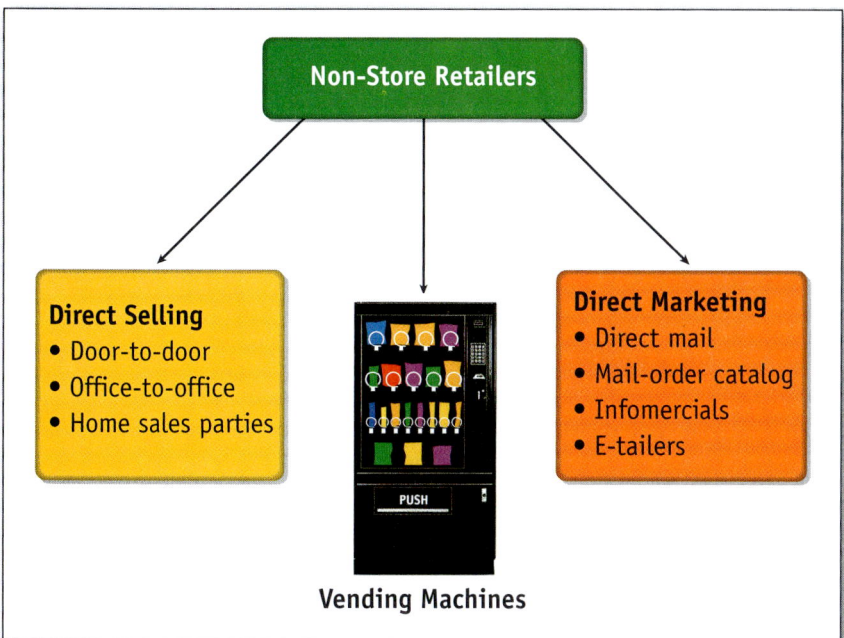

DIRECT APPROACH Non-store retailers use methods other than the traditional storefront to reach their customers. *At what type of non-store retailer have you recently shopped?*

Section 3-1 Review

RESPOND to what you've read.

1. How were early retail stores less convenient for shoppers than those of today? _____

2. How do category killers "kill" their competition? _____

3. Why might some retail customers prefer not to shop at discount stores? _____

4. What is a non-store retailer? Why is e-tailing a type of non-store retailing? _____

Quick TALK With a partner, list three online and three bricks-and-mortar retail businesses that sell CDs. *Discuss where each of you will most likely buy your next CD and why.*

Section 3-2

The E-Tail Experience

AS YOU READ...

YOU WILL LEARN
- what components form a successful e-commerce business.
- which major merchandising issues e-tailers must address as they enter the e-commerce marketplace.
- what methods customers use to pay for online purchases.
- how a customer's private information remains secure during an online transaction.

WHY IT'S IMPORTANT

When you understand the complexities of running a secure and reputable e-commerce site, you can better plan your business and purchases.

KEY TERMS
- hyperlink
- Electronic Funds Transfer
- smart card
- eWallet
- e-cash
- Secure Sockets Layer
- digital certificates

PREDICT

What are some of the challenges unique to setting up an online business?

THE NATURE OF E-TAILING

Setting up and maintaining an online business comes with a set of unique challenges. Suppose you want to create an online store that sells caps, sweatshirts, and jewelry. Where will you purchase the products you intend to resell? How will you display this merchandise online? How will you price each product? What payment methods will you offer customers? How will you fill and ship the orders?

You also have to consider concerns about the security of information on the Internet to protect both yourself and your customers. Creating an engaging retail experience online for customers will go a long way toward ensuring your success as an e-tailer.

Product Merchandising

Choosing the best product assortment and presenting your products in an appealing manner are key to selling on the Internet. How you present the products and how your customers navigate the site depend on how you categorize your products. Will you categorize them by price, type, size, or color? All of these considerations are merchandising functions. *Merchandising* activities include acquiring products for sale, setting prices, displaying products, and making them available for purchase.

Online merchandisers use hyperlinks as merchandising cues to present their products and motivate consumers. A **hyperlink** connects you to another location on the same Web site or a related Web site. Examples of merchandising cues are cross-sells, upsells, recommendations, and promotions. A *cross-sell* hyperlink takes you to an item associated with the item you're currently viewing. An *upsell* refers you to a location that presents a similar but more upscale and more expensive item. A *recommendation* takes you to a product that might interest you based on products you've purchased before. A *promotion* refers you to a "hot" product or sales item the site is currently offering. **Figure 3.3** shows how merchandising cues can be used to hook a consumer's interest.

As e-tailers become more experienced with the ins and outs of running their Web sites, they become increasingly adept at finding ways to capture the attention of customers. Many retailers doing business online, such as Pier 1 Imports and the Home Shopping Network, use software technologies like RichFX to enhance the presentation of their products with zoom photos, color change interfaces, and video clips. Zoom photos allow shoppers to see small details on products. Color change interfaces help customers see what the product would look like in a different color. Video clips provide short video showcases that zoom in and around the product from every angle.

52   Unit 1   The History, Nature, and Impact of E-Commerce

Figure 3.3

Merchandising Cues

DIRECT APPROACH This is a product page for a Grundig® Portable World Band AM/FM/SW Receiver. It contains merchandising cues. *Can you find the cross-sell? Do you see the recommendation product?*

Setting Up an Online Purchasing Process

Before you can sell goods to customers, you must either have them available in stock or have the ability to get them from a manufacturer quickly once you've received orders. Retailers generally purchase their products from vendors or manufacturers using *purchase orders*—written documents that serve as an offer to buy certain items from a supplier at a specified price.

A purchase order generally contains the following information:

- quantity
- description
- unit price
- total cost
- supplier's name and address
- date needed
- shipping method (optional)

The Web allows new forms of online collaboration between retailers and their suppliers. A collaborative planning, forecasting, and reorder system can help retailers find the products they need at the best prices. Companies such as Office Depot, Burlington Coat Factory Warehouse, and Williams-Sonoma use an online application from MarketMax Inc. that allows their retail locations to work with their suppliers via the Web to plan merchandise assortments and in-store displays. Many large companies such as Motorola use Internet tools to create, distribute, and manage Requests for Information or Invitations to Bid (collectively referred to as e-RFx tools). These tools help the companies collect bids from suppliers and manage that information to keep manufacturing processes on target.

Payment Options

After you've created an inviting environment for shoppers and they've ordered the items they want to buy, you'll want to ensure they can easily and quickly pay for their purchases. The checkout process should be flexible, affordable, and secure for the buyer and the seller.

E-tailers currently offer a variety of ways in which consumers can pay online. The following are a few of the most popular methods used today.

CREDIT CARDS Consumers use credit cards to pay for approximately 95 percent of all purchases on the Internet. Online shoppers tend to feel more secure with Web sites that can process Visa, MasterCard, or American Express. To provide this service, you will need a merchant account, which costs between $15 and $30 per month. Retailers also pay the credit card companies between $0.25 and $0.50 for each transaction and about 2 percent of the total transaction. Credit card transactions are processed electronically using the Automated Clearing House (ACH) Network.

DEBIT CARDS When customers use debit cards for their online purchases, they are authorizing the withdrawal of money from their bank accounts. Since debit cards only allow their owners to spend the amount of money they have in the bank, there is less risk of nonpayment. It is a quick transaction between the merchant and the customer's personal bank account.

ELECTRONIC FUNDS TRANSFER Electronic Funds Transfer (EFT) provides electronic payments and collections for online sales. It is safe, secure, efficient, and less expensive than paper check payments and collections. When you purchase a product online using EFT, you can pay for it by having money transferred from your checking account to the checking account of the seller. You may also use EFT to pay routine bills online. If you were to log on to your customer account, for example, at Sprint.com, you could pay your bill by authorizing an EFT from your checking account.

SMART CARDS AND eWALLETS A **smart card** looks like a credit card but has a microchip embedded in it loaded with data that can be programmed for various applications. Smart cards can be designed to be inserted into a slot and read by a special reader, to be scanned from a distance when driving through a tollbooth, or to be plugged into your computer workstation so that you can make financial transactions over the Internet. For example, if you live and travel in the Northeast, you might purchase a prepaid E-ZPass to travel in the toll system's express lanes rather than slowing down and waiting in a toll-booth line.

Smart cards are useful for telephone calling, electronic cash payments, and other applications. More than a billion smart cards are currently in use, mostly in Europe. Cardholders can use their cards to establish their identity when logging on to an Internet access provider or an online bank, to pay for parking at meters, to give hospitals or doctors personal

---

CONNECT
What are some methods of payment you've used or seen friends or family members use?

CONNECTION MATH

Returning Point
Many e-tail businesses neglect to factor in the cost of returned merchandise when calculating their expenses and how much profit they expect to make. The e-commerce company Forrester Research estimated online sales in 2004 at $185 billion. According to the e-tail trade association, Shop.org, an estimated $9.25 billion worth of merchandise was probably returned. *Based on this information, what percentage of merchandise you sell online can you expect to be returned? How much will this cost you for every $100 you make?*

Figure 3.4

eWallets

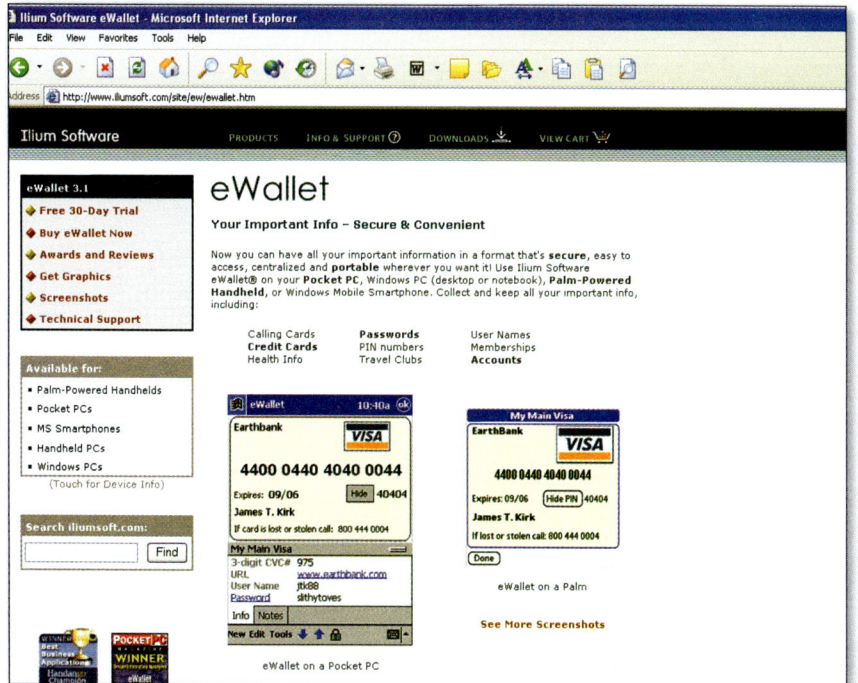

WHAT'S IN YOUR eWALLET?
An eWallet is a convenient and useful way for you to store credit card, bank account, or other sensitive financial information on a portable computing device such as a handheld computer. *In what ways might an eWallet be preferable to more conventional ways of storing important financial information?*

data without filling out a form, and to make purchases at e-commerce sites. Smart cards with electronic value are known as stored-value cards. Much like smart cards, **eWallet** is a software application that stores a customer's data for easy retrieval during online purchases. The eWallet utility encrypts your personal information and is accessible via a portable, handheld PC device. **Figure 3.4** shows an example of an eWallet.

E-CASH E-cash is a legal form of computer-based currency that allows for the purchase of items by credit card, check, or money order, providing secure online transactions and processing. E-cash provides rapid, secure, and reliable real-time payment processing worldwide. When you place an order on a merchant's Web site and use e-cash for the payment, the merchant securely transfers your order information over the Internet to your bank to request transaction authorization.

When the bank authorizes the transfer of funds from your bank account to the merchant's bank account, the merchant fulfills your order by shipping the product or transferring it digitally to you. One of the global leaders in online payments is PayPal, which has more than 30 million account members in 38 countries around the world. Once registered, PayPal members with e-mail addresses can send and receive payments online.

E-CHECKS While approximately eighty million people in the U.S. don't have credit cards, most have a checking account. E-checks, or electronic checks, provide a handy way to get payment from customers who do not own or use credit cards. Customers must register with a third-party account server before they are able to write e-checks. This process usually takes more time and requires the customer to enter more information in the checkout process.

METHOD OF PAYMENT? Although alternative means are available, a majority of both traditional and online consumers use their credit cards to pay for their purchases. *Why do you think this is the case?*

Order Fulfillment and Customer Service

Part of your customers' e-tail experience consists of receiving the goods they ordered quickly and efficiently. Were the correct goods delivered? Did the order arrive on time? If there was a problem with the order, were customer service agents available to help?

You need to consider how to warehouse your products and what methods to use to deliver them. You also need good service agents to keep track of orders, deal with customers' problems, and build good customer relationships. These aspects of e-commerce will be covered in detail in Chapters 13 and 14.

Security Issues and Concerns

One issue that may keep customers from making purchases on your Web site is security. Any personal information a customer enters on the Internet, such as a Social Security number or even a birth date, could potentially be stolen by a hacker. Another concern is disreputable merchants who fail to fulfill orders or refund customers' money. How can you assure your customers that their credit card numbers will be secure and that you are a legitimate business?

Security concerns include someone stealing personal information you've entered on the Internet and merchants who fail to fulfill your order or give you a refund.

The use of **Secure Sockets Layer** (SSL) helps encrypt and protect the information that customers enter into Web pages when making a purchase. This protocol is built into most browsers and is supported by most Web servers. SSL is only half of the solution, though; your customers also want to feel confident that you run a reputable business. **Digital certificates**, which are issued by a trusted third party, verify to customers that a company is what it claims to be. Digital certificates

QUESTION
What are some security concerns you have about buying items on the Internet?

can be purchased from many companies, including VeriSign and Netscape, and require documentation to obtain.

Both SSL and digital certificates require *encryption*—a computer program that scrambles customer information into a secret code so that unintended parties do not have access to their information. *Asymmetrical encryption* provides this protection and also issues a public and private key, or code. The public key can be used by anyone to encrypt anything, but only the private key can decrypt it. When a customer logs on to an encrypted Web page, the server gives the customer's browser a public key. The digital certificate provider confirms that the key belongs to a valid certificate used by the company at the domain where the transaction is taking place. You will learn more about encryption techniques in Chapter 6.

Advantages and Disadvantages of E-Tailing

Whether as an online buyer or seller there are both advantages and disadvantages to e-tailing you should take into account. (See **Figure 3.5**.) What's good for you as a shopper is not necessarily good for you as an e-tailer.

The main advantage for both is convenience. The Internet is open 24 hours a day, seven days a week. As an online shopper, you can buy whatever you want whenever you want without getting up from your computer. As an e-tailer you don't have to keep store hours and you're never closed. Anyone can shop on your Web site any time anywhere in

Figure 3.5

E-Tail Advantages and Disadvantages

Advantages

- E-tailing increases a business's customer base.
- Web sites attract new customers.
- Customers can shop from their homes.
- A business's online store is never closed.
- Customers can shop 24/7, regardless of weather, traffic jams, or distance from the retail store.
- A well-designed, easy-to-use, and frequently updated Web site is a valuable channel for any business.

Disadvantages

- Customers are reluctant to release personal information on a Web site.
- Customers are concerned about the security of their credit-card accounts.
- Transactions may be interrupted on the Internet.
- Customers are unable to examine merchandise or try on clothing.

TO E-TAIL OR NOT? The top concern that customers have about buying online is safety. A store's Web site must stress the security of the site before customers will feel comfortable buying online. *What can e-tailers do to reassure customers?*

the world. You don't have to pay rent for your store, keep your shelves stocked, or hire clerks.

Shopping on the Internet offers consumers greater choice than ever. You can comparison shop to find the best price for a CD or a used car without driving to different stores or even picking up the phone. For e-tailers, however, this translates into greater competition. There are literally millions of Web sites on the Internet and it's easy to get lost in the shuffle.

Security is not only a concern for online shoppers, but for e-tailers as well. Just as consumers have to be concerned about their personal information being stolen, e-tailers have to be concerned about people purchasing products using stolen or fake information. (Internet security is further discussed in Chapter 6.)

A disadvantage for both online shoppers and e-tailers is the inability to personally inspect items on the Internet. If you're in the market for clothing, gourmet food, or designer jewelry, you want to touch, taste, try on, or otherwise try out an item before you buy it. This is why e-tailing will never fully replace the experience of shopping at a traditional bricks-and-mortar retail store.

Section 3-2 Review

RESPOND to what you've read.

1. What is a cross-sell? What product might be cross-sold to a customer purchasing a coffeemaker? Why?

2. How could a company enhance the presentation of its products on its Web site? How would this influence its sales?

3. Why might a Web site's sales suffer if it only permits customers to pay using smart cards and EFTs?

4. How can a Web site assure customers of security?

Quick TALK Many people are still reluctant to shop online. **With a partner, brainstorm a convincing argument to persuade a friend or family member who is reluctant to shop online that e-commerce is safe and efficient.**

Name _____ Date _____

Worksheet 3-1

Internet Influence

Businesspeople everywhere are wondering how e-commerce will influence traditional bricks-and-mortar retailing. Will they always be in competition? Will they work together? How will the Internet influence a business and its marketing functions?

1. Interview a local business manager to determine how the Internet has influenced the manager's merchandising functions. (If students have trouble finding a local business manager to interview, invite someone to the school and have the entire class interview the person to answer this question.)

   _____
   _____
   _____
   _____
   _____
   _____
   _____
   _____
   _____
   _____

2. Imagine you have been hired as a consultant to help this business launch or enhance an Internet component. What would you suggest? Why?

   _____
   _____
   _____
   _____
   _____
   _____
   _____
   _____
   _____
   _____

3. Write a one-page paper summarizing your recommendations.

4. Create a multimedia presentation using your recommendations. Make the presentation to your class and, if possible, to the business manager you interviewed.

Name _____ Date _____

Worksheet 3-2

Merchandise Motivation

Presenting merchandise in an appealing manner on the Web is essential to successful selling. With the increasing number of e-tailers, motivating customers to purchase through visual cues is becoming increasingly complex and sophisticated.

1. Using the Internet, find examples of cross-sell, upsell, recommendation, and promotion hyperlinks. Note the Web address for each example.

   _____
   _____
   _____
   _____
   _____
   _____
   _____
   _____
   _____
   _____
   _____

2. Choose an online retailer that you frequent or are interested in visiting. What additional merchandise could you promote using the visual cues described in this chapter?

   _____
   _____
   _____
   _____
   _____
   _____
   _____
   _____
   _____
   _____
   _____

3. Prepare a brochure with key terms and helpful hints for someone who is new to online shopping. Include information about merchandising cues, navigation techniques, payment options, and security features.

Chapter 3 Review and Activities

Chapter Summary

Section 3-1

- Retailing has evolved from street vendors, small, family-owned shops, and the general store to big-box retailers and e-tailers.

- Convenience, variety, and new technology allow retailers to make the shopping experience accessible to a variety of consumer needs.

- The major categories of retailers include specialty stores, department stores, discount stores, services retailers, and non-store retailers.

- E-tailing, because it involves buying and selling via the Internet rather than through bricks-and-mortar storefronts, is a form of non-store retailing.

Section 3-2

- The components of a successful e-commerce business include merchandising, a purchasing process, payment options, order fulfillment and customer service, and security issues and concerns.

- Choosing the best product assortment and presenting these items in an attractive manner are important considerations when selling on the Web. E-tailers must design their Web sites using merchandising cues to motivate customers to buy.

- E-tailers need to set up an online purchasing process to serve their customers.

- Credit cards, debit cards, electronic funds transfer (EFT), smart cards, eWallets, e-cash, and e-checks are popular payment methods used by those who purchase items online.

- The use of Secure Sockets Layer (SSL) and digital certificates, both requiring encryption, provide security to customers who shop on the Web.

Chapter 3 Review and Activities

Review of Key Terms

a. e-cash
b. e-tailing
c. hyperlink
d. services retailers
e. Secure Sockets Layer
f. smart card
g. wholesalers
h. retailers
i. eWallet
j. digital certificates
k. non-store retailers
l. Electronic Funds Transfer

Match each term to its definition.

_____ 1. A connection to another location on the same Web site or to a related Web site.
_____ 2. The transfer of money from your checking account to the checking account of the seller.
_____ 3. Verification issued by a trusted third party that a company is what it claims to be.
_____ 4. Looks like a credit card but has a microchip loaded with data embedded in it.
_____ 5. A software application that stores a customer's data for easy use during online purchases.
_____ 6. Businesses that sell goods using means other than traditional storefronts.
_____ 7. The buying and selling of retail goods on the Internet.
_____ 8. A legal form of computer-based currency.
_____ 9. Helps protect the information that customers enter into Web pages when making purchases.
_____ 10. Businesses that sell specialized skills, such as medical care or financial planning, rather than goods.
_____ 11. Businesses that sell merchandise to the general public.
_____ 12. Businesses that sell goods to other businesses rather than directly to the public.

Applying Technology to Academics

English Language Arts—Reading

Review a print or online catalog from a non-store retailer and answer the following questions:

- What process is required to place an order?
- What forms of payment are accepted?
- Are shipping and handling charges required? What are they?
- Is there a satisfaction guaranteed policy?
- What is the procedure to follow to return an item?

English Language Arts—Writing

Write a letter to a friend recommending a product from the catalog and explaining how to use the catalog. In the letter, include information about all the questions you answered above. If possible, use a word processing program to write the letter.

English Language Arts—Speaking

Choose one of the payment options available to online consumers. Imagine you are an e-tailer and want to offer customers this particular option. Conduct research on the Internet to determine how to go about doing so. Prepare a short presentation about the option and, with a partner, discuss what you have learned.

Calculation

You sell electronics online. Last month, your business had 22 credit card transactions that totaled $37,456. The credit card company charged you $0.25 per transaction and 2 percent of the total transactions. How much did it cost you last month to provide this service to your customers?

Chapter 3 Review and Activities

Critical Thinking

1. As a consumer you shop in a variety of stores. Where do you like to shop? Explain why.
2. Do you think cash is becoming old-fashioned? Explain why or why not.

COMPETITIVE EVENT

Assume the role of sales manager for a locally owned business that manufactures and sells herbal skin and bath products. The company has enjoyed success with local distribution, but the owner feels it is time to sell these products online. The owner has asked you to create and present ideas for a Web site.

EVALUATION

You will be evaluated on how well you meet the following performance indicators:
1. Create a vision for your Web-site proposal.
2. Explain the impact of the Internet on purchasing.
3. Discuss trends in e-commerce.
4. Identify strategies for protecting business Web sites.

INTERNET ACTIVITY

To engage in online activities, visit the *E-Commerce* Web site at **ecommerce.glencoe.com**.

BusinessWeek online

TOPIC: Displaying Merchandise Online

Product display is an essential tool for attracting customers' attention. While bricks-and-mortar retailers can strategically display items throughout their stores, e-tailers are limited to a flat screen. Still, there is a variety of methods e-tailers can use.

ACTIVITY Go to the Student Center at **ecommerce.glencoe.com**. Click on *BusinessWeek* Activities and open Chapter 3. Here you'll investigate what kinds of product displays work and don't work online.

ecommerce.glencoe.com

Chapter 3 Retailing on the Internet   63

Chapter 4

Global E-Commerce

Section 4-1
Going Global

Section 4-2
The Impact of E-Commerce on International Trade

PREREADING STRATEGIES

Before you read this chapter, finish these statements in your journal:

- Based on this chapter title, I predict…
- The most important thing to remember about globalization and e-commerce is…
- Some questions I have about globalization and e-commerce are:

BusinessWeek online

Go to *BusinessWeek* online. Read an article about the economic situation in a country other than the U.S. Summarize the main point of the article and list some basic economic information about the country, such as what type of currency it uses, and what its major industries are.

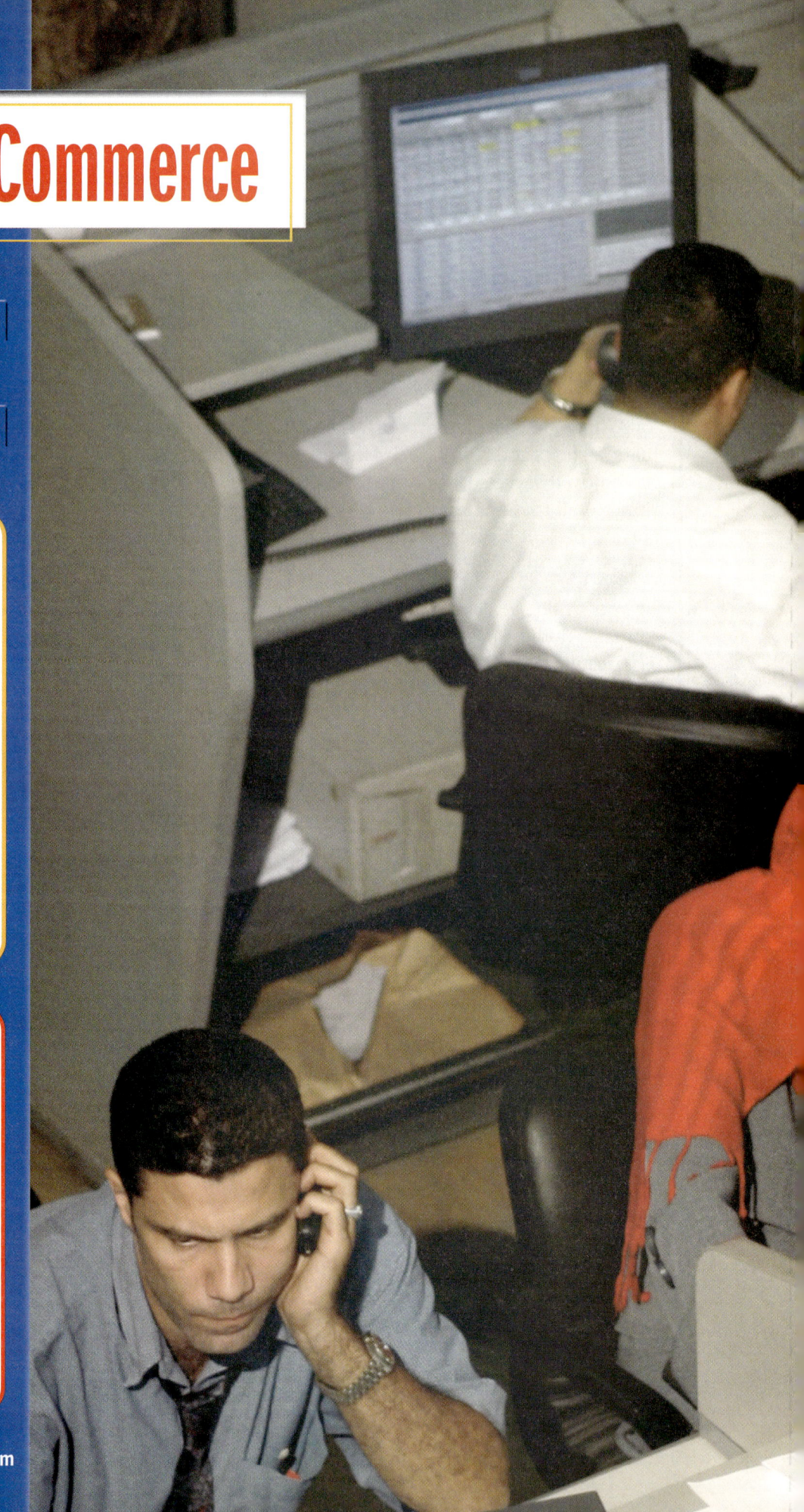

Facing New Challenges

The development of new technology is a constant process of problem solving. The solution to old problems often leads to new ones. The invention of the automobile made transportation much faster and easier, but led to traffic jams and air pollution. The Internet has made it much easier to conduct business, but created new challenges.

Drill Down

1796 Saving the World
Edward Jenner creates the smallpox vaccine, which virtually eliminates a disease that had claimed millions of lives.

1898 Diesel Motors
Rudolf Diesel receives a patent for the internal combustion engine.

1903 A Close Shave
King Camp Gillette and William Nickerson invent a double-edged safety razor, which is smoother, safer, and more hygienic for shaving.

1921 Tell the Truth
John Larson develops the polygraph machine, or lie detector.

1946 Microwave Ovens
Percy L. Spencer invents the first microwave oven and makes the creation available for commercial use.

1981 Personal Computers
IBM releases the first personal computer. The company expected to sell only 250,000 units in the first five years, but sold more than 3 million.

Get the Big Picture
List some of the challenges of starting an international e-business. How can companies overcome these obstacles and successfully globalize their businesses?

POWER READ

Be an active reader and use these reading strategies:

PREDICT what the section will be about.
CONNECT what you read with your own life.
QUESTION as you read to make sure you understand the content.
RESPOND to what you've read.

65

Section 4-1

Going Global

AS YOU READ...
YOU WILL LEARN
- what globalization means for e-commerce.
- what motivates many U.S. businesses to expand into global positions.
- why cultural considerations must be taken into account when businesses go global.

WHY IT'S IMPORTANT

Companies engaged in the global market enjoy the potential for increased revenues, but also face greater challenges in certain areas of their business. Ensuring respect for the various cultures involved is essential to success in the global arena.

KEY TERMS
- globalization
- culture
- machine translation
- human translation
- Web globalists

PREDICT

What is globalization, and how does it relate to e-commerce?

THE GLOBAL MARKETPLACE

When was the last time you heard about a hot new electronic gizmo from Tokyo? How about a political upheaval in a developing country? It's likely that news from the international scene comes to you daily, if not hourly, in today's media-rich world. Our connection to the global community is continual and instantaneous. The Internet, along with all of its technological advances, plugs you into business, political, and cultural news from just about anywhere at any time.

Challenges of Global E-Commerce

With increased technological access to the international scene, the business world is experiencing dramatic changes. **Globalization**—or enhancing connectivity and interdependence among the world's markets and businesses—is a trend you'll hear more about in the coming years.

In the e-business world, globalization specifically refers to the process of adapting a business Web site to meet the needs of users in various countries. Recent research conducted by Forrester Research indicates that worldwide online trade has grown exponentially in just four years. Although the *Computer Industry Almanac* indicates that the United States currently represents 43 percent of online users, this percentage of the market will most likely decline as the Web gains increasing popularity worldwide, particularly in Europe and the Far East. With so much international online momentum, U.S. businesses cannot afford to ignore non-English-speaking markets.

Before the Internet, companies that wanted to do business in foreign countries had to think in terms of physical location. They needed to project the costs and work involved in moving people to new places, investing in buildings and equipment, and processing orders manually in different languages and currencies. The Internet and advancing technologies have changed all this. Companies may now construct cyber-locations in several countries, translate their Web sites into the appropriate languages, and use software technologies to process complex multicurrency sales transactions in seconds. Going global still requires planning, research, and money, but in many ways the Internet has made the process far easier. In some sectors today, international expansion is necessary for success.

So what is involved in going global? It might sound like a simple task of translating text into another language, but the job is much more complex. You will need to consider various currencies, customs, government regulations, pricing strategies, technologies, and customer-support strategies.

CULTURE As you think about expanding your business into a new country, you'll also want to consider local customs, business practices, attitudes, and values. **Culture** is a way of life that includes behaviors, beliefs, values, and generally accepted symbols for a group of people. Culture can describe a religion, race, gender, or a geographic location. Cultural differences exist not only between different nations, but also within them. Think about your own values and beliefs. Are they different from those of people from other countries? Are they different from those of people who live north or south of you in the United States? Because these cultural differences can be substantial, the practice of selling to a Japanese consumer, for example, will likely differ from selling to an American consumer.

Customers from different cultures may respond differently to the same simple design element on your Web site. Suppose your Web site's design features a mostly white background. Do you think a citizen of China might react differently to that color than someone from the United States? While in the United States the color white is generally associated with virtue, order, and cleanliness, you might be surprised to learn that in China white is traditionally used for periods of mourning. To prepare a Web site that does business globally, be aware of the following issues:

- Creating *content* for a new international market goes far beyond translating a Web site's English text word-for-word. You'll need to consider your writing style, phrasing, marketing concepts, word choices, use of humor, and product names—any of which might not translate well. Adding localized content—ideas, phrases, and language specific to the culture—will help to make the content reader-friendly and accessible.

- *Design* choices such as color and layout can make or break a site's success, depending on your target audience. Graphics or photos that may appeal to U.S. customers might not have the intended impact on a shopper in Mexico.

- You'll also want to consider how to set up your site so customers in different countries feel comfortable *navigating* it. In cultures where people are fascinated with technology they prefer sites that are complex to navigate, while in other cultures people prefer sites in which navigation is simple and direct.

Once you've decided to build a Web site that is suitable for customers in Mexico, Japan, and Germany, it is critical that the site's gateway, or front door, is easy to find. Companies generally structure their foreign language sites in one of two ways:

- They provide links on their English home pages as gateways to foreign-language pages that reside within the same domain. In this case, the foreign-language pages act as subpages of the main English sites.

- They obtain a new domain name ending with the country's extension suffix and place all their foreign-language pages on that domain. For example, Nestlé provides a gateway that enables customers to select various countries' Web sites from a pull-down menu. (See **Figure 4.1** on page 68.)

CONNECT

Would you consider doing business with a company outside the United States? What might convince you to buy products featured on an international Web site?

CONNECTION
SOCIAL STUDIES

A Lagging Fad
Around the world, expect to encounter ambivalence about the Internet even today. A TNS Interactive e-commerce survey on global usage of the Internet found that Argentina, Australia, Belgium, Bulgaria, and Canada combined only accounted for 36 percent of total online use. In those countries, only 7 percent of those using the Internet shopped online. *Why do you think these populations don't shop online?*

Figure 4.1

Nestlé International Gateway

INTERNATIONAL LINKS Nestlé offers a pull-down menu that enables visitors to access its international Web sites. *How is the Japanese Web site different from the international home page?*

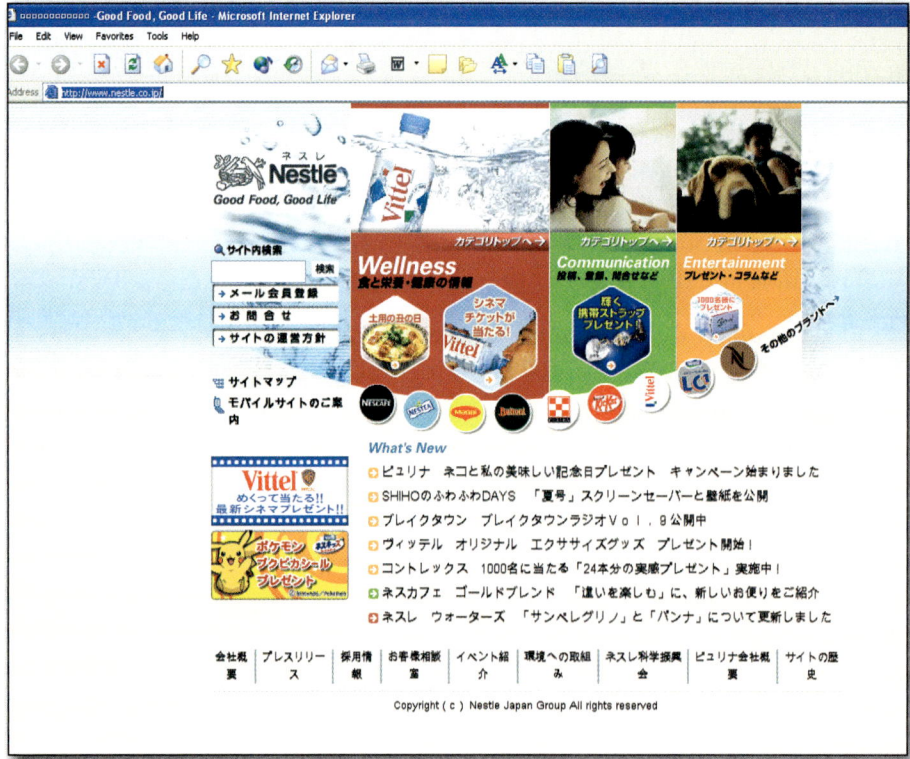

LANGUAGE Even though English is the dominant language on the Internet, online companies cannot assume that all customers who visit their Web sites are fluent in English. GlobalReach.com reports that while there are approximately 242 million English speakers using the Internet, there are also more than 438 million non-English-speaking users. Nearly half of them live in Asia or Europe. Visitors to Web sites that are translated into two or more languages will spend twice as much time surfing those sites than non-translated ones, according to GlobalReach.com's research studies. Today more than one-third of all e-commerce Web sites are in a language other than English. Which languages do Internet users speak? Take a look at **Figure 4.2** to see the breakdown.

With more than 438 million non-English speakers surfing the Web, e-commerce companies need to offer potential customers a selection of languages to increase their chances of selling their products and services. As business owners prepare their Web sites for audiences who speak different languages, they can approach translation issues in two ways. **Machine translation** may not be entirely accurate, but it can quickly translate English text into another language using a software application. You may have visited a site such as AltaVista to translate sentences or words into another language. While this method might do for a small amount of content, it is likely too time-consuming and inaccurate to handle a large corporate site. Machine translation also does not account for regional dialect or language trends. Human translation becomes necessary in these cases.

Figure 4.2

Online Population by Language

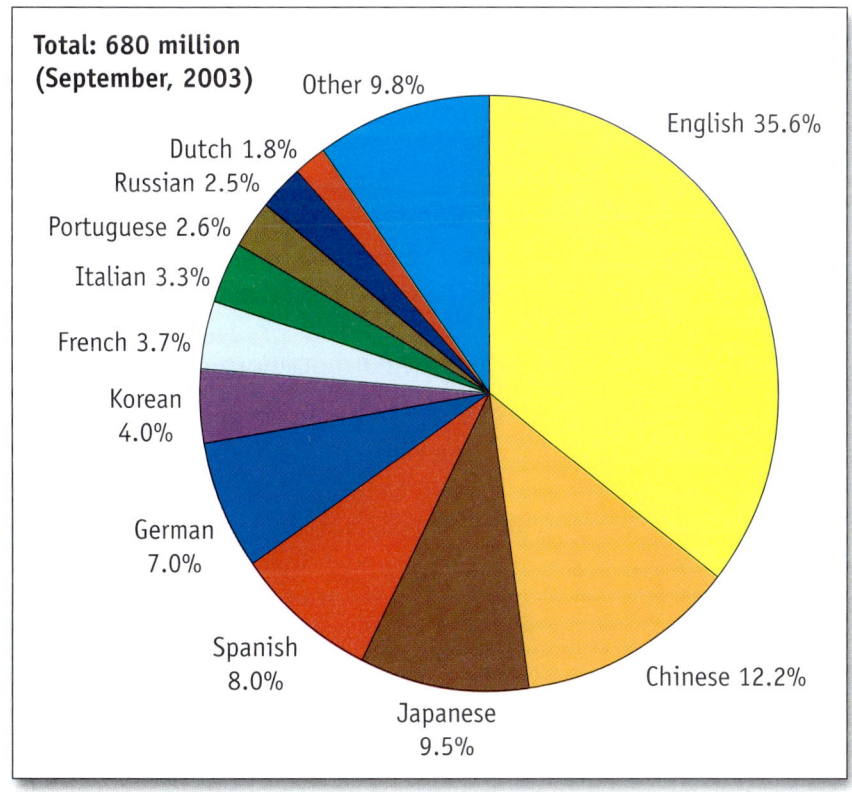

NATIVE TONGUES Internet users from around the world can access the Web and browse many sites in their own language. *Which languages are used most often by Internet users? If you were planning to expand your Web site to a new population, which might you consider based on this chart?*

Figure 4.3

Web Content by Language

LANGUAGES ONLINE Web sites appear in a variety of languages, but one language clearly dominates. *Why do you think this is the case?*

LANGUAGE	PERCENTAGE OF TOTAL WEB SITES
English	68.4%
Japanese	5.9%
German	5.8%
Chinese	3.9%
French	3.0%
Spanish	2.4%
Russian	1.9%
Italian	1.6%
Portuguese	1.4%
Korean	1.3%
Other	4.6%
Total Web pages:	313 billion

Human translation of text requires the services of a multilingual person. Such services are essential because product descriptions, marketing lingo, and purchasing instructions must be clear, concise, and targeted to the citizens of a particular country. Currently, Web sites are predominately in English, as shown in **Figure 4.3**, and translating a Web site can be expensive.

Companies such as SDL International and Bowne Global Solutions (BGS) provide translation services using a network of native speakers who understand local dialects, word usage, and customs. They also offer consultation services with **Web globalists**—consultants who advise companies on how to design their Web sites and market and sell their products effectively to international audiences. They help companies to determine whether online selling in a global market will be profitable based on both their product and service lines and their organizational structure.

COMMUNICATIONS SYSTEMS If your Web site takes orders in Japanese, but the database you use to fulfill them can't read the characters, you've got a problem. To conduct business globally, you not only have to deal with customers who speak a foreign language, you also have to use communications systems based in that language. All servers, databases, management systems, and software applications must be internationalized to handle different language character sets as well as different date formats, time zones, and currencies. Companies such as GlobalSight Corporation and Idiom Technologies, Inc. now make software products specifically designed to help develop and manage the technical challenges of multilingual Web sites.

CURRENCY Now that you've considered cultural and linguistic differences, think about the currency issues you have to face if you're selling products to customers in Japan, Germany, and Mexico. From yen to pesos, dollars to euros, you must present the pricing of your products in the local currencies and factor currency *fluctuations* into your e-business's pricing structure.

Just as languages are *translated* into one another, currencies of different countries are *converted* into one another. Different currencies have different values compared to each other. The price at which one currency can buy another currency is its *exchange rate*. For example, one American dollar is worth a certain number of Mexican pesos or Japanese yen. Exchange rates differ from day to day and country to country. One day the American dollar might be up compared to the Mexican peso, but down compared to the Japanese yen. If your e-commerce Web site operates globally, you need to keep track of exchange rates to conduct business with all your potential customers.

There are numerous services that can help you with this. WorldPay, for example, allows customers to view prices and pay for products in more than 150 currencies. Paymentech processes approximately 3 billion multicurrency transactions annually.

PRICING Since the value of the dollar in other currencies can change from day to day, merchants in the international arena must price their products to make a profit. Web sites that support *real-time transaction processing* must be continually updated with accurate information about the currency of the countries in which their buyers are located. The profit realized from these sales is dependent on the accuracy of the business's foreign-currency conversions. If the software used on a Web site cannot continually update its currency conversions, the company should not use real-time transaction processing. More traditional methods to complete sales transactions might be necessary, such as fax, telephone, and postal delivery.

> **QUESTION**
> Why would a U.S. business want to enter the global market?

DUTY-FREE SHOPS Duty-free shops, found worldwide, allow travelers in foreign lands to purchase goods imported from their home countries without paying excise taxes to the local government. *Are there similarities between the duty-free model and e-commerce?*

Section 4-1 Review

RESPOND to what you've read.

1. What is culture? How does culture impact the way businesses market and sell their products and services online? _____
_____
_____
_____
_____

2. Describe two ways to translate Web-site content into another language. Which factors should be considered when having a Web site translated? _____
_____
_____
_____
_____

3. How can dealing with customers who speak a foreign language affect your ability to communicate with them over the Internet? _____
_____
_____
_____
_____

4. Why is it important for companies doing business globally to become familiar with international regulations and import restrictions? _____
_____
_____
_____
_____

Quick TALK With a partner, brainstorm some American advertising phrases that could lose or change their meaning if a machine translated them into another language (e.g., "This is a steal!" or "You can't miss with this deal."). *Discuss the differences between their literal and slang meanings.*

Section 4-2

The Impact of E-Commerce on International Trade

BUSINESS WITHOUT BOUNDARIES

International e-commerce has taken its seat firmly at the table of the business world. As you can see in **Figure 4.4** on page 74, Forrester Research estimates that by 2006 worldwide online trade reached $12,837.3 billion in revenues—a number that will have rippling effects throughout the world community. As the volume of global trade grows online, it affects economies, governments, and people.

Have you considered how e-commerce has changed the boundaries of traditional commerce or the impact of e-commerce on competition among companies? How are governments dealing with issues of online taxation and consumer safety? How are low-income countries being impacted by the growth of e-commerce? These are all issues that accompany today's surge of online traffic, communication, and commerce.

Removing Geographic Barriers

Globalization has removed the geographic barriers faced by traditional businesses. Once confined to shopping in physical stores, customers now have the ability to shop online almost anywhere in the world at any time. The removal of geographic barriers has created new advantages and challenges including:

- Increased competition for all businesses.
- Expanded shopping alternatives to the online buyer.
- Greater opportunities for small firms to compete with large firms.
- Cross-cultural issues when marketing products to specific audiences.

CHALLENGES OF A GLOBAL E-MARKETPLACE Movie studios such as Universal and Paramount traditionally release their films in the United States before releasing them in foreign countries. By the time a film is released in Europe, it is often offered to U.S. consumers on DVD. Universal Studios does not want its European customers to skip the theater experience, however, and instead buy the DVD from a Web site like DVDPlanet.com. A shift from selling theater tickets to selling only DVDs would cost the studios millions of dollars in lost revenues.

In addition to revenue concerns, studios are also concerned with the cultural impact of their films around the world. They often edit movies to make them suitable for various markets.

AS YOU READ...

YOU WILL LEARN
- what rewards and challenges are experienced by businesses in the global market.
- how trade policies work in a global e-commerce marketplace.
- how quotas are designed to affect trade.
- which taxation concerns and regulations are common to international trade.

WHY IT'S IMPORTANT
As businesses seek to sell goods in the global market, it is essential they be aware of the rules and regulations unique to this segment of e-commerce.

KEY TERMS
- protectionists
- imports
- tariffs
- quotas
- export
- free trade
- online dispute resolution

PREDICT
How is a quota applied to imports?

Figure 4.4

Worldwide Online Trade Growth

THE FUTURE OF ONLINE TRADE Use of the Internet for trade and commerce is increasing rapidly. *By what amount was total online trade forecasted to grow worldwide by 2006?*

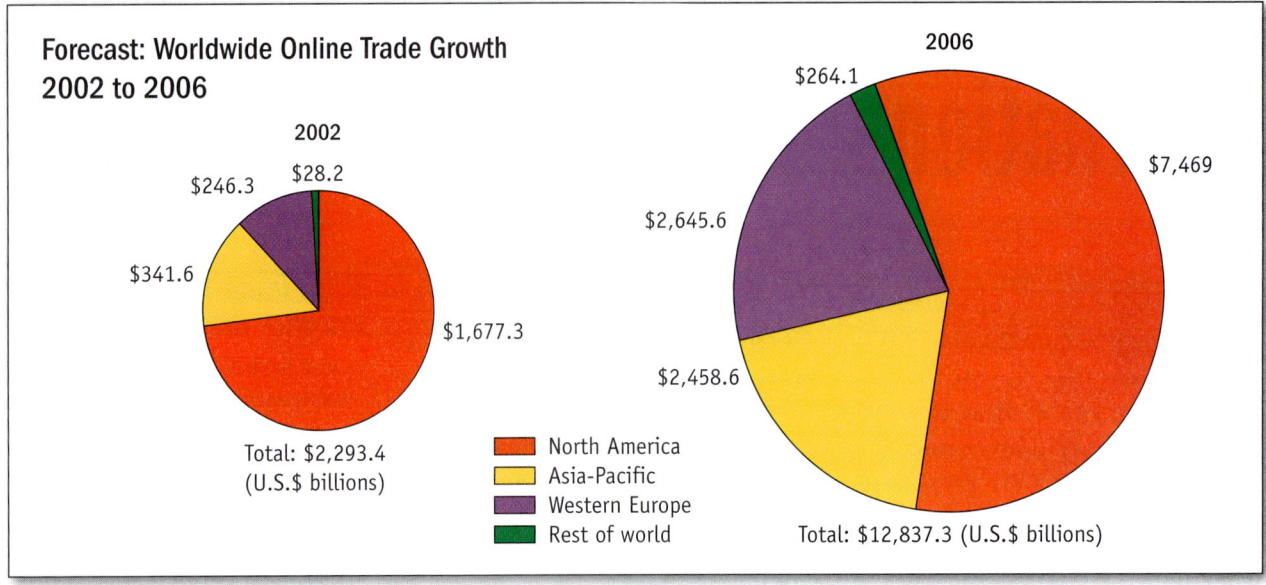

Source: "Global Online Trade Will Climb to 18% of Sales," Forrester Research, Inc., December 2001.

CHANGING THE COMPETITIVE LANDSCAPE As more and more businesses appear on the Web each day, the competitive landscape continues to shift. Some factors improve this environment for business owners, while others have a negative effect. Three factors that improve most businesses' ability to compete are:

- Businesses can get closer to their customers and reduce the need for intermediaries between them and their customers.

- Companies using the Internet can increase their efficiency and the size of their markets.

- Using Web sites to reach customers often reduces the need for a sales force, allowing new companies to enter the market at lower costs than in the past.

Conversely, e-commerce has had a negative impact on the ability of some businesses to compete in the following ways:

- Because the Internet makes "comparison shopping" much easier, consumers are in a stronger position than businesses to set prices.

- Since all companies can market to a global community, it is harder for one company to maintain a competitive advantage in a specific market.

- Because doing business on the Internet can drastically reduce overhead costs, some online companies engage in price competition, forcing bricks-and-mortar companies to lower their prices and cut their profits.

Web Site Success

SELLING STEREOS AROUND THE SPHERE
Sony.net

A few months ago you purchased a Sony stereo receiver that you adore. The quality and design won you over, and you rave about this powerful music maker to anyone who will listen. Some friends in Spain are thinking of buying the same receiver, but they want to research the product's specifications before making a final decision. They'd like to be able to buy the unit online, too. Thanks to the global Sony.net Web site, they can easily accomplish both tasks with just a few clicks of a mouse. Sony's well-organized, easy-to-use site links to more than three dozen country-specific sites, from Albania to the Ukraine.

Thinking Critically
What must a company consider when preparing its Web site to do business with countries that use different languages?

EXAMINING TRADE POLICIES As economies increasingly engage in international e-commerce, governments and individuals are examining their policies and positions on international trade. Those who favor government protection for domestic producers are known as **protectionists**. Governments and organizations that support protectionist positions believe, for example, that the interests of U.S. makers of steel (the domestic producer) should be given consideration and a competitive advantage over steel importers from Germany (foreign producers). They believe this position protects U.S. jobs and supports a more vibrant domestic economy.

Imports, or foreign goods, are goods that are manufactured in one country, then shipped to another country for sale. To provide a competitive advantage to domestic producers, importers are often required to pay **tariffs**—taxes or fees that various governments place on selected imported products. For example, to protect car manufacturers at home, a government might place a tariff on cheaper imported cars. In effect, tariffs raise the price of imported goods, ensuring that domestic producers can maintain prices that will make a profit. Other trade agreements impose quotas.

Quotas are prescribed quantities or maximum amounts of imports allowed into a country. For example, the Softwood Lumber Agreement between the United States and Canada in 1996 set export quotas for Canada's largest lumber-producing provinces and imposes punishing surcharges on above-quota shipments.

Instead of encouraging the use of imported products, many countries prefer domestic companies to produce enough products to meet local demand and then export the excess to foreign countries.

An **export** is a product sold in a country other than the one in which it was produced. The United States, for example, is a leading exporter of agricultural products and automobiles. Exported products can help a

CONNECT
When shopping online or at a local store, is what you buy ever influenced by where it was made? Why or why not?

nation's economy by generating additional revenue from sales abroad. An increase in exports could increase the need for additional jobs in local communities.

As opposed to protectionism, some governments and organizations favor **free trade**—trade among countries without barriers such as tariffs or quotas. One organization that takes this position is the World Trade Organization (WTO), which monitors trade among its member countries and assists in mediating and resolving issues and disputes. The WTO helps member organizations overcome cultural and political differences that may present obstacles. It also promotes the development and practice of ethical business behavior worldwide.

The ultimate goal of the WTO is to encourage its members to work toward achieving free trade. The WTO promotes the concept that the global economy benefits when countries import goods that they are unable to produce domestically and export excess goods that they produce well. The WTO believes this can be achieved without the additional burden of taxation.

Since the Internet ushers in a new class of transactions among citizens of various countries, issues such as trade taxation, quotas, and policies are prominent in the minds of many government officials and online businesses. These policies are re-examined and revised continually to meet the changing state of economies and address the needs of the global marketplace.

E-COMMERCE, INFRASTRUCTURES, AND DEVELOPING COUNTRIES As suggested by the *E-Commerce and Development Report 2003,* prepared by the United Nations Conference on Trade and Development, the world economy is becoming an information and communications technology (ICT)–based economy. The Internet has removed the borders that have traditionally determined the location of service providers and goods producers. The benefits of this new economy are apparent in several developed countries such as the United States and the United Kingdom, while other developing nations have not yet seen the payoff. In those countries, limited communications infrastructure to support Internet connectivity remains a barrier to growth.

E-commerce requires an infrastructure of computers, routers, and hardware; satellite, wire, and optical communications; network channels; electronic payment models; support services; and people who facilitate business processes and e-commerce transactions. Many countries lack the funding, knowledge, and people to initiate development and manage the resulting challenges. Regulating activities online is a sophisticated and complicated task. It can take hundreds, if not thousands, of highly skilled, technically competent employees to manage the process.

Since the late 1990s, an increasing number of developing countries have followed the example of more technologically developed nations and launched their own national ICT strategies. These cover a range of policy initiatives, such as raising public awareness, building infrastructure, deregulating telecommunications, educating an information-technology labor force, and changing legislation. While many developing countries have shown little or no interest in the Internet

QUESTION
What are some reasons developing countries might have for not developing a technological infrastructure?

and its economic potential, others think they need it and think it's important in a global society.

Thailand, for example, believes in the importance of e-commerce and international competition. The government has accordingly issued a policy framework to address global e-commerce, including legal issues, payment systems, human resources, infrastructure concerns, and public-awareness efforts.

REGULATION AND TAXATION It is important to educate yourself about national and regional regulations and import restrictions before launching into a new international market. For example, in most countries items such as animals, explosives, and weapons require special permits and taxes. Local laws often ban the sale of some products completely.

Along with government-imposed regulations, companies engaged in global commerce must deal with different taxation methods. Buying goods overseas can create complex tax situations. In these cases, taxation discussions involve different countries and governing bodies.

The sale of an iPod from a Web site in San Francisco to a consumer in Canada would involve U.S. and Canadian sales-tax laws as well as possible tariff taxes. Given these complexities, it is no wonder that regulating Internet sales becomes a monumental, if not impossible, task. Nevertheless, the U.S. Treasury Department has claimed that it will not be passive in controlling such problems.

DISPUTE RESOLUTION One of the main challenges facing international e-commerce is **online dispute resolution** (ODR), the process of resolving cross-border disputes in the electronic-business environment. What happens when you buy a CD from a Web site based in Japan, but you receive the wrong CD and the company charges you triple the price indicated on the site? You may encounter problems regarding cultural and language differences, determining the applicable law, and enforcement of a satisfactory resolution.

◄ **E-COMMERCE INCREASES COMPETITION, RAISING THE STAKES AND LOWERING THE COSTS** Phone service over the Internet is generally free of federal regulation. However, bricks-and-mortar phone services, which are regulated by the federal government, pay certain fees to the government as part of that regulation. *How do you think these differing approaches to phone service regulation affect the marketplace?*

Chapter 4 Global E-Commerce

Some companies now hire third-party negotiators to settle online disputes. On eBay, when a problem with a transaction arises, buyers attempt to contact sellers and negotiate a solution. If a resolution cannot be agreed upon, a third-party company called SquareTrade, Inc. helps settle the problem. SquareTrade, which is accessible from a link on eBay's Services page, employs a Web site, rather than e-mail, as the main tool for negotiation and asks the parties to try the Web-based negotiation before requesting human mediation.

The site provides forms for the parties to fill out which clarify and highlight both what is dividing the parties and the desired solutions. While parties do have an opportunity to describe concerns in their own words, the forms and the form summaries that parties have to fill out reduce the amount of complaining and demanding that occurs, which lowers the amount of anger and hostility between the parties.

When Web-based negotiation fails, SquareTrade provides a human mediator for a fee of $20. A human being facilitates the conversation, but a Web interface is still used.

Section 4-2 Review

RESPOND to what you've read.

1. How do the import and export of goods differ? _____

2. What is the World Trade Organization, and what is its main goal? _____

3. What potential barriers exist for developing countries' entry into e-commerce? _____

4. What is a protectionist? How might a protectionist in the U.S. feel about importing lumber from Canada and why? _____

Quick TALK With a partner, discuss the role and necessity of online dispute resolution negotiators in the global marketplace. *Brainstorm some generic questions a negotiator might ask a customer who is disputing an online sale.*

Name _____ Date _____

Worksheet 4-1

Cultural Diversity

Successful international e-businesses understand the specific market segments, called *target markets,* in which they wish to concentrate their efforts. Target marketing requires a solid understanding of local cultures and is crucial to businesses focusing their efforts internationally.

1. Imagine you work for a company that is planning to offer its products online to attract customers from around the world. Select one nation your company plans to target, and construct a target-market profile of the country and its culture.

   _____
   _____
   _____
   _____
   _____
   _____
   _____

2. Using your computer, create an advertisement for the products your company wishes to market abroad. Be sure to research the language, design, and colors that would be appropriate for the local culture.

3. Present your target-market research, profile, and the advertisement you have created to the class.

4. Some people choose marketing research as a career field. Consider the traits and skills needed by someone in this area (e.g., curiosity, patience, analytical skills, objectivity, business knowledge). Would you be interested in a career in marketing research? Explain why or why not.

   _____
   _____
   _____
   _____
   _____
   _____
   _____
   _____
   _____
   _____

Name _____ Date _____

Worksheet 4-2

Can the Internet Improve World Trade?

International e-commerce is growing at a remarkable pace. By changing the boundaries of traditional business and impacting competition worldwide, it is affecting world economies, governments, and people.

1. Using newspapers, magazines, or the Internet, locate two recent news articles related to e-commerce. What are the issues described in the articles?

   _____
   _____
   _____
   _____
   _____
   _____

2. How can international e-commerce improve world trade? What do you think will happen to trade barriers such as quotas and tariffs?

   _____
   _____
   _____
   _____
   _____
   _____
   _____

3. Some developed countries, like the United States and Britain, have sophisticated Internet communications structures in place. Other countries do not. How will this difference impact international e-commerce? What do you think could be done to help the underdeveloped countries improve their infrastructure?

   _____
   _____
   _____
   _____
   _____
   _____
   _____

4. Write a paper describing what you think international e-commerce will look like in ten years. Create a visual to complement your paper.

Chapter 4 Review and Activities

Chapter Summary

Section 4-1

- Globalization in e-commerce is the process of adapting a business Web site to meet the needs of users in different countries.

- Worldwide online trade will grow dramatically in the next few years as companies construct cyberlocations in countries around the world.

- Cultural differences exist between nations and within them. It is important to know and understand how your content, design, and navigation might be received by different populations.

- Businesses conducting global e-commerce must not only take into account differences in culture and language, but differences in currency, exchange rates, and technological communications systems.

Section 4-2

- Businesses involved in the global marketplace experience increased competition, provide new markets to buyers and sellers, expand shopping alternatives, and allow small firms to compete with large firms. However, there is more competition, more pressure to deliver low prices, and more emphasis on instant processing of business transactions for the businesses.

- Issues of taxation, quotas, and other trade policies are continually examined and revised to meet the changing state of our economies and needs as a global marketplace.

- Quotas are designed to restrict trade by prescribing specific quantities of imports allowed into a country.

- The current trade taxation, quotas, and policies that world governments have in place need to be examined and refined in order to meet our changing needs as a global marketplace.

Chapter 4 Review and Activities

Review of Key Terms

a. human translation
b. online dispute resolution
c. export
d. culture
e. Web globalists
f. quotas
g. free trade
h. globalization
i. imports
j. machine translation
k. tariffs
l. protectionists

Match each term to its definition.

_____ 1. Taxes or fees placed on imported products.
_____ 2. The use of a software application to change one language into another.
_____ 3. Those who favor government protection for domestic producers.
_____ 4. Trade among countries without barriers like tariffs or quotas.
_____ 5. Can be related to a religion, race, gender, or geographic location.
_____ 6. The process of increasing interdependence and connectivity among the world's markets and business.
_____ 7. Specific quantities of imports allowed into a country.
_____ 8. The process of resolving cross-border disputes in the electronic business environment.
_____ 9. The use of a multilingual person to change one language into another.
_____ 10. People who advise a company on how to design its Web site for an international audience.
_____ 11. Goods that are manufactured in one country and then shipped to another for sale.
_____ 12. A product sold in a country other than the one in which it was produced.

Applying Technology to Academics

English Language Arts—Reading

Choose a country and research the rules of business and social etiquette you need to understand to conduct e-business there. Include the following:

- Cultural do's and don'ts
- Verbal and nonverbal communication
- Business negotiations

Create a visual (e.g., a poster, PowerPoint presentation, three-dimensional display) of your work.

English Language Arts—Writing

Locate the Web site of a U.S. company that markets its products in at least one other language. How is the site structured? Is the gateway easy to find? Write a one-page paper describing the structure used, the countries the company is marketing to, and how user-friendly the site is.

English Language Arts—Speaking

Read the labels of all the products you have around you—your backpack, clothing, shoes, electronic items, etc.—and make a list of different countries in which they were made. In a group, compare your list with those of your classmates and discuss which countries the United States seems to import from the most.

Calculation

Research the exchange rate of the currency used in the country you chose for the *English Language Arts—Reading* Activity. Imagine you will be taking a trip to that country for which you have saved $1,350. Using the current exchange rate, convert your U.S. dollars into the country's currency.

Chapter 4 Review and Activities

Critical Thinking

1. Describe the needed job skills in the international e-commerce world. Explain why they are important.
2. Discuss the differences between free trade and protectionism. Which policy do you think the United States should follow, and why?

COMPETITIVE EVENT

Imagine you're the business manager for a midsize manufacturer of women's sweaters. The sweaters, which feature unique, hand-knit designs, are highly sought after because each style is produced in a limited quantity. In fact, several styles have become collector's items. Currently, the sweaters sell for $175 to $250 apiece, depending on the design. The company sells its sweaters on its Web site and at a few select upscale retailers. Up to now, the company has only sold to U.S. and Canadian customers, but the president has asked for your input regarding whether the company should accept international orders. Discuss this issue with the company president (in this case, your teacher), outlining the factors to consider before engaging in global e-commerce.

EVALUATION

You will be evaluated on how well you meet the following performance indicators:
1. Describe the impact of e-commerce on international trade.
2. Identify the impact of cultural and social factors on world trade.
3. Explain the nature of international trade.
4. Discuss trends in e-commerce.
5. Address people properly.

INTERNET ACTIVITY

To engage in online activities, visit the *E-Commerce* Web site at **ecommerce.glencoe.com**.

BusinessWeek online

TOPIC: E-Commerce in Brazil

The bustling South American country of Brazil is not just a hot vacation spot. With the most computer users in South America, Brazil is a huge market for e-commerce, and a prime center for Internet development.

ACTIVITY Go to the Student Center at **ecommerce.glencoe.com**. Click on *BusinessWeek* Activities and open Chapter 4. There you'll learn more about Brazil and its potential as a major player in the world of e-commerce.

SELF-ASSESSMENT

What Is Your Potential?

Perhaps you thought the only way to learn a subject was through a textbook and a dry lecture. Think again. Dr. Howard Gardner, a professor at Harvard University, found that humans are more complex than I.Q. tests. Instead he developed eight intellectual pathways to explain how people learn and develop. This is your chance to analyze your intelligence potential.

PART I

Complete each section by placing a "1" next to each statement that you feel accurately describes you. If you do not identify with a statement, leave the space provided blank. Then total the column in each section.

Section 1

____ I enjoy categorizing things by common traits.
____ Ecological issues are important to me.
____ Hiking and camping are enjoyable activities.
____ I enjoy gardening.
____ I believe preserving our national parks is important.
____ Putting things in hierarchies makes sense to me.
____ Animals are important in my life.
____ My home has a recycling system in place.
____ I enjoy studying biology, botany, and/or zoology.
____ I spend a great deal of time outdoors.

TOTAL for Section 1 _____

Section 2

____ I easily pick up on patterns.
____ I focus in on noise and sounds.
____ Moving to a beat is easy for me.
____ I've always been interested in playing an instrument.
____ The cadence of poetry intrigues me.
____ I remember things by putting them in a rhyme.
____ Concentration is difficult while listening to a radio or television.
____ I enjoy many kinds of music.
____ Musicals are more interesting than dramatic plays.
____ Remembering song lyrics is easy for me.

TOTAL for Section 2 _____

Section 3

____ I keep my things neat and orderly.
____ Step-by-step directions are a big help.
____ Solving problems comes easily to me.
____ I get easily frustrated with disorganized people.
____ I can complete calculations quickly in my head.
____ Puzzles requiring reasoning are fun.
____ I can't begin an assignment until all my questions are answered.
____ Structure helps me be successful.
____ I find working on a computer spreadsheet or database rewarding.
____ Things have to make sense to me or I am dissatisfied.

TOTAL for Section 3 _____

Section 4

____ I learn best interacting with others.
____ The more the merrier.
____ Study groups are very productive for me.
____ I enjoy chat rooms.
____ Participating in politics is important.
____ Television and radio talk shows are enjoyable.
____ I am a "team player."
____ I dislike working alone.
____ Clubs and extracurricular activities are fun.
____ I pay attention to social issues and causes.

TOTAL for Section 4 _____

Section 5

____ I enjoy making things with my hands.
____ Sitting still for long periods of time is difficult for me.
____ I enjoy outdoor games and sports.
____ I value non-verbal communication such as sign language.
____ A fit body is important for a fit mind.
____ Arts and crafts are enjoyable pastimes.
____ Expression through dance is beautiful.
____ I like working with tools.
____ I live an active lifestyle.
____ I learn by doing.

TOTAL for Section 5 _____

Section 6

____ I enjoy reading all kinds of materials.
____ Taking notes helps me remember and understand.
____ I faithfully contact friends through letters and/or e-mail.
____ It is easy for me to explain my ideas to others.
____ I keep a journal.
____ Word puzzles like crosswords and jumbles are fun.
____ I write for pleasure.
____ I enjoy playing with words like puns, anagrams, and spoonerisms.
____ Foreign languages interest me.
____ Debates and public speaking are activities I like to participate in.

TOTAL for Section 6 _____

Section 7

____ I am keenly aware of my moral beliefs.
____ I learn best when I have an emotional attachment to the subject.
____ Fairness is important to me.
____ My attitude affects how I learn.
____ Social justice issues concern me.
____ Working alone can be just as productive as working in a group.
____ I need to know why I should do something before I agree to do it.
____ When I believe in something I will give 100 percent effort to it.
____ I like to be involved in causes that help others.
____ I am willing to protest or sign a petition to right a wrong.

TOTAL for Section 7 _____

Section 8

____ I can imagine ideas in my mind.
____ Rearranging a room is fun for me.
____ I enjoy creating art using varied media.
____ I remember well using graphic organizers.
____ Performance art can be very gratifying.
____ Spreadsheets are great for making charts, graphs, and tables.
____ Three dimensional puzzles bring me much enjoyment.
____ Music videos are very stimulating.
____ I can recall things in mental pictures.
____ I am good at reading maps and blueprints.

TOTAL for Section 8 _____

Continued on next page

SELF-ASSESSMENT

PART II

Now carry forward your total from each section and multiply by 10 below:

Section	Total Forward	Multiply	SCORE
1		X10	
2		X10	
3		X10	
4		X10	
5		X10	
6		X10	
7		X10	
8		X10	

PART III

Now plot your scores on the bar graph provided:

	Sec 1	Sec 2	Sec 3	Sec 4	Sec 5	Sec 6	Sec 7	Sec 8
100								
90								
80								
70								
60								
50								
40								
30								
20								
10								
0								

PART IV

Here is a key to your strength(s):

SECTION 1	This indicates your naturalist strength.
SECTION 2	This indicates your musical-rhythmical strength.
SECTION 3	This indicates your logical-mathematical strength.
SECTION 4	This indicates your interpersonal strength.
SECTION 5	This indicates your bodily-kinesthetic strength.
SECTION 6	This indicates your verbal-linguistic strength.
SECTION 7	This indicates your intrapersonal strength.
SECTION 8	This indicates your visual strength.

Source: Modified from Walter McKenzie, Surfaquarium Consulting

IN BRIEF

As you think about your career possibilities, keep your results to this quiz in mind. Perhaps it will help focus your career and educational goals in this course and in life.

TYPE OF INTELLIGENCE	MEANING
Bodily-Kinesthetic	Movement-Smart
Interpersonal	People-Smart
Intrapersonal	Introspection-Smart
Logical-Mathematical	Number-Smart
Musical-Rhythmical	Sound-Smart
Naturalist	Nature-Smart
Verbal-Linguistic	Word-Smart
Visual-Spatial	Art-Smart

BusinessWeek It's Your Project

SPRINGBOARD: POP QUIZ

Can you pick the false headline? Which headline did not appear in *BusinessWeek*?
- **a.** "This Volvo Is Not a Guy Thing"
- **b.** "Hispanic Nation"
- **c.** "Investors Need to Wake Up"

This lab will require research. Check out *BusinessWeek* online for useful articles and resources.

What Is True?

No one said this was going to be easy...but it's your chance to invent your activity. This is your opportunity to become an expert in what is true. The idea is to prove whether or not your question is true. Welcome to your journey... you're in charge. The thinking is up to you.

Your project in Unit 1 is to discover, determine, and answer what is true about a topic of your choosing. This requires you to think on a higher level. Find the truth behind one of these topics listed below or come up with one on your own:

- Does democracy work?
- Do morals influence behavior?
- What obligations do you have to preserve the natural world?
- Is "reality TV" really real?
- Is speech ever unprotected?
- Will gene modification benefit future generations?
- Does history repeat itself?
- Do colors influence emotions?
- Is the Internet a wasteland of information?
- Are mind and body interconnected?

Step 1 Start with *BusinessWeek*

Orient your lab by conducting research in the *BusinessWeek* online archives. Use keywords in relation to the topic of your selection. The articles may offer some insight as you progress through this lab.

Step 2 Investigate and Engage

Ask yourself these questions:
- What do you already know about this question's topic?
- How does this topic relate to technology?
- How should you show the connection between your question and technology?
- What is it you want to learn in this project?

Step 3 Identify the Obstacles

- What facts or ideas back up your question?
- Write out the problem(s) you're trying to solve.

Step 4 Conduct Research and Seek Solutions

By now, you should have a central question you are going to investigate and solve. With that question in mind, ask yourself these questions:
- How are you going to find the answer?
- Where will you start looking for answers?

Step 5 Select Information and Analyze Data

- How do you know what information is relevant and accurate to your question?
- Translate the information into meaningful statements.

Step 6 Connect to the Real World

Create a way to showcase your project—a PowerPoint presentation, a paper, a role play, a speech, a Web page, an audio-visual presentation, and so on. It's up to you.

As you prepare your final project, be sure to emphasize:

- How does this project connect to the real world? Perhaps it connects to an individual, a family, a community, the world, and so on.
- How does this project prepare me for the real world? What can I apply directly to the real world?

Step 7 Self-Evaluation

You're used to teachers evaluating you. This time it's your chance to evaluate yourself. Create your own self-assessment for the project. Use it as a tool for practicing reflection. It might be a checklist, a conference with your teacher, a daily reflection log, an evaluation, or a small group discussion. Decide what works best for you. Be honest with yourself; determine your strengths and weaknesses during the project's journey.

UNIT 2
YOU AND E-COMMERCE

In This Unit . . .

Chapter 5
Building a Career in E-Commerce

Chapter 6
Ethical, Legal, and Social Responsibilities in E-Commerce

E-COMMERCE Online

To learn more about careers and security issues in e-commerce, visit the *E-Commerce* Web site at **ecommerce.glencoe.com**.

BusinessWeek Byte

Privacy Matters: Perils and Promise of Online Schmoozing

They say everyone is connected to everyone else in the world by no more than six degrees of separation. Now a new set of dot-coms such as Friendster, Ryze, and Tribe.net are putting that notion to the test, allowing individuals to create an electronic Web of friends, families, and business contacts, who in turn, are connected to their other friends, families, and business contacts. The idea: By linking each of us into a broad but still relevant network, I can find whatever I'm looking for—a date, a new job, or a used TV.

Social-networking sites find themselves in a Goldilocks-style dilemma: If they share too much information, the services become a spammers' paradise. Share too little, and they defeat the power of social networking, where you can discover and communicate with people you may not know but with whom you share something in common. The amount of information shared has to be just right.

Source: Excerpted with permission from Jane Black, "Privacy Matters: The Perils and Promise of Online Schmoozing," *BusinessWeek*, February 20, 2004.

To read this *BusinessWeek* article in its entirety, go to **ecommerce.glencoe.com**. Click on the Student Center, find *BusinessWeek* Articles, and go to Unit 2.

THINK ABOUT IT

What steps can you take when using a site such as Friendster to safeguard your privacy?

BusinessWeek It's Your Project

Do you ever feel like your privacy is being invaded? The Unit 2 project, "Who Has the Right to Know?" on page 134, asks you to consider the extent of your legal right to privacy.

 ecommerce.glencoe.com

Chapter 5
Building a Career in E-Commerce

Section 5-1
Basics of the E-Commerce Workplace

Section 5-2
Searching for an E-Commerce Job

PREREADING STRATEGIES
Before you read this chapter, finish these statements in your journal:
- Based on this chapter title, I predict...
- The most important thing to remember about searching for an e-commerce job is...
- Some questions I have about finding a career in e-commerce are:

BusinessWeek online
Go to *BusinessWeek* online. Browse the site and links to other sites for job opportunities. In class, discuss how your search went and the types of jobs you found.

Tools for Advancement

Building a career is a slow process that takes initiative to build your skills and look for the next opportunity. Technology is now part of just about every aspect of career planning.

Drill Down

▸ **1790 First U.S. Patent**
William Pollard of Philadelphia issues the first U.S. patent for a machine that spins cotton.

▸ **1844 New Communication**
"What hath God wrought" is the message that Samuel F. B. Morse delivers by way of telegraph to the Supreme Court in Washington, D.C.

▸ **1888 Capturing the World**
George Eastman introduces the portable box camera.

▸ **1891 Moving Stairs**
At the Coney Island amusement park in Brooklyn, New York, Jesse W. Reno introduces moving stairs at a 25 degree angle. The conveyor belt transports passengers. At the 1900 World's Fair, it is called the escalator.

▸ **1939 Using Binary Code**
Two Iowa State College colleagues complete the prototype for the first digital computer.

▸ **1975 Successful Buddies**
High school friends Bill Gates and Paul Allen form Microsoft to write computer software.

▸ **1994 Online Ads**
The *New York Times* publishes its want ads on the Web.

Get the Big Picture

Computer technology has not only created new types of jobs, but new ways to look for jobs. What type of career would you like to pursue? Pick an online job site and see what kinds of jobs it offers in your chosen field.

POWER READ

Be an active reader and use these reading strategies:

PREDICT what the section will be about.

CONNECT what you read with your own life.

QUESTION as you read to make sure you understand the content.

RESPOND to what you've read.

Section 5-1

Basics of the E-Commerce Workplace

AS YOU READ...

YOU WILL LEARN

- how your career will influence your lifestyle.
- the difference between a job and a career.
- career strategies for discovering e-commerce jobs.
- how to prepare for careers in e-commerce.
- common traits that employers desire in e-commerce workers.
- the importance of maintaining a healthy work environment.

WHY IT'S IMPORTANT

Everyone desires a lifestyle that provides job satisfaction, recognition for accomplishments, time for leisure activities with family and friends, and material possessions for a comfortable life. The career you select will be a major factor in determining your lifestyle.

KEY TERMS

- job shadowing
- internship
- mentorship
- multimedia
- Occupational Safety and Health Administration
- ergonomics

PREPARING FOR A FUTURE IN E-COMMERCE

When you look down at your dinner plate, do you see primarily one color or do you see many colors? Eating foods with a variety of colors indicates you're getting a range of nutrients (and this is a good thing). This analogy can also be applied to your lifestyle choices. Once you enter the real world of work, will you spend all your time at the job, eat at your desk, leave work for another job, and then come home to crash on the sofa? Or will you work at one job you like, take a leisurely lunch and leave on time to have the rest of the evening to yourself? All of these decisions have to do with what kind of lifestyle you envision for yourself. Balance between your professional and personal life is important.

Influences on Your Lifestyle

Take a moment to complete **Figure 5.1** individually or as a class. Estimate your real-world living expenses. These are the most important ways your career will influence your lifestyle.

- **Income.** The money you earn from working at a job will let you purchase goods and services that satisfy your needs and wants and define the type of lifestyle you will live.

- **Achievement and Recognition.** Knowing you are doing a good job and being recognized by others for that good job contribute toward your sense of accomplishment. This has a strong influence on your lifestyle.

- **Social and Leisure Time Activities.** Your job can provide you with opportunities for forming friendships and doing things with others that you enjoy.

The job you choose and the career path you follow will have a strong impact on the lifestyle you will enjoy. When you are satisfied with your job, you are more likely to succeed.

Careers versus Jobs

The line of work you select will influence your lifestyle. When you have identified a job that interests you, think of it in terms of where it will lead. A *job* is work that a person does for pay. It is usually a stepping-stone to a higher position as you advance toward your career goal. A *career* is a series of jobs built on a foundation of interest, knowledge, training, and experience.

Your career can be thought of as a path. As you grow in work experience and skills, you advance along your career path by being promoted to higher-level jobs with more responsibility. Your forward movement in your career is an indication of your success in the world of work.

How to Approach the E-Commerce Workplace

When you think of your ideal workplace, what does it look like? Who are the people you're working with? Perhaps you're working alone. Jumping into a career in e-commerce, you might encounter creative types, risk takers, decision makers, and self-starters.

PREDICT

Where do you predict you will be living in ten years?

Figure 5.1

Lifestyle Worksheet

EXPENSES	+	AMOUNT
Rent Payment	+	
Renter's Insurance	+	
Water/Sewer Bill	+	
Electric Bill	+	
Gas Bill	+	
Telephone Bill	+	
Car Payment	+	
Car Insurance	+	
Cost of Gas	+	
Groceries	+	
Clothing	+	
Entertainment	+	
TOTAL A		

DEFINING A LIFESTYLE for yourself means deciding where you want to live, how you want to live, and what career plan will fund these opportunities. If you don't already have a sense of the lifestyle you want, now is the time to start envisioning your future. *As a class, complete the expenses portion of this figure and discuss the outcomes. Are you surprised with the results? Share your impressions.*

1. Take Total A and multiply by 33 percent to get a rough estimate of the dollar amount of payroll withholding taxes you will pay (25 percent tax bracket).

2. Add the payroll withholding taxes to Total A, this will give you a new Total B.

3. Divide Total B by 160 (average hours worked in a month, based on 40 hours a week). The answer will tell you how much money you would need to earn per hour to pay your bills.

Source: Minot Public Schools, Career and Technical Education, Minot, North Dakota

CONNECTION MATH

What's in It for Me?
According to U.S. Census figures, a student who does not graduate from high school can expect average lifetime earnings of $936,000. A student who does graduate from high school can expect average lifetime earnings of $1,216,000. *How much is your high school diploma really worth?*

CONNECT
Be as specific as possible in describing the lifestyle you will be living.

SELECTING A CAREER STRATEGY You will have many opportunities to explore and experience careers in e-commerce. Your teacher or guidance counselor can assist you with your career exploration activities. If you want to sample several different e-commerce careers, a **job shadowing** experience is ideal for you. You will be able to observe several workers engaged in different aspects of e-commerce to learn the requirements of the job as well as determine your interest in this career field. You will not be paid to shadow someone. This is more of an informational interview for you to gain access into the world of work. To job shadow someone you might write a letter or contact a company that interests you. Express your interest in learning more about the kinds of jobs people have at that company. At this stage you are testing the waters to see if particular jobs really do interest you or not. The job shadow might last a few hours or a few days. Work out the details with the company to get your questions answered.

If you find a career you like through a shadowing experience, you can apply for an **internship**. With an internship, you work with employees over a period of time sampling the career tasks by completing hands-on projects. Internships are listed on company Web sites, in the bookstore or library (tell the clerk or librarian that you're looking for "a book with a current listing of internships"), or in the guidance counselor's office. Start your search early because the internship application process might start early. You could take an internship in the summer or part-time during the school year. Internships are typically unpaid, but there are some that pay. Be sure to ask your school and the company if you can earn school credit for interning.

Another strategy you can use to explore your career options is a **mentorship**. In a mentorship you will be paid for your work. You will also be assigned a mentor who will guide you, answer your questions, and help you explore and prepare for this career.

You have many strategies available for exploring and experiencing your career interests. By using these strategies you will learn the knowledge, skills, and attitudes required for success in e-commerce. Now is a good time to begin your career exploration and to plan educational experiences that will lead to an exciting entry-level job or postsecondary education in e-commerce.

TRAINING AND EDUCATION FOR A CAREER IN E-COMMERCE During high school is a perfect time to begin thinking about and preparing for an e-commerce career. By selecting appropriate courses, you begin developing the basic knowledge, job skills, and attitudes possessed by successful workers in this dynamic career field. Selecting the right high school courses will prepare you for either an entry-level job in e-commerce after graduation from high school or for admission to a technical school, college, or university where you will receive the education and specialized training needed for an e-commerce career. The key to success is to start planning the training and education you will need today.

All workers in e-commerce need effective communication skills—especially reading and writing skills. As an e-commerce worker, you may need to read technical manuals, news and information releases, tables and charts, as well as correspondence. You may need to write copy for product announcements and advertisements, letters in response to customer inquiries, or text for your company's Web page. Creative communication skills are essential for success in e-commerce. Good grades are required in English courses and language arts elective courses, such as creative writing and technical communications.

Everyone in e-commerce careers works with technology, computers, and the Internet. Elective courses at your school in computer applications, Web-page design, **multimedia** presentations, and programming will help you develop the skills used in e-commerce jobs. Multimedia combines text, graphics, motion, and sound to convey a message. Usually a computer software program is used to create the multimedia presentation.

Many workers in e-commerce use math skills every day on the job. These workers know, understand, and are able to apply algebra, statistics, probability, logical reasoning, and basic math skills when solving problems and making decisions on the job. Doing well in your math courses and taking electives in statistics and probability will help you prepare for a career in e-commerce.

Depending on which e-commerce career you like, you may need to take courses in art or business. The art electives would help you develop artistic skills used by workers in the creative design jobs in e-commerce. Business and marketing courses will help you develop the knowledge, skills, and attitudes needed in management, finance, and marketing e-commerce jobs. This business course that you are currently taking is helping you to learn about the many career options in e-commerce.

MULTIMEDIA You might have to create and give a presentation. *How does a multimedia presentation grab your audience's attention?*

The E-Commerce Workplace Environment

Technology has changed the workplace and given workers options on when and where they may complete their job tasks and work assignments. Not all e-commerce companies have a traditional bricks-and-mortar building. Some e-commerce businesses are completely based on computers in workers' homes.

As an e-commerce worker, you may be able to work at your home when it is convenient for you. Technology brings the workplace to you instead of you traveling to the workplace. You can complete your work activities on the computer in your home.

The Internet allows businesses to be open 24 hours a day, 7 days a week. This may allow you the flexibility of setting a work schedule that is convenient for you and blends in with your other commitments. You and your fellow workers at an e-commerce business may not need to work the same hours.

QUESTION

What factors do you think will most influence your lifestyle?

EQUAL OPPORTUNITY
Competition is a way of life. You are competing with others to get your foot in a company's door. *What separates you from the other job applicants?*

Some e-commerce companies like Alloy.com maintain a traditional business office and warehouse. If you choose to work for this type of e-commerce business, you will go to its office or warehouse facility to work a traditional 8:00 A.M. to 5:00 P.M. schedule. With these types of companies you will be working in e-commerce but in a traditional business environment.

If you start your own e-business company or work for a company that sells its products on Internet auction Web sites like eBay, your workplace will most likely be a computer in your home. You will be able to work when it is convenient for you to complete the business activities necessary for selling the merchandise.

E-commerce is an equal opportunity employer. It does not discriminate against any workers and has a diverse workforce. Workers with disabilities, limited work availability, and busy schedules can be successful in e-commerce.

An e-commerce career offers flexibility in where you will work and when you will work. Your computer connects you to the rest of the world where you can buy and sell products and services, find relevant information, and complete the activities needed to make your business successful.

ERGONOMICS You want to work in a healthy, comfortable, and productive environment. Often workers experience discomfort from working long hours at a computer. These discomforts, if left untreated, may become serious injuries, including musculoskeletal disorders (MSDs) such as Carpal Tunnel Syndrome, upper body strain, eyestrain, and headache.

A recent **Occupational Safety and Health Administration** (OSHA) press release states that 1.8 million workers a year report work-related musculoskeletal disorders to their employer. More than 600,000 of these injured workers take time off from work because of their injuries, costing employers over $13 billion in lost productivity.

Ergonomics is the study of the physical, environmental, and emotional areas of work. It looks at how workers perform tasks by studying their capabilities and limitations, and the tools, equipment, and furniture they use. Its goal is to increase worker productivity by achieving a healthy and comfortable fit between the worker and the computer workstation. Recommendations from ergonomic studies, which you could use at school or home as well as in the workplace, include the following:

- varying the work tasks throughout the day to avoid fatigue.

- eliminating glare on the monitor screen to make it more comfortable and easier to read the displayed information.

Unit 2 You and E-Commerce

- taking occasional breaks and doing stretching exercises to help restore energy, loosen tight muscles, and increase productivity.
- organizing your work area to reduce stress.

Common Traits for Those Working in the Field

Individuals who are most successful working for an e-commerce company have several common traits. They enjoy the creativity of problem solving and are not afraid to take risks and make decisions. These workers are Internet literate and are committed to keeping pace with the quickly changing Internet. They are self-starters who show initiative by willingly assuming additional responsibilities. The e-commerce workplace is ideal for the creative employee who is flexible, wants to be challenged, and exhibits a strong work ethic. It is a work environment that encourages and fosters competition.

The e-commerce environment is constantly changing. E-businesses open and close their doors about as often as restaurants in a metropolitan area. Each business seeks a market segment to control over the competition. Therefore, workers in today's e-commerce world need to be flexible to adapt to their ever-evolving job descriptions and responsibilities. At times job assignments can be vague and unclear, and even chaotic, but exciting and challenging.

Section 5-1 Review

RESPOND to what you've read.

1. What is the difference between a job and a career? _____

2. What are some strategies you can use to learn about careers in e-commerce? _____

3. What can you do when working on a computer to make your workstation safe and healthy for yourself? Explain how each suggestion will help you remain safe and healthy. _____

Quick TALK

Visualize yourself five to ten years from now working for an e-business. Working with a partner, describe your e-commerce job and your desired lifestyle. *How will your job enable you to live that lifestyle?*

Section 5-2

Searching for an E-Commerce Job

AS YOU READ...

YOU WILL LEARN

- about growing career opportunities in e-commerce.
- sources available to provide you with job and career information.
- what to include in an electronic résumé and portfolio.

WHY IT'S IMPORTANT

Most likely you will work for 50 years during your lifetime. Assuming an average work week of 40 hours, this amounts to over 100,000 hours that you will spend in the workplace. Therefore, it's important you select a career that is stimulating and enjoyable.

KEY TERMS

- Occupational Outlook Handbook
- electronic résumé
- digital portfolio

PREDICT

What resources are available for you to learn about careers in e-commerce?

GETTING THE E-COMMERCE JOB YOU WANT

When it comes to computer and information systems jobs, about two in five are in the service sector. While most of these service firms using computer and information managers are insurance companies and financial services, service firms also include protective services and health care. In order to get to the managerial level, you would need strong technical knowledge, work experience, and formal training. Before you climb to the top of the career ladder in computer science, you need to explore the entry-level positions that are available.

Trends in E-Commerce

According to the **Occupational Outlook Handbook** (OOH), a career reference manual written and updated by the U.S. Department of Labor, Bureau of Labor Statistics, the number of computer and data processing specialists' jobs is expected to grow 86 percent in the first decade of the twenty-first century. This will be the fastest growing of all career fields.

The growth of the Internet and the rapid growth of electronic commerce will increase opportunities in careers specializing in developing and maintaining Web sites, creating and maintaining corporate Intranets, designing sophisticated computer networks, developing hardware and software products, and developing and maintaining site security systems. These career fields will increase faster than the average for all occupations through the year 2010.

With the Internet and electronic businesses creating tremendous volumes of data, there is a growing need to store, manage, and retrieve data effectively and efficiently. There will be above average increases in job openings (at Internet service providers, for example) for those who have data management skills.

Online sales of products and services are expected to grow rapidly, creating additional job opportunities for Internet sales managers, Web masters, and technical support workers. These e-commerce retailing jobs will grow much faster than similar retailing jobs in traditional retail stores.

The U.S. Department of Labor projects that each of the e-commerce jobs listed in **Figure 5.2** will have a large number of job openings through 2010. Jobs in these career fields offer growth opportunities for you.

Figure 5.2

Future Jobs in E-Commerce

E-COMMERCE FUTURE Given the outlook for the information technology industry presented in this table, what are your conclusions based on the total employment and the job openings? *Is this an optimistic snapshot of the industry?*

Occupation	Total employment in 1,000s		2000-2010 increase in total employment		2000-2010 average annual job openings in 1,000s		Median annual earnings	Education/ training category
	2000	2010	Number in 1,000s	Percent	Due to growth and total replacement needs	Due to growth and net replacement needs		
Computer programmers	585	680	95	16.2	45	22	$57,590	Bachelor's degree
Computer support specialists	506	996	490	97.0	89	51	$36,460	Associate's degree
Computer systems analysts	431	689	258	59.7	60	30	$59,330	Bachelor's degree
Computer and information systems managers	313	463	150	47.9	43	20	$78,830	Degree plus work experience
Network and computer systems administrators	229	416	187	81.9	37	20	$51,280	Bachelor's degree
Training and development specialists	204	244	40	19.4	24	8	$40,830	Bachelor's degree
Graphic designers	190	241	51	26.7	33	7	$34,570	Bachelor's degree
Network systems and data communications analysts	119	211	92	77.5	19	10	$54,510	Bachelor's degree
Database administrators	106	176	70	65.9	15	7	$51,990	Bachelor's degree
Technical writers	57	74	17	29.6	7	3	$47,790	Bachelor's degree

Career Opportunities in E-Commerce

The Computing Technology Industry Association lists six categories for tech-related jobs: Internet Site Designer, Project Manager, Internet Network Specialist, Internet Security Specialist, Internet Application Developer, and Internet E-Commerce Specialist.

The Technical Careers Chart shown in **Figure 5.3** highlights some entry-level e-business and Internet jobs that would start your journey down a career path in e-commerce. For each career path, job tasks, skills the worker would use, and education and training required are listed. Do any of these jobs match your interests? What additional information would you like to know about this job? Use the Internet to research your chosen job to find additional career facts and information.

Figure 5.3

Internet and E-Business Types of Jobs

E-COMMERCE JOBS Positions in the field of e-commerce offer an array of duties that require different skill levels and training requirements. *What jobs are good for people with creativity?*

Entry-Level Job Title	Internet Application Developer	Internet E-Commerce Specialist	Internet Site Designer
Alternate Job Titles	• Web Developer • Internet Systems Developer • Internet Developer	• E-Commerce Designer	• Site Designer
Essential Function	• Develops, implements, maintains, prepares, and tests Web applications	• Develops and applies e-commerce marketing goals • Develops e-commerce Web site • Formulates and manages e-commerce sites	• Applies human factors to Web site design • Creates and implements Web content • Monitors Web activity
Sampling of Skills	• Works well with team • Uses programming standards • Customizes Web page hit counters • Converts media specifications to computer language	• Ability to sell one's ideas • Business sharpness • Works well on a team • Self-motivated • Online check processing	• Ability to sell one's ideas • Artistic • Creativity • Self-motivated
Knowledge and Training	• Less than two years of experience in the field • Knowledge of difference in client versus server scripting • Knowledge of development tools and data manipulation tools	• Two to four years experience in the field • Knowledge of script development tools • Knows Internet access options and costs • May require a bachelor's degree	• Two to four years experience in the field • May require a bachelor's degree or a technical certification

Additionally, if you are interested in marketing or writing, there are excellent job opportunities in Internet marketing and content production. For example, the Direct Marketing Association's commissioned study found that online direct marketing sales reached more than $136 billion in 2005 and more than 730,000 jobs are forecasted in this career field. A career opportunity in e-commerce marketing might require you to work at an advertising, marketing, or Internet agency where you would perform a variety of job tasks, such as conducting statistical analysis of online marketing efforts. If an Internet marketing career interests you, taking business and marketing courses in high school will help you prepare for your first job in this field.

CONNECT

Why do you think family members would be good sources of information on careers in e-commerce?

Internet Network Specialist	Internet Security Specialist	Project Manager
• Internetworking Professional	• Internet Security	• Associate Project Coordinator • Lead Program Manager • Project Leader
• Analyzes and applies core protocol analysis methods • Deploys and manages the TCP/IP network • Monitors network traffic and congestion • Tests performance and transfer times	• Develops countermeasures to security issues • Identifies security threats • Performs various phases of security audits • Supports security policy	• Defines, monitors, and revises project scopes • Evaluates and mitigates project risks • Oversees communication and reporting on projects
• Creativity • Decision making • Detail oriented • Rapid response to problems	• Creativity • Conscientious • Detail oriented • Problem solving • Works without supervisor	• Ability to sell one's ideas • Applies analytical thinking • Creativity • Decision making • Detail oriented • Flexibility • Judgment • Leadership
• Less than two years experience in the field • Knowledge of IP addressing rules and classes	• Less than two years experience in the field • Requires a bachelor's degree or higher • Understands network security issues and solutions	• Two to four years experience in the field • Knowledge of human motivation theories • May require a bachelor's degree or a technical certification

Source: The Computing Technology Industry Association

In a content-producing career you might work for a television or radio station writing news stories, for an e-business writing press releases, or for a publishing company writing text for books, magazines, or newspapers. If creative or technical writing jobs interest you, taking creative writing and technical writing courses in high school will help you prepare for a career in content producing.

DEVELOPING SEARCH TACTICS FOR YOUR JOB

Thinking about a career in e-commerce? You can find the OOH online, in your library, or in a guidance counselor's office. By looking up a career, you can learn the type of tasks a worker performs on the job, the training, skills, and experiences needed for success, the working conditions, yearly salary, and projected number of yearly job openings.

How to Look for a Job

You can find career information by reading professional and trade magazines in the career field that interests you. Talk to friends or family who work in these career fields to learn more about the type of activities and tasks they perform on the job. You will not find "e-commerce jobs" in the newspaper's classified section. Instead you will have to look under such headings as customer service, health care, hospitality, information technology, manufacturing, publishing, retail, or sales and marketing. All the jobs under the e-commerce umbrella do not logically fit in one category. E-commerce jobs are part of many different industries.

The fastest growing source of career information is on the Web. As the Web grows, the amount of career information it provides is continually expanding. You can use the Internet for a variety of job-seeking tasks, including job openings, salary and benefit information, education and experience required, future trends, and career paths.

Web Site Success

IT PROVIDES A LESS SCARY PROCESS
Monster.com

After graduation you're thinking a summer job in Boise or Toronto sounds adventurous. They are miles apart, but each appeals to you for some reason. Monster.com, an online recruitment portal, makes it easier than ever for employees and job seekers. With 20 million résumés in its databases and companies in 21 countries looking for the right candidate, you can create your own profile for someone to pick you. Monster.com provides career advice, like résumé tips, interviewing strategies, salary information, networking etiquette, and relocation tools.

Thinking Critically
More and more résumé experts advise job seekers to make their résumé scannable. *Name at least three techniques you should use in order to make your résumé scannable.*

Creating an E-Résumé and a Digital Portfolio

A résumé is a structured, written summary of your job objective, education, employment history, and job qualifications. It is your advertisement, designed to create an interest in an employer to learn more about you by inviting you in for a personal interview. Your résumé should focus attention on your best features by matching your strengths with the requirements of the job. An effective résumé is your key for getting the job of your choice.

There are different types of résumés you can create to market yourself. The traditional paper printed résumés include chronological, functional, and combination. A chronological résumé lists your work experience in a timeline approach, from present to past. If you do not have a lot of workplace experience, this might not be the best approach for you. A functional résumé groups your work experience and skills together but focuses more on your skills exhibited. You can use this one to highlight your skills if you don't have extensive work experience. Lastly, the combination résumé highlights your work experience and skills, each given equal weight.

Along with a traditional résumé, create an **electronic résumé**, or e-résumé, to send to potential employers via e-mail or the Internet. Look at the electronic résumé in **Figure 5.4** as one example you could emulate. Most Fortune 500 companies today encourage applicants to submit

QUESTION

What sources will you use to learn about careers?

Figure 5.4
Electronic Résumé

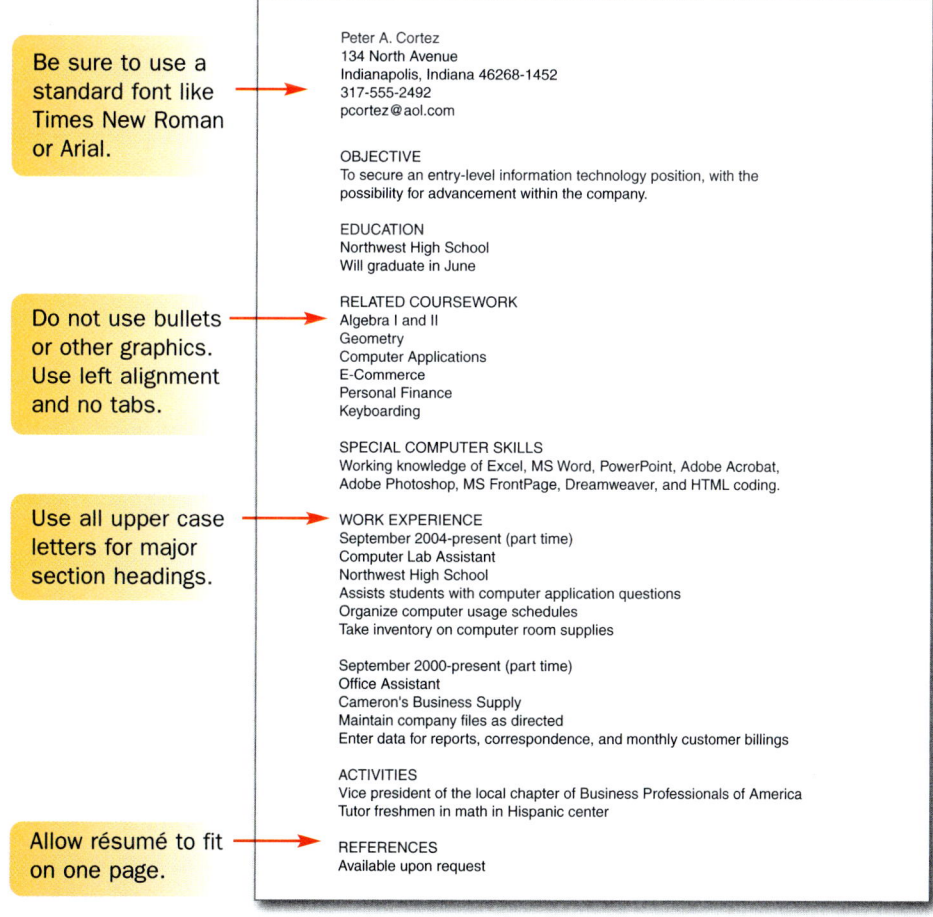

- Be sure to use a standard font like Times New Roman or Arial.
- Do not use bullets or other graphics. Use left alignment and no tabs.
- Use all upper case letters for major section headings.
- Allow résumé to fit on one page.

ELECTRONIC RÉSUMÉS With the competitive job market, a snail mail résumé on clean, white bond paper doesn't always make it to the hiring manager's desk on time. An electronic résumé can be your answer to speedy delivery. *What are some reasons for completing an e-résumé?*

Chapter 5 Building a Career in E-Commerce

an electronic or scannable résumé first. If a prospective employer later requests an interview with you, a traditional résumé can be submitted with your completed application.

An electronic résumé contains the same information as a traditional résumé except the style and format must be computer-friendly. Your electronic résumé eliminates graphics, boldface print, underlines, italics, and tab settings. Most scannable résumés are converted to plain text using ASCII format so any scanner on any computer can read it.

Your electronic résumé should also include a 15 to 20 keyword summary that defines your skills, education, experience, and job qualifications. Select nouns for your keywords that an employer will understand and use when searching résumés for a candidate to fill a job opening. Checking want ads in the newspaper will give you an indication of the "buzzwords" an employer will look for when scanning a résumé. This keyword list follows your name and address on the résumé. There are many different formats available. Go to the bookstore or library, or complete an online search to find résumé formats.

Applicants seeking jobs in the field of e-commerce may be asked to provide employers with a **digital portfolio**. This hypermedia document is a collection of screens or pages linked together with navigational buttons for easy access. Each screen emphasizes a personal or employment strength that would be important for a person seeking a job in e-commerce. Your digital portfolio would include any honors and awards you have received, any relevant tasks, activities, or jobs you have completed, and any job-related skills you possess. It is an opportunity for you to highlight your strengths to a prospective employer.

Section 5-2 Review

RESPOND to what you've read.

1. What makes an effective electronic résumé? _____

2. Which jobs are expected to have the highest growth rate over the next ten years? _____

3. What sources are available to you when looking for a job in your career field? _____

Quick TALK With a partner, brainstorm words that you could list in the keyword summary of your electronic résumé to define your skills, education, experience, and job qualifications. *Why did you select these words to describe yourself to a potential employer?*

Name _____ Date _____

Worksheet 5-1

Describing Your Future Lifestyle

Think about the ideal lifestyle you want as an adult. In the space below, create a representation of your desired future lifestyle by drawing a diagram or picture, creating a collage, or writing a short essay. Based on the lifestyle you have described, answer the questions below.

1. Describe the education and training you will need to achieve this lifestyle. _____
   _____
   _____
   _____

2. Describe the type of interests, hobbies, or activities you would like to pursue in your leisure time. _____
   _____
   _____
   _____

3. Describe the achievements and recognition you hope to receive in your lifetime.
   _____
   _____
   _____

Name _____ Date _____

Worksheet 5-2

Learning About an E-Commerce Job

Check your local or a major newspaper's help-wanted ads and pick out three e-commerce jobs that interest you. Then answer the following questions:

1. Why did these jobs interest you? Explain. _____
   _____
   _____

2. What similarities are there among the three jobs? _____
   _____
   _____

3. What differences are there among the three jobs? _____
   _____
   _____

4. Select one of the jobs advertised and describe the qualifications needed for that job. _____
   _____
   _____

5. How would you prepare for that job? Use your imagination to determine the kinds of knowledge, training, and experience needed for the job. _____
   _____
   _____

6. Using the Internet, research this e-commerce job. Prepare a written career report that includes the type of tasks a worker performs on the job; the training, skills, and experiences needed for success; the working conditions, yearly salary, and projected number of yearly job openings.

7. Create a multimedia presentation to accompany an oral summary of your career report to the class.

Chapter 5 Review and Activities

Chapter Summary

Section 5-1

- The career you select will be a major factor in determining your lifestyle. Your income, achievement, and recognition satisfaction, and social and leisure time activities affect your lifestyle.

- A job is work that you do for pay. A career is a series of jobs that you move through as you gain knowledge, training, and experience to advance along a path toward greater responsibilities.

- Use job shadowing, internships, and mentorship as real-world opportunities to prepare yourself for a career in e-commerce.

- Take academically challenging courses in high school to prepare for life after graduation.

- Individuals working in e-commerce are often creative problem solvers, risk takers, and decision makers. A keen knowledge of technology and the Internet helps a person maneuver through the industry. While workers consider themselves self-starters, they are also team players. Workers are able to adapt to change, which helps in e-commerce's dynamic environment.

- To reduce your risk of computer-related injuries, you should vary your work tasks, eliminate glare on your monitor screen, take occasional breaks, do stretching exercises, and keep your work area organized.

Section 5-2

- Careers expected to have higher than average job growth include software and hardware support specialists, system administrators, retail and wholesale salespersons, customer service representatives, and programmers.

- Job applicants often use the Occupational Outlook Handbook, family and friends, the Internet, professional and trade magazines, and newspaper advertisements as resources for finding a job.

- An effective résumé contains information about your career objective, education, employment history, and job qualifications. An electronic résumé is a scannable list of 15 to 20 keywords that a human resources department uses to match a candidate's qualifications with the job requirements.

Chapter 5 Review and Activities

Review of Key Terms

a. job shadowing
b. internship
c. mentorship
d. multimedia
e. ergonomics
f. Occupational Safety and Health Administration
g. Occupational Outlook Handbook
h. electronic résumé
i. digital portfolio

Match each term to its definition.

_____ 1. A short-term unpaid work experience in which you receive training and hands-on experience in a specific career area.

_____ 2. An online career reference written and updated by the U.S. Department of Labor that provides information about the type of tasks, training, skills, experiences, working conditions, yearly salary, and projected job openings for a career.

_____ 3. A visual presentation that combines text, graphics, motion, and sound to convey a message.

_____ 4. An electronic collection of your goals, achievements, honors, and reflections that demonstrate career growth over a period of time.

_____ 5. A work experience where you are paid for your work and receive professional guidance and training from an experienced worker in your selected career field.

_____ 6. The department of the U.S. government responsible for ensuring the safety and health of the work environment.

_____ 7. The study of the physical, environmental, and emotional areas of work.

_____ 8. Learning about a job by following a competent worker to witness firsthand the work environment, the skills needed, and the tasks performed in a career area of interest.

_____ 9. A computer-friendly document that uses keywords to provide an employer with information regarding a job candidate's professional experience, education, and job qualifications.

Applying Technology to Academics

English Language Arts—Reading

Using the Internet, search the online version of the Occupational Outlook Handbook for an e-commerce career of interest to you. For the career selected, identify the tasks a worker performs, the training, skills, and experience required, the working conditions, the average yearly salary, and projected job openings for this career.

English Language Arts—Writing

Identify a career in which you have an interest. Assume you are applying for an internship in that career field, and create an electronic résumé based on your qualifications for this internship using a word-processing software program. Go to ecommerce.glencoe.com for a sampling of online résumé templates you can use.

English Language Arts—Speaking

Using the information you learned from searching the Occupational Outlook Handbook, prepare an oral presentation for your class about the career you researched. Prepare three visuals that can be used to enhance your presentation. You might prepare these in PowerPoint, Word, Adobe PhotoShop, Adobe Illustrator, or QuarkXpress.

Graphing Software Program

Select three e-commerce career paths of interest to you. Using the Internet, look up the salary for an entry-level job, an intermediate-level job, and a top-level job in each of the three career fields. Using a graphing software program, create a chart that displays the salary information you found.

Chapter 5 Review and Activities

Critical Thinking

1. Refer to Figure 5.2 on page 101. Explain why computer and information systems managers make more than twice what graphic designers do?
2. Refer to Figure 5.2. Determine what position you are most likely to start at upon entering the e-commerce workforce?

COMPETITIVE EVENT

You're to assume the role of employee of a small needlepoint shop. Your shop is located in Honolulu, Hawaii. The store is well known for its hand-painted canvases of the beautiful tropical flowers native to Hawaii. Local artists paint the canvases. The prices for these canvases range from $50 to $100, depending on the size of the canvas. The store's customers are primarily residents of Honolulu and tourists. The store's owner (judge) would like to increase the store's sales. To help with that goal, the store's owner (judge) has asked you about what it would entail to create a Web site for the store, complete with an online catalog. You are to make your recommendations about the proposed Web site to the store's owner (judge).

EVALUATION

You will be evaluated on how well you meet the following performance indicators:
1. Identify ways that technology impacts business.
2. Create and post a basic Web page.
3. Describe current business trends.
4. Handle customer inquiries.
5. Make oral presentations.

INTERNET ACTIVITY

To engage in online activities, visit the *E-Commerce* Web site at **ecommerce.glencoe.com**.

BusinessWeek online

TOPIC: E-Résumés

While the days of paper résumés aren't necessarily behind us, modern businesses—and modern job-seekers—are increasingly enjoying the benefits of e-résumés. Easily customizable and cost-effective, e-résumés improve upon every feature of a standard résumé except one: tangibility.

ACTIVITY Go to the Student Center at **ecommerce.glencoe.com**. Click on *BusinessWeek* Activities and open Chapter 5. There you'll learn more about e-résumés—what they are and how they're created.

Chapter 6

Ethical, Legal, and Social Responsibilities in E-Commerce

Section 6-1
Internet Law and Ethics

Section 6-2
Privacy Issues and the Internet

Section 6-3
Internet Security

PREREADING STRATEGIES

Before you read this chapter, finish these statements in your journal:

- Based on this chapter title, I predict...
- The most important thing to remember about ethical, legal, and social issues in e-commerce is...
- Some questions I have about ethical, legal, and social responsibilities pertaining to e-commerce are:

BusinessWeek online

Go to *BusinessWeek* online. Find an article about an online security issue, such as antivirus software or piracy. Read the article, then write in your journal about the consequences of unethical behavior on the Internet.

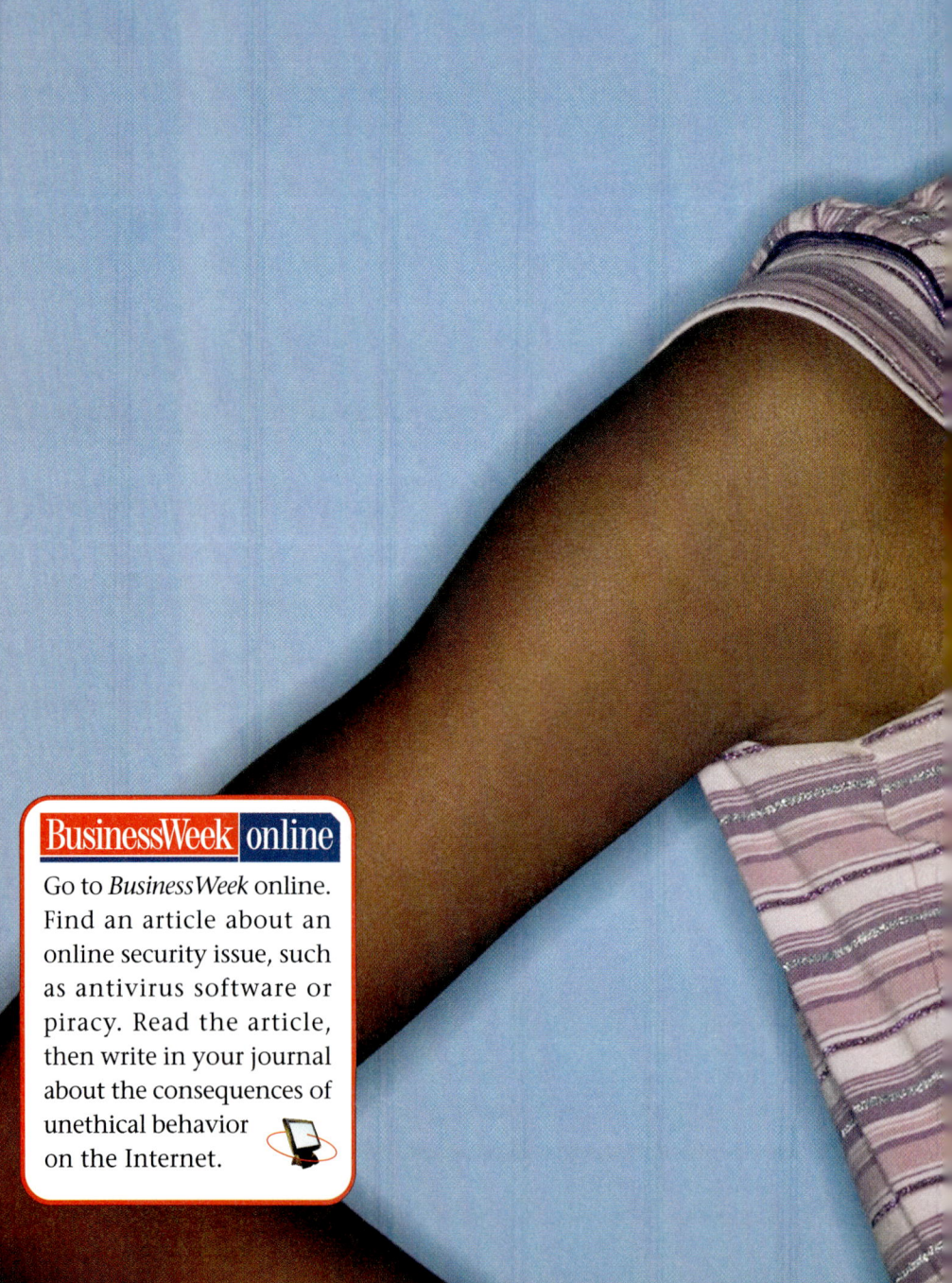

112   ecommerce.glencoe.com

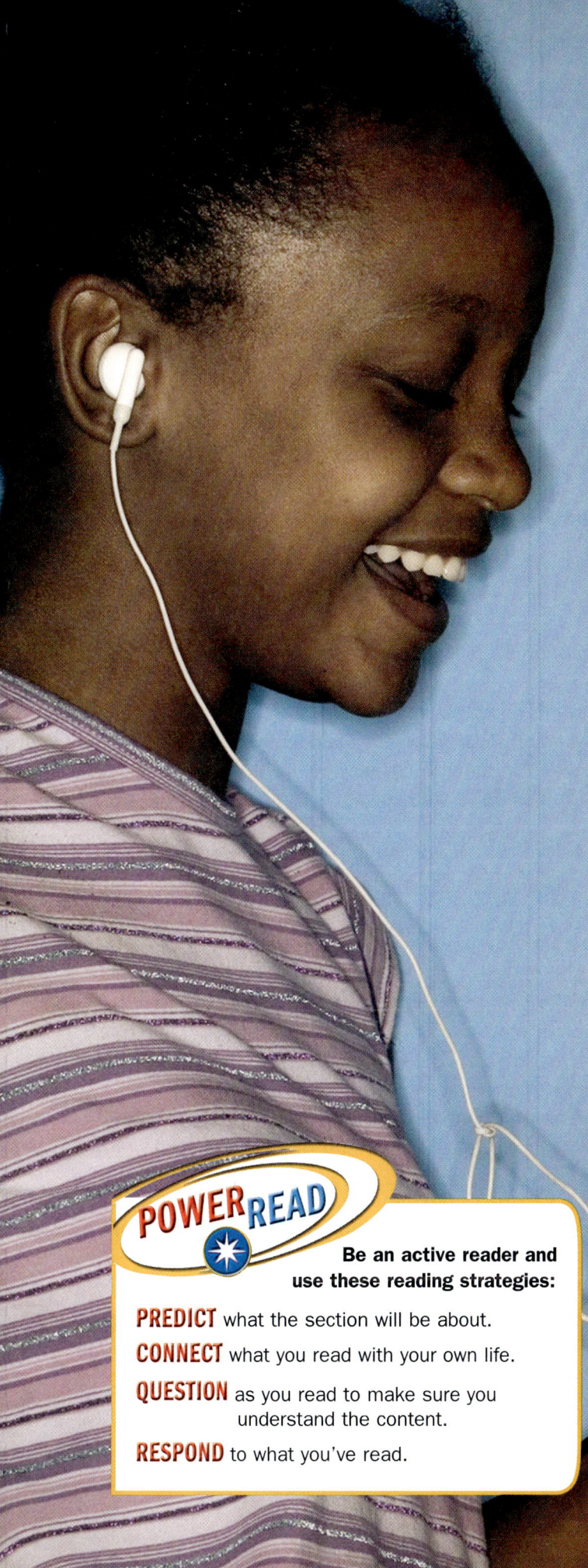

Safeguards Against Crime

Every new invention creates new opportunities for illegal activities and requires new laws to combat them. Protecting privacy and intellectual rights on the Internet poses unique challenges.

Drill Down

▶ **1858 Crime Prevention**
Edwin T. Holmes installs the first electric burglar alarm in Boston, Massachusetts.

▶ **1865 Money Makers**
The U.S. Treasury Department estimates that 50 percent of all U.S. paper currency in circulation is counterfeit.

▶ **1925 Safe Deposit**
The Safe Cabinet Company tests the first bank vault.

▶ **1984 Going Mobile**
Motorola introduces the first cell phone, available for $4,000.

▶ **1993 White House on the Web**
The White House enters the World Wide Web by creating its own Web site and e-mail accounts for the president and vice president.

▶ **1999 Presidential Hack**
Hackers shut down White House Web site for the first time for six hours with a "denial of service attack."

▶ **2004 Phone Bug**
Researchers invent the first virus program capable of spreading from cell phone to cell phone.

Get the Big Picture
What types of criminal activities has the Internet created or made easier to pursue?

POWER READ

Be an active reader and use these reading strategies:

PREDICT what the section will be about.
CONNECT what you read with your own life.
QUESTION as you read to make sure you understand the content.
RESPOND to what you've read.

Section 6-1

Internet Law and Ethics

AS YOU READ...

YOU WILL LEARN

- why laws are a necessity in the online world.
- how cyberlaw is designed to protect companies and people.
- why easily recalled domain names are an asset to businesses.
- how patent and copyright laws protect against the misuse of original work.
- about trademark laws that are unique to e-commerce.

WHY IT'S IMPORTANT

Laws are necessary in all segments of society, and e-commerce is no exception. Those in the online world should recognize that copyright, patent, and trademark laws protect much of the material found on the Internet. In addition, they should be aware that there are other types of laws they must follow.

KEY TERMS

- cyberlaw
- copyright
- trademark
- patent

LEGALITIES OF THE INTERNET

The Internet landscape may sometimes seem like the Wild West—an uncharted territory with surprises at every pass. Users may often feel invisible in a Web of computer systems, networks, satellites, and cables. Since there are no real storefronts, face-to-face encounters, or handshakes to seal a deal, it may seem like a place where anything goes. While most transactions and behaviors on the Internet are ethical, some individuals and groups attempt to use the Web for disreputable or illegal purposes. What laws apply to e-commerce? What ethical responsibilities do consumers and sellers have as they participate in transactions?

Cyberlaw

Just as the real world requires rules and regulations to keep things running smoothly, the Internet has given birth to a host of laws that govern how we interact online. **Cyberlaw**—an evolving legal framework that governs Internet activities—covers topics ranging from copyright infringement to e-mail privacy, identity theft, and interstate e-commerce. In fact, courts around the world are creating Internet law right now. Judges face the challenge of applying the legal frameworks of the physical world to online disputes, which is not always an easy task. Let's take a look at a few of the most common legal issues faced by Web site creators and users today.

COPYRIGHT Although information sharing is one of the primary uses of the Web, not everything posted online is available for reuse or downloading. Copyright protects most text, photos, sounds, software, logos, images, and videos on the Web. **Copyright** is a form of protection provided by U.S. law that grants exclusive rights for the original works of an author. These works may include software, articles, books, music, art, or performance pieces. Copyright protection is available for both published and unpublished works.

For works created after 1978, copyright generally lasts for the life of the author plus 70 years. If you held copyright protection on a piece of your artwork, it would have begun at the time the work was created—that's when it became your property. Only you, the author, or those you grant rights to, can rightfully claim copyright.

As you create your Web site, you may be tempted to look at other sites for design ideas, graphics, or text. Remember that under copyright law, you may not use or copy someone's work unless you have been given permission to do so.

To protect your property online, you should place a copyright notice on your Web page and advise Web browsers about what they may and may not do with your work. A *copyright notice* should contain the word "Copyright" or the symbol ©, the year in which the work was created, and the name of the copyright owner. At the bottom of most Web pages, you will find a copyright notice. For example, the Sun Microsystems Web site posts the following notice:

Copyright 1994–2004 Sun Microsystems, Inc.

The Microsoft Web site uses this format:

©2004 Microsoft Corporation. All rights reserved.

In addition, most sites include more detailed information on copyright and how their Web materials may or may not be used. Take a look at the copyright information provided in **Figure 6.1** from the Web site for the San Francisco Museum of Modern Art.

Simply posting a copyright notice does not necessarily prevent your work from being copied or used elsewhere. If you have valuable content you want to protect, consider using passwords for access, or do not post the material on your Web site at all.

If you wish to sue someone for using your work, your copyright must be registered with the Library of Congress. The registration process is inexpensive and may be done at any time up to three months after the infringement has occurred.

PREDICT

Why are disputes over domain names more common today than they were five years ago?

Figure 6.1

Copyright Information

San Francisco Museum of Modern Art Web Site
COPYRIGHT

SFMOMA is committed to protecting the copyrights and other intellectual property rights of creative artists and others. SFMOMA has taken all reasonable steps to ensure that all reproductions on this Web site are produced according to applicable laws. Copyrights and other proprietary rights in the materials on this site may subsist in individuals and entities other than SFMOMA. You hereby acknowledge and agree that, subject to valid third-party rights, SFMOMA retains all intellectual property rights, including without limitation copyrights, trademarks, service marks, patents, and other proprietary rights, in and to the texts, images, marks, data, and all other content contained in this Web site (collectively "Materials"). SFMOMA expressly prohibits the reproduction, distribution, transmission, sale, transfer, creation of derivative works, modification, public display, public performance, publication or any commercial exploitation of any Materials on this Web site, except for the purposes of fair use as defined in the copyright laws, and as described below.

PROPERTY RIGHTS Many Web sites include copyright information to remind Web-site users of what they may and may not lawfully use. *Which types of works are protected by copyright on the San Francisco Museum of Modern Art's Web site?*

SFMOMA
sfmoma.org Copyright Statement
© San Francisco Museum of Modern Art

CONNECT

Which breakfast cereal do you eat most often? Why is it important for the cereal's manufacturer to ensure the product name is trademarked?

DOMAIN NAME DISPUTES Today's businesses realize that having a domain name that is the same as their company's or product's name can be extremely valuable. As a result, disputes over domain names have become more common.

Imagine you've decided to open a Web-site-design business called Blue Sky Design. You'd like your domain name to reflect your company's name: BlueSkyDesign.com. If you discover someone else owns the domain name, you may decide to choose a different name or attempt to buy the domain from its current owner.

When disputes over domain names occur, the issues can be settled in court by presenting legal arguments about why registered domain names should be cancelled or transferred to another party. Often, the courts rely on trademark law to settle these claims. In the following section, you will learn about trademarks and how they relate to the world of e-commerce.

TRADEMARK The expansion of the World Wide Web has led to increased trademark-infringement claims and lawsuits. Have you noticed the ® symbol next to a product name such as McDonald's McGriddles® Breakfast Sandwiches? This symbol is used to identify a trademark. A **trademark** is the use of a word, phrase, symbol, product shape, or logo by a manufacturer or business to identify its goods and to distinguish them from others on the market. Similar to a trademark, a service mark is used on services to identify their uniqueness and is also protected by trademark law. A registered trademark (®) is one that has been *registered* with the federal and/or state government and can be secured by the U.S. Patent and Trademark Office for a fee.

Though trademarks do not have to be registered, this action provides more legal protection. The ™ symbol signifies that application has been made but trademark registration has not yet taken place.

When a business uses a name or phrase trademarked by another company or person to identify its own product, the trademark has been *infringed*. Trademark infringement can cause confusion about the product's source or sponsorship. Imagine if you decided to offer McGriddles

▶ **KNOCK-OFFS** Despite the efforts of Harry Potter author J.K. Rowling, imitations of her best-selling series continue to invade the market. *Tanya Grotter and the Magic Double Bass* sells for less than half the cost of a Harry Potter book in Moscow. *Harry Potter and the Leopard-Walk-Up-To-Dragon* was published in China by an author falsely claiming to be Rowling. **Why do you think patent- and copyright-infringing material continues to be produced?**

Figure 6.2

Patented Business Process

© 2004 Amazon.com, Inc. All rights reserved.

COMPETITIVE ADVANTAGE Amazon.com patented its 1-Click payment process in 1999. *What advantage do you think this patent gives Amazon over its competitors?*

sandwiches at your sandwich shop. They could be easily confused with McDonald's McGriddles breakfast sandwiches—a federally registered trademark. McDonald's could sue you for trademark infringement and force you to stop using the McGriddles name.

PATENT Have you ever had a great idea for a new product or business process and wanted to protect the idea as your own? A patent is a property right granted to the inventor of a product or process by the U.S. Patent and Trademark Office. Generally, a new patent lasts 20 years from the date on which its application was filed. A **patent** excludes others from "making, using, offering for sale, or selling" the invention or idea in the United States or "importing" the invention into the United States.

Patents may be granted for new products such as a biodegradable cotton material or new business processes such as Amazon.com's 1-Click system. The 1-Click system, shown in **Figure 6.2**, expedites online orders by allowing repeat customers to bypass certain data-entry forms by using the information already stored in their accounts.

Since 1998 an increasing number of patents have been issued to software and Internet companies that have devised novel ways of doing business. These patents, which usually combine software with business methodology, are commonly referred to as *business method patents* or *Internet patents*. A company that develops or acquires such a patent can block others from using the business method for approximately 17 years.

Ethical Considerations on the Internet

Because the Internet is constantly changing, adherence to an ethical framework is more important than ever. While there may not be written laws governing every e-business situation, an ethical person will do the right thing in these circumstances. Self-regulation, codes of personal and business conduct, and good behavior provide the foundation for an online business framework that the world can trust.

QUESTION What does a copyright notice say about a Web site's contents?

TAXING THE INTERNET As you learned in Chapter 4, e-businesses in the United States must charge sales tax on purchases by customers who live in the same state where the selling company is based or has a physical location. Many state and local government officials wish to see this policy expanded so that all Internet transactions are taxed, thus ensuring a larger sales tax revenue base.

The U.S. government is reviewing not only the implications of Internet sales tax and collection, but also the taxation of Internet use. When you key in a Web address on the Internet, you are routed through various volunteer computers that act as hosts in delivering your message. It would be almost impossible to track this kind of access and to tax each connection point. It remains to be seen whether the government will require users to pay taxes on their Internet connections in the future.

Section 6-1 Review

RESPOND to what you've read.

1. What options do people or companies have if the domain names they want are already taken?
   _____
   _____
   _____

2. How long does a copyright stay in effect? _____
   _____
   _____
   _____

3. How are domain-name disputes generally settled? _____
   _____
   _____
   _____

4. What does it mean when a trademark has been infringed? _____
   _____
   _____
   _____

Quick TALK With a partner, brainstorm possible ways in which Internet usage could be fairly taxed as a source of revenue for the government. *Discuss which of your solutions is the best and why, and present your findings to the class.*

Section 6-2

Privacy Issues and the Internet

ONLINE PRIVACY

In the world of e-business, the information that can be collected about customers digitally amounts to a competitive advantage. When you shop online at a site such as LandsEnd.com or Barnes&Noble.com, you are asked to provide personal data, including your address, phone number, e-mail address, and credit-card information. You may enter this information under the assumption that it will not be shared with other businesses.

Many online companies, however, make no guarantee they will use this information only to process your order or request. They might also sell it to other businesses that will use it to market their own products and services. Selling personal data has become big business on the Internet.

Privacy Policy

Reputable online companies understand how important it is to respect the privacy of their customers. To earn your trust, these businesses post privacy policies on their Web sites. A **privacy policy** is a written, legally binding statement informing Web-site users about how their personal information will be managed and maintained.

Many online companies also give registered users the opportunity to have their names removed from e-mail lists that distribute customized information or targeted advertising. This is designed to guarantee your e-mail address will not be used, either by the Web site or the companies it has an agreement with, to distribute promotional materials.

PROTECTING THE PRIVACY RIGHTS OF THE INDIVIDUAL

The federal government has taken various steps to protect consumers' rights on the Internet. Using its authority under Section 5 of the Federal Trade Commission Act, which prohibits unfair or deceptive business practices, the Federal Trade Commission (FTC) has brought a number of cases to enforce the promises in privacy statements.

When it comes to protecting personal data, businesses are not the only entities held to this responsibility. The U.S. government must also protect the personal data of its citizens. The *Privacy Act of 1974* was designed to regulate the collection, maintenance, use, and dissemination of personal information by federal executive-branch agencies. Broadly stated, the Privacy Act is designed to balance the government's need to collect, maintain, use, and disclose information with the individual's right to be protected against invasions of privacy.

The monitoring of e-mail is not only an issue in fighting terrorism and crime, it is also a concern in the business world. The federal

---

AS YOU READ...

YOU WILL LEARN
- why privacy policies are important to those who conduct business online.
- why online companies collect personal information from their customers.
- what efforts have been made to protect children in the online world.
- what government legislation protects individuals' rights and privacy online.

WHY IT'S IMPORTANT

All Internet users, including minors, need to be assured of their privacy and the safety of their personal information online. Since 1997 Congress has been working to craft legislation specifically designed for this purpose.

KEY TERMS
- privacy policy
- Electronic Communications Privacy Act of 2000
- spam

PREDICT

How would a company be able to make money with its customers' personal information?

CONNECT

Have you ever entered personal information onto a Web site form? Were you concerned about confidentiality? Why, or why not?

CONNECTION
SOCIAL STUDIES

Who's Watching You on the Web?

According to the organization Working to Halt Online Abuse, instances of cyberstalking—using the Internet to harass or threaten someone—are on the rise. Statistics show that 63 percent of victims are between 18 and 30 years old. *How can you protect yourself from becoming a victim of online harassment, threats, or impersonation?*

---

government has enacted legislation supporting the right of businesses and organizations, including school districts and state agencies, to monitor the use of e-mail and the Internet on their computers. This applies to students as well as employees.

The **Electronic Communications Privacy Act of 2000** allows companies to monitor an employee's e-mail if either the sender or the recipient has given consent or the organization can demonstrate why the company or its employees could be harmed by the electronic communication. Either party to the e-mail, or the business, has the right to submit e-mail to legal authorities when either has been the victim of harassment, hostile communication, or defamation.

Individuals who post on Web site bulletin boards or make comments in chat rooms about companies or individuals, depicting them in a manner that is unfounded or untrue, have been sued, and, in some cases, arrested. You do not have the right to falsely accuse your company of illegal business practices or slander coworkers using e-mail, chat rooms, or bulletin boards. Courts have held that these are illegal acts punishable by fine and/or imprisonment.

CHILDREN'S ONLINE PRIVACY PROTECTION ACT As the nation's consumer-protection agency, the FTC plays a vital role in safeguarding the privacy of all consumers. The FTC enforces the *Children's Online Privacy Protection Act of 1998,* also known as COPPA, to prevent the collection of personally identifiable information from children without their parents' consent.

The goal of COPPA is to give parents control over what information is collected from their children online and how such information is used. The act applies to operators of commercial Web sites directed to children under 13 who collect personal information from them; to operators of general-audience sites who knowingly collect personal information from children under 13; and to operators of general-audience sites who have a separate children's area and collect personal information from children under 13.

PRIVACY PROTECTION GROUPS With the rapid increase of online shopping, it is more important than ever that consumers are adequately educated about privacy and Internet usage. National advertising campaigns have been launched informing citizens of the need to protect their personal information, and new citizen privacy-protection groups have been organized.

The *Electronic Frontier Foundation (EFF)* is a nonprofit organization of lawyers, volunteers, and contributors who are working to protect your digital rights. The EFF's activities include providing legal counsel to defend citizens' rights; increasing public awareness about digital rights, such as electronic voting (e-voting); holding educational seminars and classes; and providing informational resources on its Web site.

SPAM If you ever go on the Internet, you're probably familiar with spam. **Spam** is unsolicited "junk" e-mail sent by companies to potential customers. Like junk mail in the postal system, spam makes up more than half of all messages sent on a daily basis. Unlike junk mail, the companies that send spam are not limited by the cost of postage. As a result, they are able to send out millions of messages to Internet users around the world. These companies, called *spammers,* have purchased customers' e-mail addresses as part of a mailing list.

Internet users, businesses, and ISPs have to spend valuable time sorting through unsolicited e-mail, which advertises everything from herbal remedies to get-rich-quick schemes, or risk clogging their systems. Billions of dollars in productivity are wasted in the constant effort to combat the problem.

As an Internet user, spam can be a major nuisance because of its sheer volume. It can also carry viruses and other dangers that damage your computer system. There are ways you can combat spam yourself. IPOs, such as AOL, provide options for you to block and report spam. You can also discourage spammers by forwarding any spam you receive back to them. One of the best ways to deal with spam, as with any e-mail you receive from someone you don't know, is simple. Don't open it.

QUESTION

Why is it necessary to have separate laws governing privacy for children?

Section 6-2 Review

RESPOND to what you've read.

1. When is it permissible for a company to monitor an employee's e-mail? _____

2. Why was the Privacy Act of 1974 created? _____

3. What is the intended goal of the Children's Online Privacy Protection Act of 1998? _____

4. What is spam? Why is it considered harmful? _____

Quick TALK In some cases, law-enforcement agencies monitor e-mail and other online communications in their pursuit of criminals. With your partner, discuss whether government surveillance into personal e-mail and communications is ever warranted. *Formulate an opinion, and discuss your reasons behind it.*

Chapter 6 Ethical, Legal, and Social Responsibilities in E-Commerce

Section 6-3

Internet Security

AS YOU READ...

YOU WILL LEARN

- from what dangers companies need to protect their computer databases.
- how firewalls can protect important computer data.
- what hackers can do with information they obtain from computer systems.
- how to protect computer systems of all sizes from illegal acts.

WHY IT'S IMPORTANT

Computer systems are compromised daily by those seeking confidential information for illegal purposes. Protecting against this type of activity is crucial for all computer users.

KEY TERMS

- hacker
- cybersquatters
- firewall
- Internet filter
- encryption
- virus
- worms
- logic bomb

PREDICT

Does a home computer need to have a firewall? Why or why not?

DIGITAL PROTECTION

E-businesses are connected to the Internet 24 hours a day, seven days a week. This constant connectivity can leave companies vulnerable to computer hackers, viruses, worms, and logic bombs. For this reason, e-businesses must be particularly careful to protect their customer and product data, software, servers, communication ports, trade secrets, and business processes. Let's take a look at some of the ways in which e-commerce businesses can protect these assets.

Security of Data

Maintaining the security of corporate data is one of the highest priorities of any e-commerce business. A company that conducts business on the Internet maintains a variety of databases filled with sensitive data such as customer-information files, product data, sales data, and financial information.

Though companies may spend millions of dollars to secure sensitive data, breaches can still happen. A **hacker** is a person who uses technical expertise to break into computer systems for malicious purposes. Because hackers break in electronically, they can work from a computer located almost anywhere in the world.

Another problem e-businesses face is cybersquatters. "Squatters" are people who illegally move in on someone else's property. **Cybersquatters** are people who register domain names of well-known companies or individuals with the sole purpose of selling the names back to the rightful owners to make money.

In one case, a person registered over 5,500 Web-site addresses of already existing sites by changing the spelling or capitalization of the addresses. This diverted so many users from the real sites that it cost them potentially millions of dollars. The perpetrator was ultimately identified and prosecuted.

In another case, Bill Gates, the founder of Microsoft and one of the richest men in the world, was fooled by a cybersquatter. The cybersquatter bought the name BillGates.com and tried to sell the name back to Bill Gates. Gates refused to pay and took the case to court. The court ultimately ruled in Bill Gates's favor.

The loss of sensitive company data can be devastating to a business. Almost 80 percent of companies that lose their database files are unable to reopen after the loss. While insurance can be purchased to reduce the risk of physical loss of merchandise, database files can rarely be insured.

In addition to the perils an e-business can face due to security breaches, the online consumer encounters new threats. The failure of e-businesses to adequately protect consumer data has led to an increase in *identity theft*. Identity theft is the practice of running up bills or committing crimes in someone else's name. Since the Internet acts as a conduit for the spread of personal data, high-tech criminals can use this information to open credit-card accounts, raid bank accounts, or secure loans in your name illegally. The information provided in **Figure 6.3** on page 124 outlines the credit-card protection offered by Gap online.

FIREWALLS Companies use a number of applications and tools to protect the data stored on their computers. One important safeguard is installing firewalls on their computer servers. A **firewall** is a combination of hardware and software used to block potential hackers from gaining access to a computer system. Any computer that is always connected to the Internet needs a firewall.

Most firewalls use some kind of packet filtering. The *packet-level firewall* is a software firewall used mostly in home settings. A packet-level firewall analyzes network traffic at the transport protocol layer to see if it matches the rules that define acceptable data flows. This limited set of rules determines whether communication is allowed based upon the information contained within the Internet and transport-layer headers and the direction in which the packet is headed.

A better level of security is afforded by an *application-level firewall*, which evaluates network packets for valid data at the application layer before allowing a connection. An *application layer* is how an application interacts with network operating systems. This is a more advanced security system with a more complex set of rules, allowing traffic to pass through only after a rigorous hurdle of tests.

In addition, some businesses install a *bastion host*, which is a heavily fortified computer server that handles all incoming requests over the Internet. Think of a bastion host as the walls of a castle fortified with

Web Site Success

HACKING AWAY AT THE HACKERS
Symantec.com

Is it really necessary to protect your computer from viruses and other Internet dangers? Symantec's sky-high sales numbers show that most consumers think it is. The Symantec.com site both enables consumers to browse and purchase its antivirus and firewall products and serves as a source of information about the latest online threats. Its Security Response section describes the virus or worm by name, categorizes its threat level, and gives instructions on how to respond to the risk.

Thinking Critically
Each year Symantec's antivirus software products see strong spikes in sales. *What factors might be contributing to the company's continued success?*

Figure 6.3

Credit-Card Safeguard

BUYER PROTECTION Many online retailers offer credit-card safeguards and other consumer protections. *What protection does Gap online offer to keep your credit-card information secure? If you're making a purchase online at The Gap, how can you determine if you are accessing its secure server?*

Our Security Safeguard

At gap.com we don't just stand behind our merchandise. We also do our best to provide a safe and convenient online shopping experience. When you make a purchase at gap.com, we save your credit card information for an even faster checkout next time.

We safeguard the integrity of our customers' credit card information. If your credit card or debit card ("credit card") is used in an unauthorized manner on our site (as determined by your credit card company), we will cover whatever amount your credit card company doesn't—up to $50.

If you're not quite comfortable shopping online, or would prefer to make your purchase via telephone, just give us a call at 1-800-GAP-STYLE, option 1. Our Customer Service representatives are standing by 24 hours a day, seven days a week. They'll be happy to take your gap.com order right over the phone.

Credit Card Safety

Protecting the safety of your credit card information is important to us.

We use Secure Sockets Layer (SSL) technology to protect the security of your credit card information as it is transmitted to us.

SSL is the gold standard in Internet encryption technology, which is a fancy way of saying that it's a highly sophisticated method of scrambling data as it travels from your computer to our website's servers.

If someone steals your credit card information or uses your credit card in an unauthorized manner on our site (as determined by your credit card company)—through no fault of your own—we'll cover up to $50 of the charges your credit card company doesn't. What's more, we'll do what we can to make the whole process as hassle-free as can be.

What about my bank or credit card company?

Under the Fair Credit Billing Act, your bank or credit card company cannot hold you liable for more than $50 of unauthorized or fraudulent charges on a credit card. Your liability for unauthorized use of a debit card may be higher, but most debit card issuers voluntarily apply the $50 limit to their cards, as well. If your bank does hold you liable for any amount up to $50, gap.com will cover the liability imposed on you, up to $50. Gap.com will cover this amount **only** if the unauthorized use of your credit card occurred at gap.com and through no fault of your own.

Remember: If unauthorized use of your credit or debit card does occur, you must notify your card provider, in accordance with the agreement you have with the company.

How do I know my credit card information is secure?

To make sure you are accessing our secure server before you submit personal financial information, look at the lower left-hand corner of your browser. If you see an unbroken key or a closed lock (depending upon your browser), then SSL is active. To double-check for security, look at the URL or Location line of your browser. If you have accessed a secure server, the first characters of the address in that line should change from "http" to "**https**."

Some browser versions and some firewalls don't permit communication through secure servers like the ones we use to process orders at gap.com. If for any reason you cannot access the secure server, please feel free to place your order with us by phone at 1-800-GAP-STYLE, option 1. Our Customer Service representatives will be happy to assist you.

guns and soldiers. In the event of attack, the bastion host is the only affected server, and internal servers with sensitive data and processes are protected inside the castle walls. All communication that passes through a firewall system is monitored using screening criteria that protect the integrity of the computer system.

INTERNET FILTERS An **Internet filter** is another tool used by organizations to secure computer systems. It is a software program that limits access to Web sites on the Internet. If your school has installed an Internet filter on its computers, the program limits or blocks access to certain Web sites and Internet destinations.

There are several reasons why an organization might want to use an Internet filter. A company can prevent virus infection by blocking access to Web pages coded to inflict damage. Internet filters, like antivirus software, are updated frequently with alerts to block the latest types of destructive Web-page programming. A company can also use an Internet filter to limit the number and types of Web sites employees are able to visit.

Internet Policy Consulting reports that 10 billion hours in productive workplace time are lost each year due to Internet surfing. Using average salary costs, businesses lose approximately $250 billion in wages due to non-work-related Internet surfing. **Figure 6.4** outlines some of the ways in which employees use the Internet during the workday.

ENCRYPTION What other tactics can be employed to protect e-businesses and their customers? **Encryption** is the scrambling of data from

Figure 6.4
Monitoring Employee Internet Usage

Internet Usage Statistics

30% to 40% of Internet use in the workplace is not related to business.

70% of all traffic on Internet adult-content sites occurs during the nine-to-five workday.

37% of workers say they surf the Web constantly at work.

People are spending more time surfing the Internet at work than they are at home, mainly because home connection speeds pale in comparison to the faster connections companies give their employees.

77.7% of major U.S. companies keep tabs on employees by checking their e-mail, Internet, phone calls, or computer files, or by videotaping them at work.

63% of companies monitor workers' Internet connections, and **47%** store and review employee e-mail.

27% of companies say that they've fired employees for misuse of office e-mail.

PERSONAL TIME? It is a well-known fact that many workers use their work e-mail and Internet access recreationally. *What percentage of U.S. companies monitor how their employees use the Internet, e-mail, and phones while at work?*

> **SECURITY ENFORCEMENT**
> Service providers such as SBC Yahoo! are asking subscribers to upgrade their security measures and to follow procedures such as those described in Figure 6.5. Customers who delay or refuse to modify their behavior may risk disconnection from their ISP. *Why do you think ISPs are taking such a tough position on Internet security?*

plain text into code once it is sent from a computer. (Encryption was first discussed in Chapter 3). This makes the data unreadable to everyone except the recipient and helps to combat computer hackers who develop tools to monitor data being passed over Internet connections. E-commerce sites use encryption to secure credit-card data transmitted by customers. The next time you make an online purchase, take a look at the following two items to make sure your connection is encrypted:

- The "http" portion of the Web-page address will read "http**s**." The "s" indicates that the page is "secure."
- A small gold padlock will appear in the bottom-right-hand corner of your browser window. The padlock is a universal Web symbol indicating a secure connection.

Pretty good privacy (PGP) is the most popular program for encrypting and decrypting e-mail messages. It requires both the sender and recipient to have an encryption key, which is a code used to translate the message. It can also be used to send an encrypted digital signature that verifies the sender's identity and lets the recipient know that the message was not altered while traveling over the Internet.

If a hacker intercepts the message, it will read as scrambled words and numbers without the key to decipher it. Computer databases that store customer information, including credit-card information, are traditionally encrypted as well. If an attacker is able to steal any information found in these databases, it would be useless data and nearly impossible to interpret.

ANTIVIRUS SOFTWARE Perhaps the greatest threat to computers and data on the Internet are viruses, which are often delivered using business-communication channels such as e-mail and Web sites. A **virus** is a program written to inflict harm on a computer system and to interfere with its normal operation. Computer Economics estimates that computer viruses cost businesses billions of dollars per year. Virus infections cause a loss of productivity and sales, as well as a need for expensive security systems and the cleaning of infected systems.

> **CONNECT**
> Have you or a friend received an e-mail that you thought may have contained a virus? What did you do with that e-mail? Would you handle it differently today?

Some virus attacks have been motivated by anger over job layoffs or personal grudges against individuals or companies. Fortunately, there are ways in which people and businesses can protect themselves from viruses distributed over the Internet. Norton and McAfee are two companies that produce *antivirus software* designed to monitor and eliminate viruses. Once a virus is detected, the user is notified either not to open the e-mail message or not to click on a Web site's hotlink. If the virus has already attacked your computer, you are instructed on how to remove the virus and fix any damage that was done.

There are several types of viruses on the Internet. Trojan viruses are the most common. Named after the story of the Trojan horse used by the Greeks to gain access to the city of Troy, a *Trojan virus* refers to a program that appears safe but actually contains something harmful. These viruses are typically distributed through e-mail messages that request unsuspecting receivers to click on attached files or Web links. When you click on it, you unleash a dangerous program that may erase your hard disk or send your credit-card number, social-security number, or passwords to a stranger. Basically, this virus lets someone take charge of your computer and use the information illegally.

When a computer has been infected by a virus, a variety of things may happen. Your computer may slow down and become inoperable. A virus may search your computer hard drive for a particular type of file such as a word-processing document and automatically erase the data contained in the file. It may lie dormant for a period of time and then begin to selectively delete files from your hard drive. **Figure 6.5** outlines protective measures you can take to fight damaging viruses.

Worms operate a little differently. **Worms** are programs that replicate themselves exponentially, causing malicious actions against the

QUESTION

What is identity theft?

Figure 6.5

Antivirus-Protection Actions

- Never click on links within an e-mail that comes from someone you do not know.

- Never accept or run files you receive via e-mail from someone you don't know. If you have any doubts, write the person back, and ask for verification that they sent you a file. Some of the more recent viruses send e-mail (and files) to everyone listed in a user's e-mail address book; then, the virus deletes itself, and you have no idea what happened.

- When downloading files off the Internet, be sure they're from reputable sites.

- Schedule regular times for virus scans and updates. With more than 200 new viruses being reported each month, updates are critical.

- Back up important files regularly in case a virus causes a loss of data.

SELF-DEFENSE MEASURES
While antivirus software can protect your computer and files, the following guidelines should be followed. *What are some of the risks of not ensuring that your files are protected?*

computer's resources and files. Worms generally come through e-mail messages. If you open an e-mail message that has a worm, this program sends the worm file out to everyone listed in your e-mail address book.

Worms spread over networks and enter systems when they are connected to the Internet. Hackers attempt to create worms that affect millions of computers worldwide. If your machine is infected with a worm, it can spontaneously crash and reboot. The Windows operating system offers downloadable patches and updates to users against infection.

A **logic bomb**, also called a slag code, is a programming code added to a software program on your computer that lies dormant until a predetermined period of time passes or an event occurs, triggering the code into action. Logic bombs typically are malicious in intent and can perform actions such as reformatting a hard drive, or deleting, altering, or corrupting data. It's important to update your computer-protection software and screen your e-mail regularly. According to some accounts, there are as many as 80 new viruses, worms, and logic bombs created *per day*.

Section 6-3 Review

RESPOND to what you've read.

1. What are hackers? How can they damage computer systems? _____
_____
_____

2. What is encryption? Why is it necessary? _____
_____
_____

3. What is PGP? _____
_____
_____

4. Why would a hacker want to destroy computer databases and hard drives? What type of people take part in this? _____
_____
_____

Quick TALK With a partner, brainstorm a list of reasons you could use to convince a friend or family member to get virus protection for their home computer. *Anticipate some arguments that person might give, and plan your rebuttals.*

128   Unit 2   You and E-Commerce

Name _____ Date _____

Worksheet 6-1

Cybercrime

Cyberlaws are evolving, just as the Internet is evolving. Often, the U.S. Congress can't keep up with the changes taking place online. The judicial system faces the challenge of trying to apply current laws to the cyberworld, and hackers are working harder to create new problems for online users.

1. Research two or three cyberlaws that are NOT mentioned in the chapter. Write a short summary of each.

   _____
   _____
   _____
   _____

2. Research a cyberlaw from another country. What is its purpose? When was it enacted?

   _____
   _____
   _____
   _____

3. Explain identity theft.

   _____
   _____
   _____
   _____

4. Passwords are often used to help safeguard computers and online shoppers. What are some appropriate guidelines for creating passwords?

   _____
   _____
   _____
   _____
   _____

5. Create a one-page flyer or information sheet outlining ways to protect yourself from identity theft.

Name _____ Date _____

Worksheet 6-2

Fact or Fiction?

Computer viruses cost businesses billions of dollars per year. Almost as problematic, however, are computer-virus hoaxes. Hoax warnings are scare alerts started by people who want to cause harm and passed on by innocent users who think they are helping the community by spreading the warning. Unfortunately, entire computer systems have crashed due to these warnings.

1. Research computer hoaxes. Choose one that interests you, and describe it.

   _____
   _____
   _____
   _____
   _____

2. How can people identify a computer hoax?

   _____
   _____
   _____
   _____
   _____

3. What are some ways to reduce computer hoaxes?

   _____
   _____
   _____
   _____
   _____

4. Create a public-service announcement (either for the radio or newspaper) about computer hoaxes. Provide information you think the public needs to know to protect itself.

   _____
   _____
   _____
   _____
   _____
   _____
   _____

Chapter 6 Review and Activities

Chapter Summary

Section 6-1

- Laws are necessary in the online world because some individuals and groups attempt to use the Internet for dishonest purposes.

- Copyright infringement, e-mail privacy, identity theft, and interstate e-commerce are some of the reasons why cyberlaws exist.

- A domain name that is the same as a company's or product's name can be extremely valuable.

- Copyright and patent laws protect authors, writers, inventors, and others against the misuse or theft of their original work.

- Not everything posted online is freely available for reuse or downloading. Trademark laws protect much of the intellectual property found on the Web.

Section 6-2

- Everyone who is using the Internet, including minors, needs assurance of their privacy and the safety of their personal information online.

- Online companies collect personal information from their customers to process orders and provide them with services. However, some online companies sell your personal data with or without your permission.

- The Privacy Act of 1974, the Electronic Communications Privacy Act of 2000, and the Children's Online Privacy Protection Act of 1998 are designed to protect the personal data of U.S. citizens.

Section 6-3

- Companies' computer databases need protection from hackers, viruses, worms, and logic bombs.

- Firewalls block potential hackers from gaining access to computer systems.

- Hackers can use the data they obtain to steal consumers' credit-card numbers or even their identity.

- Firewalls, Internet filters, encryption, and antivirus software are tools that can be used to protect computer systems of all sizes from illegal acts.

Chapter 6 Review and Activities

Review of Key Terms

a. spam
b. Electronic Communications Privacy Act of 2000
c. hacker
d. trademark
e. worms
f. firewall
g. copyright
h. patent
i. Internet filter
j. virus
k. logic bomb
l. privacy policy
m. cyberlaw
n. encryption
o. cybersquatters

Match each term to its definition.

_____ 1. Software and hardware used to block potential hackers and unauthorized intruders from gaining access to a computer system.
_____ 2. The scrambling of data to make it unreadable.
_____ 3. The use of a word, phrase, symbol, product shape, or logo that distinguishes one business's products from another.
_____ 4. A software program that limits access to Web sites on the Internet.
_____ 5. Someone who uses technical expertise to break into computer systems for malicious purposes.
_____ 6. A written, legally binding statement informing Web-site users of how their personal information is being managed.
_____ 7. Programs that replicate themselves exponentially.
_____ 8. Allows companies to monitor an employee's e-mail.
_____ 9. Programming code that lies dormant until a set period of time passes or an event occurs.
_____ 10. The grant of a property right to the inventor of a product or process.
_____ 11. A program written to inflict harm on a computer system.
_____ 12. Unsolicited "junk" e-mail.
_____ 13. A form of protection provided by U.S. law for exclusive rights over the individual works of an author.
_____ 14. An evolving legal framework that governs Internet activities.
_____ 15. People who register domain names of well-known companies or individuals to sell the names back to the rightful owners.

Applying Technology to Academics

English Language Arts—Reading
Using the Internet, access the U.S. Department of Justice's cybercrime Web site. Review the site. What are the major sections? What do you think the purpose of this site is? Read one of the articles, then explain it to a classmate.

English Language Arts—Writing
Most people find spam very annoying, and the federal government is considering regulating it. Research your own Internet provider's acceptable use policy regarding spam. Write a paragraph describing the policy.

English Language Arts—Speaking
Working with a group, research two common computer viruses. Determine the type of virus, how it operates, who created it, and when it first was activated. Then, share your findings orally with the class.

Graphing Software Program
Research the cost of four different antivirus software packages. Check the price in regular bricks-and-mortar stores and on the Internet. Create a graph depicting your findings. Is it less expensive to purchase the software online or in a store?

Chapter 6 Review and Activities

Critical Thinking

1. Describe how an ethical framework when doing business on the Internet is more important than ever before.
2. How do you feel about the U.S. government monitoring the e-mail of American citizens? Explain what you believe regarding the privacy of your personal information.

COMPETITIVE EVENT

Imagine you're the sales manager of an online bookstore specializing in new and used mystery books. The bookstore does a brisk business with an efficient order processing and fulfillment process. An employee (judge) has mentioned reading about legal and regulatory issues affecting online businesses. The employee (judge) has asked you to explain how those legal and regulatory issues affect your business. Discuss some of those issues with the employee (judge).

EVALUATION

You will be evaluated on how well you meet the following performance indicators:
1. Describe legal considerations in e-commerce.
2. Explain the nature of trade regulations.
3. Explain the nature of personnel regulations.
4. Explain the nature of tax regulations on business.
5. Explain the nature of a business's reporting requirements.

INTERNET ACTIVITY

To engage in online activities, visit the *E-Commerce* Web site at **ecommerce.glencoe.com**.

BusinessWeek online

TOPIC: Internet Security

The news is often full of reports of hackers finding and exploiting holes in computer systems. For the home computer user, these attacks can result in loss of data and tremendous inconvenience. For a business engaged in e-commerce, however, an attack on a computer network can have devastating consequences.

ACTIVITY Go to the Student Center at **ecommerce.glencoe.com**. Click on *BusinessWeek* Activities and open Chapter 6. There you'll learn more about security risks online and what can be done to prevent and defeat them.

BusinessWeek It's Your Project

SPRINGBOARD: POP QUIZ

What right does the Fourth Amendment to the Constitution guarantee?
- **a.** The right to remain silent.
- **b.** The right to free speech.
- **c.** The right to be safe from unreasonable searches and seizures.

This lab will require some research. Go to *BusinessWeek* online for useful articles and resources.

Who Has the Right to Know?

Rights are a two-way street. The government must protect individual rights spelled out in the United States Constitution. You have an obligation to respect the rights of others. That goes for the right of privacy, which some people have called "the right to be left alone" by the government.

What's the government? That's a tricky question. All levels of government—local, state, and federal—must protect your right to privacy. Public schools are part of the government. Are teachers and principals required to honor that right? What happens when two rights conflict—for example, when the police invade a crime suspect's privacy by searching the person's house? These are questions that you might like to explore in this project. Privacy is a complicated, fascinating subject. It comes down to this: What right do you have to be left alone by the government, by your parents, and by anyone else, and what right do they have to know what you're doing?

Design an activity that requires you to examine one area of the right to privacy. Start with the list provided. If you don't want to explore any of these issues, that's fine. Do a little research into privacy on the Web. Find an issue relating to privacy in which you're interested. Then investigate that issue, following the steps on the next page.

Under what circumstances might the government, an individual, or a company have the right to:
- search your school locker?
- write down your Social Security number?
- install "spyware" in your computer to monitor your Internet habits and behavior?
- put "cookies" in your computer, to allow sites to identify you automatically?
- make you show identification to a police officer, even if you have done nothing wrong?
- go through your backpack at an airport, where you have put some private things?
- listen in on your phone conversations?
- take credit-card information over the Internet?
- read your diary?
- search your pockets?
- withhold government documents about individuals who don't want them made public?
- make you take off your shoes at an airport?
- make you pass through a metal detector?
- look into your school records?
- make you tell them where you have been?

Courts have dealt with many of these issues. Now it's your turn.

134

Step 1 Start with *BusinessWeek*

Orient your project by conducting research in the *BusinessWeek* online archives. Use key words that relate to your topic to find articles. The articles may offer some insight as you progress through this project.

Step 2 Investigate and Engage

Ask yourself these questions:
- What do I already know about this topic?
- How does this topic relate to technology?
- How should I show the connection between the issue I chose and technology?
- What do I want to learn from this project?

Step 3 Identify the Obstacles

- What facts, information, or ideas led you to pose your question?
- Write out the problem(s) you intend to solve with your research.

Step 4 Conduct Research and Seek Solutions

By now, you should have chosen the question about the right to privacy that you intend to answer. With that question in mind, ask yourself:
- How am I going to find an answer?
- Where will I start to look for an answer?

Step 5 Select Information and Analyze Data

- How can you determine what information is accurate and what information is relevant to your question?
- Translate the information into statements that make sense and relate to your question.

Step 6 Connect to the Real World

Create a way to present your project. For example, you might give a PowerPoint® presentation or a speech. You might create a Web page or a role play, or write an essay or even a short story. It's up to you.

As you prepare your final project, be sure to address these questions:
- How does my project connect to the real world? It could apply to an individual, a family, a community, the world, or all of them.
- How does this project prepare me for the real world?

Step 7 Self-Evaluation

You're used to teachers grading you. This time you get to grade yourself. Create your own self-assessment tool. Create one that will help you reflect on your project. It might be a checklist, a talk with your teacher, a log in which to measure your progress, a small-group discussion, or something else. Be honest with yourself. If you aren't, you won't be able to determine how well you dealt with your project every step of the way.

UNIT 3

BUSINESS STRUCTURES AND THE BUSINESS PLAN IN E-COMMERCE

In This Unit...

Chapter 7
Business Structures and Economics in E-Commerce

Chapter 8
Revenue Models and the Business Plan in E-Commerce

E-Commerce Online

To learn more about business structures and the business plan in e-commerce, visit the E-Commerce Web site at **ecommerce.glencoe.com**.

BusinessWeek Byte

Architects: Ivan Seidenberg, Verizon

When it comes to e-business, most telecom companies are more talk than action. Not Verizon Communications Inc. and its CEO, Ivan G. Seidenberg. He's making the investments now to turn the promise of ubiquitous broadband connections into a reality. Over the next 10 to 15 years, he expects Verizon to spend $20 billion to $40 billion to provide super fast Net connections to every home and business in its 29-state territory. At least 10 times faster than today's broadband services, it will be able to carry everything from phone calls to high-definition television. That will support a new generation of digital interactivity.

Imagine a world in which customer service and sales are once again conducted eye-to-eye, but via the Web. "Broadband is the digital economy," Seidenberg says. And he aims to be its master builder.

Source: Excerpted with permission from "Architects: Ivan Seidenberg, Verizon," *BusinessWeek*, September 29, 2003.

To read this *BusinessWeek* article in its entirety, go to **ecommerce.glencoe.com**. Click on the Student Center, find *BusinessWeek* Articles, and go to Unit 3.

THINK ABOUT IT

In what specific ways does broadband technology benefit the digital economy?

BusinessWeek It's Your Project

Have you ever wanted to do something for your school, your town, or even the world but didn't know how to go about it? The Unit 3 project, "What's Your Goal?" on page 180, will teach you how to set and achieve your goals.

 ecommerce.glencoe.com

Chapter 7

Business Structures and Economics in E-Commerce

Section 7-1
Business Structures

Section 7-2
Components of the New Economy

PREREADING STRATEGIES

Before you read this chapter, finish these statements in your journal:

- Based on this chapter title, I predict…
- The most important thing to remember about the impact of e-commerce on the economy is…
- Some questions I have about business ownership and e-commerce are:

BusinessWeek online

Go to *BusinessWeek* online. Read an article on the current state of inflation. In your journal, explain how the inflation rate is affecting the economy and what's being done about it.

138 ecommerce.glencoe.com

Old Structures, New Technologies

From the invention of the safety lock to the advent of e-mail, new technology has revolutionized business. Companies that started as small, single-person operations have grown into giant corporations. A company's growth in a market economy depends in part on how it is structured.

Drill Down

1784 Under Lock and Key
Joseph Bramah patents the safety lock. It soon becomes standard protection for homes and property.

1839 Signed, Sealed, Delivered
The first envelopes are manufactured in New York. Previously, people folded and mailed letters without sealing them.

1862 Fantastic Plastic!
Alexander Parks invents the first man-made plastic, which manufacturers use to make everything from toys to automobiles.

1899 Hoop Metal
Johann Vaaler invents the paper clip, which he describes as a wire bent into a "rectangular, triangular, or otherwise shaped hoop."

1929 Car Radios
Paul Galvin creates the car radio. Today, radios are a standard feature in new cars.

1932 Watch the Meter
Carl C. Magee invents the parking meter. Cities around the world earn millions of dollars in revenue each year from parking fees and fines.

Get the Big Picture
Make a list of all the potential e-commerce business structures. Considering the strengths and weaknesses of each, which structure would be best for your business?

POWER READ

Be an active reader and use these reading strategies:

PREDICT what the section will be about.
CONNECT what you read with your own life.
QUESTION as you read to make sure you understand the content.
RESPOND to what you've read.

Section 7-1

Business Structures

AS YOU READ...

YOU WILL LEARN
- how to identify the basic types of business structures.
- what the key differences are between business structures.
- how common tax issues apply to each business structure.

WHY IT'S IMPORTANT

Different types of businesses require different types of structures, each with its own operational framework and tax laws. Examining the differences between these structures will help you understand how business is conducted online.

KEY TERMS
- sole proprietorship
- basic partnership
- limited partnership
- corporation
- stockholder
- shares
- initial public offering
- board of directors

PREDICT

Which type of business structure do you think is the most popular in the United States?

TYPES OF BUSINESS OWNERSHIP

E-commerce businesses are structured much the same as bricks-and-mortar businesses. How they're structured depends on the type of ownership. A business can be owned by one person, by several persons, or by a group of stockholders. Each type of ownership has different requirements regarding who pays taxes, who reaps the profits, and who is responsible for any debts or lawsuits, or *liabilities*.

You have several options when choosing how to structure your business. The first, a sole proprietorship, is the closest thing to saying, "Legally, I am my business." The second, a partnership, is like saying, "Legally, I am part of my business." The third, a corporation, is like saying, "Legally, my business and I are distinct from each other." The fourth option, a limited liability company, says, "Legally, my business and I are distinct, except in a few circumstances."

Sole Proprietorship

A **sole proprietorship** is a business owned and operated by one person. Three-quarters of all U.S. businesses are sole proprietorships. Interior decorators, caterers, and doctors frequently use this business structure. If you plan to run a small business that depends solely on your work and creative energy, and if you don't need more start-up money than you already have, this is a fine way to begin. You can also start a business as a sole proprietorship and then change its structure later.

A sole proprietorship can have only one owner. The IRS views married couples as one person for tax purposes, so a married couple can be a sole proprietor. The IRS makes no distinction between you and your business, so all the profits are yours to keep, and you have to pay taxes on them. When you file your income taxes, your business expenses and earnings will be reported on your personal income-tax return. This can work to your advantage because any losses suffered by your business will reduce the amount of taxes you owe.

If your business is making money and you have other income, taxes can get trickier. When you work for a company, your employer usually takes money out of your paycheck to cover your taxes. At the end of the year, it generally works out that you've already paid your share. If you have additional income from a home business, however, you might find yourself owing a lot in taxes at year's end.

As the owner of a sole proprietorship, if your business sells a harmful product or service, you can be sued directly. For this reason, it's important that the sole proprietor buy business insurance as a protection against wrongful lawsuits. If a court finds wrongdoing on your part, however, your business insurance won't cover your liabilities. As a sole proprietor, you're also liable for your company's debts.

Except for personal loans, it's difficult for a sole proprietor to raise money from outside sources. *Creditors*—people who lend money—will often ask for collateral before making a loan to a sole proprietor. *Collateral* is something valuable, such as a car or house, used to secure a loan and that the creditor can take if the loan is not repaid. Even loans made by the government's Small Business Administration, a good source of advice and support for the sole proprietor, are usually backed by collateral. A creditor whose loan is backed by collateral is known as a *secured creditor*. Someone who lends you money without collateral is called an *unsecured creditor*. Typically bank loans are secured, while credit-card debts are not.

Partnership

In a **basic partnership**, two or more individuals form a business and share equally in it. Law firms might be the most familiar examples of partnerships. In a law firm, a group of lawyers, often with different specialties, pool their resources and share in the business's profits and losses. To help with the workload, they might hire young new lawyers, who hope to eventually be made partners and share in the firm's profits. Though law firms usually have many partners, a partnership can be formed between as few as two individuals.

Before starting a partnership, make sure you work well with your partners. Although partners might divide responsibilities based on areas of expertise, all have equal authority to run the business and equal responsibility for any debts, taxes, or legal bills. The advantage of this is that no one partner has to bear all of the risk alone. The disadvantage is that if one partner runs up huge debts or cheats a customer, all the other partners can be held responsible.

It's not necessary that every partner invest an equal amount of money in the business. Wealthier partners might invest more money and receive a greater share of the profits but still have no more authority than the partners with a smaller investment. For tax purposes, money made or lost by the business is reported on each partner's personal income-tax statement. This can either increase or decrease your personal tax burden.

If one member of a two-person partnership decides to leave the business or dies, the partnership is dissolved. A partnership with more than two people can plan ahead by drawing up an agreement that allows the remaining partners to buy the share of the former partner.

LIMITED PARTNERSHIP In a **limited partnership**, all partners do not have equal rights. Like a basic partnership, a limited partnership consists of at least two people. A *general partner* assumes the management responsibilities for the business. A *limited partner* is not involved in the management, but puts up the money to finance the company and shares in its profits and losses. All limited partnerships must have at least one general partner and one limited partner but can have two or more of each.

If you're forming your own business, you'll probably be the general partner in this type of arrangement. Suppose you're lucky enough to know ten people who think you're brilliant and want to invest money

CONNECT

Can you name a sole proprietorship and a basic partnership in your area? What corporations can you name?

BEN AND JERRY The small business partnership formed by Ben Cohen and Jerry Greenfield in 1978—which produced the innovative Ben & Jerry's line of ice cream—became a public corporation when demand for their product soared. In 2000, the corporation was purchased by Unilever, which also sells the Dove and Breyers ice cream brands, among others. *As a sole proprietorship or partnership grows, why does the idea of turning the company into a corporation and going public make more sense?*

QUESTION

Is it necessary for members of a partnership to invest the same amount of money in a business?

in your business but who don't have the time to help you run it. By taking on the role of general partner, you get to run the company while your limited partners put up the money.

As the general partner, you have sole responsibility for the business. Although the limited partners share some of the burden for the company's debts, taxes, and legal expenses, their obligation ends at the size of their investment. Your obligations, as with a sole proprietorship, are unlimited. For the privilege of running the business alone with other people's money, you take on increased personal risk.

You're also bound by the terms of a partnership agreement to serve the interests of the limited partners. If you break any part of that agreement, the partnership can be dissolved, and your limited partners can sue you personally.

Corporation

Sole proprietorships and partnerships are considered, legally and for tax purposes, extensions of their owners. A general **corporation**, also known as a *C corporation,* is considered an entity all its own, separate from its owners. A corporation pays its own taxes, enters into contracts, can buy or sell property, and takes on all legal and debt burdens. Because a corporation is considered separate from its owners, if the business sells a harmful or faulty product, the company gets sued and not the people who own or run it.

Although the vast majority of businesses are sole proprietorships, corporations account for the vast majority of sales. (See **Figure 7.1**.) If you own part of a corporation, you're a **stockholder**. The units of ownership held by each stockholder are called **shares**. Typically, a certificate of stock ownership represents these shares, though it is rare that certificates are actually sent to investors. Owners of the corporation can also be hired as employees and paid a salary by the company.

Corporations are divided into two types: public and private. Travel agent Priceline.com is a public corporation. Investors in Priceline can buy or sell their shares whenever they want. The construction giant,

Bechtel, is a private company. Its stock is not available for sale to the general public.

As you learned in Chapter 3, public corporations offer their stock for sale on open markets such as the New York Stock Exchange (NYSE) and the NASDAQ. The Securities and Exchange Commission (SEC) regulates corporations, which are required to publicly file annual and quarterly reports, as well as other reports. These documents are available to the public on the SEC's Web site.

Public stock ownership grew quickly in the twentieth century. In 1965 just 10 percent of the U.S. population owned stock. By 1997 that number grew to 43 percent. As you learned in Chapter 3, to become a public corporation, a company needs to work with an investment bank to arrange an **initial public offering** (IPO), or first sale of its shares. This process can cost millions of dollars.

With either kind of corporation, stockholders are usually passive investors; they're not involved in the company's day-to-day management. A corporation is managed by a team of executives hired to run the company in the best interests of its stockholders. Management answers to a **board of directors**, a group of people elected by the stockholders to keep an eye on the management team. The board of directors can hire and fire members of management. When stockholders elect their board, they usually get one vote for each share they own.

To form a corporation, you must file articles of incorporation with your home state. The fee is typically between $50 and $100. The process can become complex, however, and costly if you need to hire an attorney to draft the articles of incorporation and submit them to the state. The larger and more complex your company is, the more complicated the process. When you file, you must also state how many shares of company stock will be available for investors. Some states charge higher fees for companies with larger numbers of shares.

The corporate structure is best for entrepreneurs who intend to raise money because it clearly spells out the rights of stockholders. In addition, only the money you invest in the corporation is at risk. If your

CONNECTION: LANGUAGE ARTS

Relief Plan
Small businesses leverage the Internet to help run their companies. When small to midsize companies are unprepared for hacker attacks, three out of five enterprises that experience an attack will go out of business, according to Gartner, Inc. Especially after September 11, 2001, small businesses are encouraged to have disaster-relief plans. *Discuss the various kinds of disasters that could affect businesses and the types of plans small companies should have in place.*

Figure 7.1

Types of Business Ownership

OWNERSHIP STATS Seventy-two percent of American businesses are sole proprietorships. *What is the average dollar volume in sales for a sole proprietorship compared to the dollar volume in sales for a corporation?*

Type of Business Ownership	Number of Businesses	Percent of Total Businesses	Total Volume in Sales
Sole Proprietorship	17,904,731	72%	$ 1,020,957,284
Partnership	1,338,796	5%	$ 1,829,568,091
Corporation	5,045,274	20%	$17,636,561,349
Limited Liability Company	718,704	3%	$ 344,751,557

corporation goes bankrupt or loses a lawsuit, its creditors are not allowed to seize your personal assets to pay the company's debts. A disadvantage to a corporation, however, is that you can be taxed on your business profits and taxed again on those same profits when paid to stockholders as stock dividends. This is called *double taxation.*

SUBCHAPTER S CORPORATION A *Subchapter S corporation* (or simply S corporation) is a business with 75 or fewer stockholders that is set up like a corporation but is taxed like a partnership. You should choose this business structure if you want your share of the corporation's profits and losses to be reported on your personal income-tax statement. The corporation itself pays no taxes. Its investors pay taxes on their share of the profits and can use their share of the losses to offset any other personal income they've had for the year. This is called *pass-through tax treatment.* In this case, the double-taxation problem faced by the C corporation is avoided.

The main problem with an S corporation is that it's limited to 75 stockholders. If you've reached your investor limit but still need to raise money through stock sales, you must ask your current stockholders for more money or change the company to a C corporation. Once you change the structure, you have to keep it in place for five years.

LIMITED LIABILITY COMPANY A *limited liability company* (LLC) combines the features of a sole proprietorship, a partnership, and a corporation. Like a corporation, an LLC is considered a legal entity separate from its owners. Members of an LLC enjoy the same privilege of pass-through taxation as partnerships and S corporations. They can also claim both the profits and losses of the business on their personal income taxes. Members can choose to be passive investors or active in the company's management. A partnership must consist of at least two people but, like a sole proprietorship, an LLC can be formed by only one person (except in Massachusetts and the District of Columbia).

Some states impose an annual tax or fee on LLCs in addition to the personal income tax imposed on their members. Although LLC members are not personally liable for the company's debts, there are some exceptions. Small LLC members might be asked to guarantee a business loan with personal assets such as property or a bank account. Those assets can then be seized if the business fails. If the business can't pay its taxes, the IRS can seek payment from "responsible persons," which includes all members of the LLC.

How America Incorporates Itself

Each type of business structure has its advantages and disadvantages. Since the vast majority of American businesses are small, one-person shops, it's no surprise that sole proprietorships are the most popular business structure. No matter what type of structure you choose, you are subject to both state and federal laws. Except in the case of a sole proprietorship, when you start a business in the United States, you have to register with your state. To gather information about the rules and fees for registering a business in your state, go to the appropriate local agency, such as the Division of Corporations or the

Incorporation Commission. Depending on the state, these agencies are generally administered by either the Secretary of State or the State Treasury.

There's no right answer to which form of ownership is best for your e-commerce business. The decision depends on the type of company you want to form, how much money you have to invest, and how much risk you're willing to take. To better understand aspects of business ownership, visit Nolo.com, and browse the Legal Encyclopedia and FAQs.

Section 7-1 Review

RESPOND to what you've read.

1. Why is business insurance important for sole proprietorships? How is wrongdoing on the part of business owners handled by insurance companies? _____
   _____
   _____
   _____
   _____

2. How does a limited partnership differ from a basic business partnership? _____
   _____
   _____
   _____
   _____

3. What is a public corporation? How does it compare to a private company with regard to stock?
   _____
   _____
   _____
   _____

4. What is a Subchapter S corporation? What is its greatest disadvantage? _____
   _____
   _____
   _____

Quick TALK With your partner, list five corporations in which you would both like to own stock. Independently from your partner, list the stocks in order of preference. ***Compare your list with your partner's, and discuss the reasons for the order you chose.***

Section 7-2

Components of the New Economy

AS YOU READ...

YOU WILL LEARN

- how the market economy affects e-commerce consumers.
- why economic information affects the e-commerce market.
- how e-commerce impacts state and local economies.
- which tax laws affect e-commerce.
- what economic indicators can tell us about e-commerce.

WHY IT'S IMPORTANT

The Internet has created new ways to do business and has made it much easier to start your own business. E-commerce has altered established economic traditions, such as local businesses, tax laws, and consumer buying habits. To stay competitive in today's economy, it is essential for most businesses, regardless of their structure, to have a presence online.

KEY TERMS

- market economy
- supply and demand
- inflation
- economic indicators

PREDICT

How do you think online businesses affect state and local economies?

How E-Commerce Influences Economics

The economy goes hand in hand with technology. To succeed in your endeavors, you need to understand how technology affects the way we do business. The establishment of railroads in the nineteenth century expanded both the territory and the economy of the United States by bridging distances and opening new markets. The invention of the automobile in the twentieth century increased the speed with which products could be transported and created new markets in oil, tires, and auto repair.

In the twenty-first century, computer technologies and the Internet have made it possible to conduct business around the globe instantaneously without leaving the home or office and have created new markets in video games, cell phones, e-banking, and e-tailing.

Like the telephone, the Internet is a tool that facilitates communication over long and short distances. In the United States, this tool functions as part of the market economy. When considering the Internet and how it affects the economy, it's important to first understand the workings of a market economy.

The Market Economy

In a **market economy**, also called a capitalist economy, individual buyers and sellers interact with one another to freely exchange goods, services, and money. In a market economy, prices are not set by the government, but by **supply and demand**—the relationship between the amount of a good or service that is available and how much people are willing to pay for it. For example, if a drought causes a shortage of corn crops, the price of all corn-related products, from popcorn to soft drinks made with corn syrup, will go up. But if there's a bumper crop of corn, the price of those products might go down.

The online auction company eBay provides an excellent example of how a market economy works. If you want to sell a digital camera for $100, you can list it on eBay at that starting price. If more than one buyer wants the camera, they'll compete for it by bidding against one another. In this case, you'll be able to get the best price for the camera from the buyer who is willing to pay the most. If no one bids on the camera, you'll know the demand is lower than expected, and you'll have to lower your price. You might discover someone else is offering the same camera on eBay for less. In this case, you might want to lower your price to compete with the other seller.

Figure 7.2

Doing Your Comparison Shopping Online

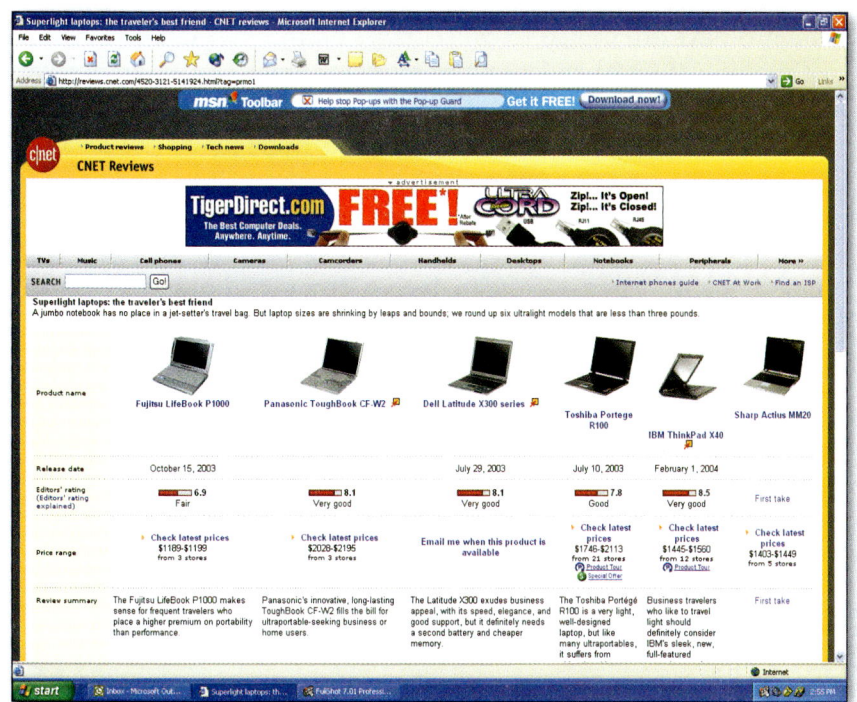

COMPARING PRICES Online shopping offers the advantage of comparing products and prices at a glance, without having to place a call or leave your desk. *What are the disadvantages of shopping online?*

CNET Networks, Inc. disclaims any responsibility for products described on their site. All product information, including prices, features, and availability, is subject to change without notice. Copyright © 1995–2004 CNET Networks, Inc. All rights reserved.

Another example of how e-commerce works in a market economy is *comparison shopping* on the Internet. Because a consumer can visit multiple Web sites, it's much easier to find the best price for a product on the Internet. Suppose you want to buy a hardcover copy of Dan Brown's popular novel *The Da Vinci Code*. Amazon.com offers a new copy for $14.97. Overstock.com has a new copy for $11.22. On BookFinder.com, you find a used copy for only $7.95. By comparison shopping online, you can open up three Web-browser windows to check the prices at all three stores at the same time. You can also compare the prices of several similar products on a single site, as illustrated in **Figure 7.2**.

The Impact of E-Commerce on Local and State Economies

The migration of commerce, from city shops to suburban malls to online stores, has changed the way in which business interacts with government. In the United States, according to the principle of *federalism*, local and state governments should have authority over any issue not specifically granted to the federal government in the U.S. Constitution. The Constitution's Commerce Clause states that the federal government is charged with ensuring that there is free trade of goods and services among every state in the union.

While the state government of California can regulate and tax a business based in California, it can't regulate or tax a New York–based

CONNECT

When have you purchased something online that you could have purchased locally? Why did you choose to make your purchase online?

Chapter 7 Business Structures and Economics in E-Commerce **147**

business that happens to be selling products to Californians. Companies that sell products online are able to compete with local businesses in every state, and yet they have to pay taxes only in the state where they maintain their headquarters. In recent years, however, some state governments have attempted to collect taxes on Internet sales.

Currently, federal regulations exempt most Internet sales from taxes. A report from the General Accounting Office of the federal government estimated that state and local sales and use tax losses for 2003 were $12.4 billion, or 5 percent of projected sales-tax revenue. If the federal government lifts its moratorium on the taxation of Internet transactions, it will be interesting to see how governments will tax online sales. If a consumer living in Aspen, Colorado, buys a product from a Connecticut company with a Web site located on a server in Delaware, which government should collect the sales tax?

Another impact on local and state economies is the loss of business for traditional bricks-and-mortar retail stores. Many "Main Street" businesses fear they will be unable to compete with online merchants and be forced to go out of business. If they close, not only will taxes be lost; jobs will be eliminated, and property values will decline. The loss of tax revenue for the local government could mean less money for vital city services such as the fire, police, and sanitation departments, libraries, and parks.

The growth of the Internet shows no sign of slowing down. In the year 2000, the estimated number of people using the Internet worldwide was 300 million—up from just 10 million in 1995. The projected number of people on the Internet in 2005 is estimated at one billion. A recent Forrester Research survey found that 43 percent of all households now own at least one personal computer.

However, traditional retailers have some key advantages. Some products, such as appliances, can't be delivered via digital download. "Main Street" shops offer products without the charges and delays involved in shipping online orders. Shops that offer personal services like haircuts are not easily replaced by the Internet.

▶ **NO SALES TAX** Five states do not collect a sales tax on purchases: Alaska, Delaware, Montana, New Hampshire, and Oregon. *Why might other states be more concerned about changes in Internet tax policy?*

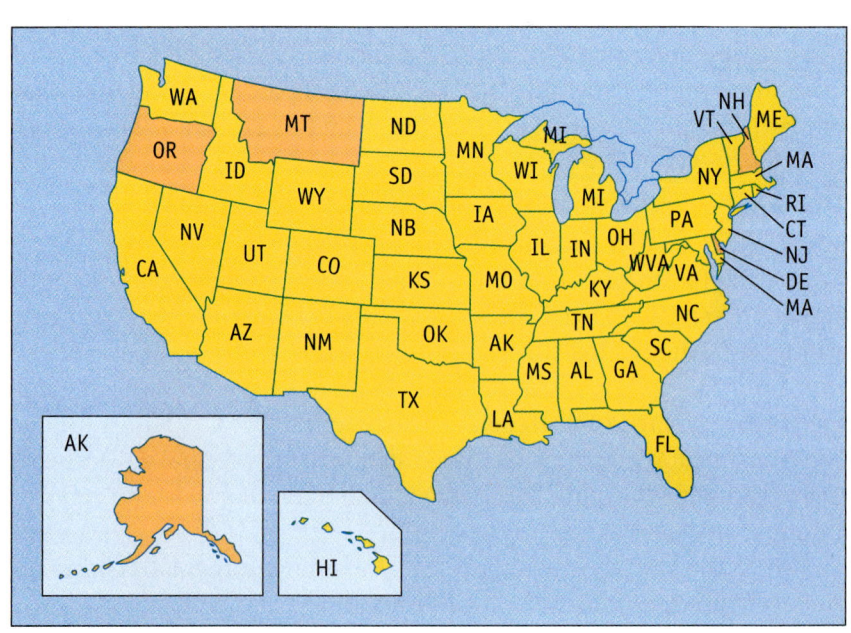

Economic Impact of E-Commerce

Although e-commerce amounts to less than 3 percent of total retail sales, it keeps prices down and, in turn, lowers **inflation**, which measures the rise in the cost of consumer goods. Economists Ethan S. Harris and Joseph T. Abate of Lehman Brothers point out that nearly every item that can be purchased at a traditional retail store is now available online for a price that averages 13 percent less, including shipping costs.

Some of the differences in savings Harris and Abate discovered in various types of goods include:

- prescription drugs are 28 percent cheaper online.
- savings on apparel average 38 percent.
- savings average 4 percent for home electronics, including shipping, and 5 percent without shipping.
- toys cost 9 percent more on the Internet than at retail stores, including shipping costs.
- hardware is 2 percent more expensive online, including shipping costs.

Online savings are the result of sellers spending less money on the costs of real estate, advertising, inventory storage, and transportation. These items account for close to 38 percent of the price of consumer goods.

QUESTION
Which types of businesses may be immune to the effects of e-commerce?

Economic Indicators

The Census Bureau of the Department of Commerce tracks many key statistics used to analyze business conditions and make forecasts. Among them are retail e-commerce sales, the unemployment rate, the

Web Site Success

JUST PUBLISH
Pyra Labs

Has this ever happened to you? You wake up in the middle of the night and think, I MUST publish something—now! Fortunately, thanks to Internet blogging, getting your voice heard is easier than ever. A *weblog*, or *blog*, is personal publishing on demand. For example, consider Pyra Labs, the company behind Blogger, a "push-button publishing tool for people." Whether you want a serious forum or a place to share your photos and innermost feelings, a weblog is a personal approach to Internet publishing that doesn't require you to know programming code. You can go to any number of blog sites to get started.

Thinking Critically
How does Pyra, and in turn Blogger, benefit from being acquired by Google?

inflation rate, consumer spending, and the balance of trade. These statistics, reported on a quarterly basis, are called **economic indicators**. According to the U.S. Department of Commerce, retail e-commerce sales in the U.S. for the second quarter of 2003 were $12.5 billion, an increase of almost 28 percent from the previous year. E-commerce sales in the second quarter of 2003 accounted for 1.5 percent of total sales. In the second quarter of 2002, e-commerce sales were 1.2 percent of total sales. In an economy as large as the United States', this is a major increase.

Making the Best of Both Worlds

As the Internet grows in popularity, all businesses, online and off, will have to find new ways to compete. Powell's City of Books, which has seven stores that sell new and used books in Portland, Oregon, began selling books on the Web in the late 1990s. Although it competes with online superstores such as Amazon.com and Barnes&Noble.com, Powells.com has a unique advantage in Oregon, where local customers can find a book on its Web site and then pick it up at one of its stores. Besides being a local favorite, its Web site has been named a favorite in *Forbes* magazine's *Best of the Web* newsletter.

Although plenty of companies exist solely on the Web, large retailers like Wal-Mart, Target, and Bed Bath & Beyond have all chosen to compete both online and off. After all, the Internet is a tool, and twenty-first-century entrepreneurs would do well to use it.

Section 7-2 Review

RESPOND to what you've read.

1. How are prices determined on the Internet? _____
   _____

2. What do economists predict for the future of e-commerce? What does this say about how traditional companies do business? _____
   _____

3. How does e-commerce affect inflation? _____
   _____

4. What are economic indicators? How often are these figures released, and why are they important?
   _____
   _____

Quick TALK Assume an older family member has limited experience with e-commerce shopping and is looking for the best-priced CD player. **With your partner, brainstorm steps your family member should take to comparison shop, using three Web sites.**

Name _____ Date _____

Worksheet 7-1

Do You Have What It Takes?

Chances are, at some time in your life, you may consider going into business for yourself. E-commerce and the Internet might make you think that starting a business is easy. While it may be exciting, owning a business is extremely challenging. Do you have what it takes?

1. Describe new and unique opportunities that might be available to you as an entrepreneur because of rapidly changing technology and the global marketplace.

   _____
   _____
   _____
   _____
   _____
   _____
   _____

2. Using the Internet, research the personal characteristics and skills necessary to prosper as an entrepreneur. Do you think you have what it takes? Explain.

   _____
   _____
   _____
   _____
   _____
   _____
   _____

3. Using the Internet, access the Small Business Administration's Web site. If you decide to open a business, this site will be very helpful. What type of information is available? Choose one topic on the site, research it thoroughly, create a presentation, and present it to your class.

   _____
   _____
   _____
   _____
   _____
   _____
   _____

Name _____ Date _____

Worksheet 7-2

What Do Statistics Really Reveal?

The Internet allows us to access a range of information quickly. Some of this is in the form of statistics, which are used to collect, process, and interpret data. The U.S. government uses statistics in a variety of ways.

1. Using the Internet, locate the Web site that reports economic indicators for the U.S. government. What type of information is available? Choose one indicator, and explain what you think it means.

   _____
   _____
   _____
   _____
   _____
   _____
   _____

2. Using the U.S. Census Bureau E-Stats Web site, locate one statistic that you find interesting. Define it.

   _____
   _____
   _____
   _____
   _____
   _____
   _____

3. The Pew Internet & American Life Project conducts research to explore the impact of the Internet on children, families, communities, the workplace, schools, health care, and civic/political life. Access the organization's site. Choose one set of data, and read it. Summarize your findings in a one-page paper, and explain it to a member of your class.

   _____
   _____
   _____
   _____
   _____
   _____
   _____

Chapter 7 Review and Activities

Chapter Summary

Section 7-1

- Businesses can be structured as sole proprietorships, partnerships, corporations, or limited liability companies.

- Since there are key differences between each type of business structure, you need to weigh your goals and tolerance for personal risk before determining the best structure for you.

- As a sole proprietor, you must report all of your business expenses and income on your personal income-tax statement. In a partnership, money made or lost by the business is reported on your personal income-tax statement. A general corporation pays its own taxes. As in a partnership, members of an LLC enjoy the privileges of pass-through taxation.

Section 7-2

- Whether through selling items online or comparison shopping using multiple Web sites, e-commerce makes a market economy function faster and more efficiently.

- E-commerce impacts state and local economies through lost sales-tax revenue for governments and lost business for traditional bricks-and-mortar retailers.

- Federal regulations exempt most Internet sales from taxes.

- By keeping down prices, e-commerce helps to keep inflation down.

Chapter 7 Review and Activities

Review of Key Terms

a. stockholder
b. board of directors
c. limited partnership
d. economic indicators
e. sole proprietorship
f. inflation
g. corporation
h. supply and demand
i. basic partnership
j. initial public offering
k. market economy
l. shares

Match each term to its definition.

_____ 1. The units of ownership held by each stockholder.
_____ 2. A structure in which all partners do not have equal rights.
_____ 3. Someone who is part owner of a corporation.
_____ 4. People elected by stockholders to keep an eye on a corporation's management team.
_____ 5. The relationship between the amount of a good or service available and how much people are willing to pay for it.
_____ 6. A business owned and operated by one person.
_____ 7. Retail e-commerce sales, the unemployment rate, the inflation rate, and consumer spending are examples.
_____ 8. The rise in the cost of consumer goods.
_____ 9. Two or more individuals form a business and share equally in it.
_____ 10. The first sale of a company's stock.
_____ 11. Where individual buyers and sellers interact with one another to exchange goods, services, and money.
_____ 12. A business structure in which the company is considered separate from the people who own and run it.

Applying Technology to Academics

ELA **English Language Arts—Reading**
Using the Internet, research a recent Internet public offering (IPO). Report the type of company, the product or service it sells, and the price of the stock the first time it was offered. In addition, track the trading price for the IPO during the next week.

ELA **English Language Arts—Writing**
Research and write a one-page paper about *Net Geners*. Describe who they are, how they learn, and how they shop. Then, compare them to *Baby Boomers*. How are they different? How are they the same?

ELA **English Language Arts—Speaking**
You and a coworker have decided to start an online auction site specializing in used CDs and DVDs. You have identified your target market and done your research. The only question is how you will structure your business. Will it be a partnership or a corporation? Choose a structure, and convince your partner that this is the right decision.

MATH **Calculation**
Choose a product that you purchase frequently. Find out how the prices for it have changed since you were born. Determine if the change is due to inflation or other factors. Calculate the percentage of increase or decrease.

Chapter 7 Review and Activities

Critical Thinking

1. Working as a team member on a class project is similar in many ways to being in a partnership. What are some advantages and disadvantages of working together?
2. Do you think sales tax should be collected for online transactions? If so, who should collect the tax? If not, why not?

COMPETITIVE EVENT

You're a management consultant for a firm that specializes in advising new businesses during their planning stages. You're appointed to help an entrepreneur with an online flower and gift business. The owner needs direction on establishing the business's ownership structure. You'll need to explain the basic types of ownership and recommend one that you feel best suits the business. After hearing your advice, the owner will decide which structure to use.

EVALUATION

You will be evaluated on how well you meet the following performance indicators:
1. Explain the different types of business ownership.
2. Describe current business trends.
3. Discuss trends in e-commerce.
4. Determine factors affecting business risk.
5. Make oral presentations.

INTERNET ACTIVITY

To engage in online activities related to the chapter content, visit the *E-Commerce* Web site at **ecommerce.glencoe.com**.

BusinessWeek online

TOPIC: Entrepreneurship

Many people consider owning a business to be part of the American Dream. Successfully realizing that dream depends on careful planning and execution.

ACTIVITY Go to the Student Center at **ecommerce.glencoe.com**. Click on the *BusinessWeek* Activities and open Chapter 7. There you'll learn more about what it takes to set up a business and what resources exist for aspiring entrepreneurs.

Chapter 8

Revenue Models and the Business Plan in E-Commerce

Section 8-1
E-Commerce Revenue Models

Section 8-2
E-Commerce Business Plan

PREREADING STRATEGIES
Before you read this chapter, finish these statements in your journal:
- Based on this chapter title, I predict...
- The most important thing to remember about revenue models and business plans is...
- Some questions I have about e-commerce revenue models and business plans are:

BusinessWeek online
Go to *BusinessWeek* online. Find an article about launching a small business. Make a list of the steps involved and describe how they might be useful in setting up an online business.

E-Commerce Road Map

When starting a business, online entrepreneurs face many of the same questions as their bricks-and-mortar counterparts. "What kinds of goods or services am I offering?" "Who is my target audience?" "Who is the competition?" Answering these questions requires some basic business planning.

Drill Down

1823 Slicker Idea
Scotland's Charles Mackintosh creates the raincoat, which quickly becomes an international best-seller.

1849 Department Stores
Henry Charles Harrod founds a grocery store in London's Knightsbridge district. Today, Harrods is considered the most famous and most extravagant department store in the world.

1889 Going Up?
Brothers Charles and Norton Otis unveil the first functional electric elevator. It gains enormous popularity in skyscrapers being erected throughout the world.

1928 Miracle Vaccine
Scottish biologist Alexander Fleming discovers penicillin, the miraculous antibiotic that would control polio.

1968 Mouse on the Move
Douglas Engelbart invents the computer mouse and introduces it at a technology conference in San Francisco.

1982 Digital Music
Sony releases the first compact disc (CD) in Japan and Europe. The technology eventually becomes the dominant format in music.

Get the Big Picture
There are more ways to make money in e-commerce than selling a product or service directly to consumers. What are some other ways to make money?

POWER READ

Be an active reader and use these reading strategies:

PREDICT what the section will be about.
CONNECT what you read with your own life.
QUESTION as you read to make sure you understand the content.
RESPOND to what you've read.

Section 8-1

E-Commerce Revenue Models

AS YOU READ...

YOU WILL LEARN
- how revenue models are defined.
- how revenue models are developed.
- how and why revenue models differ in complexity.

WHY IT'S IMPORTANT

A revenue model determines how a company makes money and how the company functions. A well-thought-out model can mean the difference between success and failure. Understanding what revenue models entail can help you understand more about how companies operate.

KEY TERMS
- revenue model
- e-zine
- blog
- affiliate program
- licensing

PREDICT

How might you decide what to charge for products you create or design, as opposed to those provided by a supplier?

WHAT IS A REVENUE MODEL?

A **revenue model** describes how your company generates income, or *revenue*. Your company's revenue model might be simple, like the one used by eBay, which allows users to sell items through its Web site in exchange for a percentage of the selling price. Your revenue model could be more complicated, like the one used by Salon.com. The Web site sells articles to readers for a subscription price, sells space on its site to advertisers, and sells books by Salon writers both on its Web site and in bricks-and-mortar bookstores. The type of revenue model you choose will depend on the type of product or service you intend to sell. **Figure 8.1** identifies the basic types of revenue models in e-commerce.

Types of Revenue Models

When you're devising a revenue model, it's good to think in general terms about your product. It might not seem like books and lawn mowers have much in common, but in the decade after it was founded, Amazon.com expanded from a bookstore into a department store that sells appliances, sporting goods, and even Segway™ Human Transporters. Amazon's sales are now eight times what they were when the company went public in 1997. All of its products, from books to computers, fall into the same general category of *consumer goods*. Consumer goods are physical items that people buy for personal or household use. Because its products are all the same general type, Amazon has been able to sell them using one revenue model.

Intuit, the maker of TurboTax, sells a service, rather than a product. For a fee, U.S. taxpayers can go to TurboTax.com and use Intuit's software to prepare their tax forms. To provide this service, the managers of Intuit had to figure out how much it cost them to develop their software, maintain the Web site, and provide customer service. Because Amazon sells consumer goods and Intuit sells consumer services, the companies operate under different revenue models.

Salon has yet another revenue model. The online magazine provides "content," including original articles about culture and politics. Readers can access these articles either by paying for a subscription or by watching an advertisement, which enables them to get a "free day pass." In addition, Salon publishes books by its authors, which it sells both through its site and to traditional bookstores. As such, Salon's revenue model covers three distinct products: the articles, which it sells to its readers; its readership base, which it markets to advertisers; and its books, which it sells both directly to consumers and to bookstores.

Figure 8.1

E-Commerce Models

E-Commerce Models	Online Applications
Product Sales	consumer goods, retail, wholesale
Advertising	search engines, Web portals, directory pages, commercial Web sites
Service	online fee-for-service, fee-for-transaction, subscriptions
Inter-Organizational	procurement, supply chain, distribution
Third-Party	e-malls, online auction, online exchange
Licensing	intellectual rights, patents, copyrighted material

PROFIT GENERATORS Although much effort has gone into developing new revenue models to drive e-commerce, the traditional models that drive bricks-and-mortar companies have proved to be the most successful for online companies as well. *Consider your own Internet use. What products or services do you buy or use online? Who is profiting, and how are they doing it?*

In all three cases—Amazon, Intuit, and Salon—the type of product determines the revenue model used.

The type of revenue model you use can also be altered to suit the needs of your company as it grows. Salon.com, for example, started as a magazine that only sold space to advertisers. The books and subscription services were added later to generate additional revenue.

PRODUCT SALES The most basic revenue model involves selling one product or related product line to customers who use your Web site as they would a print catalog. There are two types of products you can sell: products you make, which you should sell at a higher price than they cost to make; and products you buy, which you should sell at a higher price than you paid for them.

You might want to start out by selling a craft made by you, a partner, or an employee. Crafts aren't limited to items like handmade jewelry or figurines. They can include any specially made item, from gourmet salsa to designer skateboards. Don't underestimate the potential of homegrown crafts businesses. In 1977 Ben Cohen and Jerry Greenfield took a correspondence course in ice-cream making. A year later they spent $12,000 to start Ben & Jerry's ice-cream shop. Twenty-three years later Cohen and Greenfield sold their company to food giant Unilever for $326 million.

If you're not interested in making your own products, you can follow the Amazon.com model. Amazon doesn't produce books, CDs, or televisions; it buys them from wholesalers, then sells them to the public. As you learned in Chapter 3, *wholesalers* are businesses that sell products in large quantities to stores or distributors at discounted prices, rather than directly to the public. Because retailers like Amazon buy their products at a discount, they can afford to sell them for less than it might cost you to buy them at a store.

Figure 8.2

Online Advertising

BENEFITS OF ONLINE ADVERTISING Online advertising is an important way for advertisers to reach an audience and attract potential customers. *How might brokerage firms gain from placing advertising on the Yahoo! Finance Web site?*

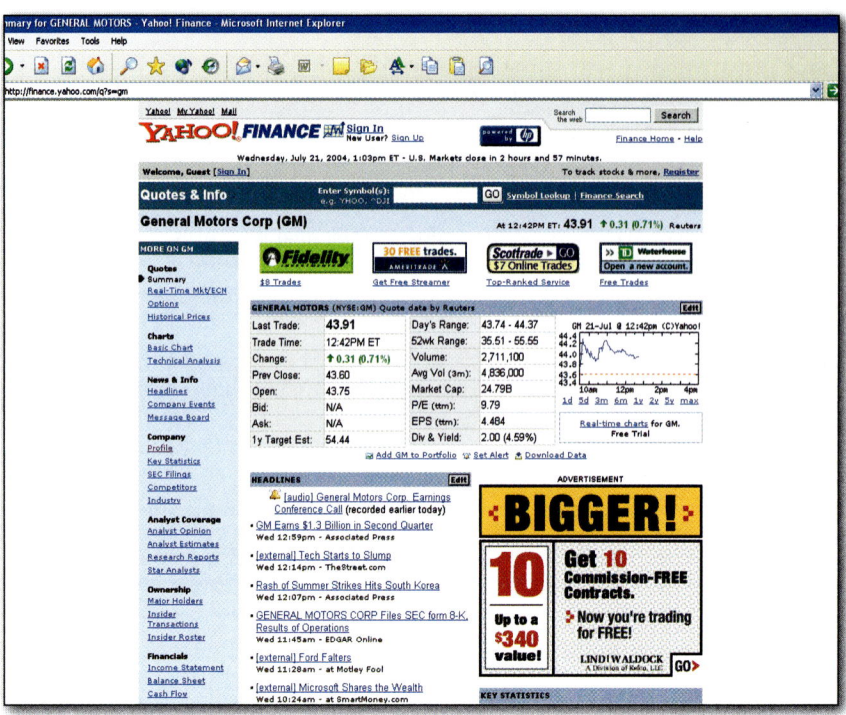

Reproduced with permission of Yahoo! Inc. © 2004 by Yahoo! Inc. YAHOO! and the YAHOO! logo are trademarks of Yahoo! Inc.

CONNECTION — MATH

No Boundaries, No Zones

Swatch, the Swiss maker of funky and fashionable wristwatches, inaugurated Internet time in 1998. Instead of the usual hours/minutes/seconds system, Swatch instilled a new measure called ".beats." Biel MeanTime (BMT) is the universal reference for Internet time. A day of Internet time begins at Central European Winter/Standard Time at midnight BMT, or at 000 .beats. One Swatch .beat is equivalent to 1 minute and 26.4 seconds. *If one .beat is equal to 0.001 day, how many .beats are in one day?*

Not only do retailers have to negotiate the best prices they can; they have to find reputable wholesalers. In the book industry, the Ingram Book Group is a well-known wholesaler. It has an inventory of one million books "available for immediate shipment." Finding a wholesaler with a large inventory and the ability to deliver products quickly will keep you from losing sales on items that are out of stock or unavailable.

AD SALES Like magazines, TV stations, and other media outlets, Web sites can sell advertising space as a source of revenue. If you've ever visited a commercial Web site, you're familiar with Internet advertising. U.S. businesses spend an average of $250 billion each year on advertising. At present, only about 3 percent of that is spent on Internet ads. Most Internet advertising money is spent on Web sites that reach a vast audience, like Amazon or Salon. Although Yahoo! Search provides a valuable service to Internet users, it does not charge for its service. Instead, it sells advertising on its site, as shown in **Figure 8.2**. Companies also pay Yahoo! for prominent placement on Yahoo!'s search results.

When you start an online business, your Web site will be virtually unknown, so you will have to seek out advertisers. Although your business may be small, it is possible to attract advertising revenue from other small businesses that sell a similar product or appeal to a similar audience. This requires some research.

First you need to define the type of goods or services you offer, then determine which kinds of businesses offer related goods and services. You'll want to visit the Web sites of these related businesses to find out where they advertise and what kinds of ads they run. For example, if you're selling designer skateboards, you could check out sites for sporting-goods stores, extreme-sports events, and teen lifestyle magazines. After you've looked at enough sites, you can contact some of them to see if they want to advertise on your site. You don't need to limit yourself to online businesses. You can also look for potential advertisers among local businesses that don't have a Web presence.

If you're just starting an online business, it can take a while to build up enough customers to attract advertisers. Although selling advertising can be a major source of revenue, you also need to advertise your own business, which can cost a lot of money. For this reason, many online businesses trade advertisements with each other. Once you've built up a large customer base, you can begin selling ad space on your own.

SUBSCRIPTIONS Most magazines, newspapers, and information services earn income by selling advertising and subscriptions. E-zines, like Salon.com, similarly generate revenue by selling subscriptions. An **e-zine**—or electronic magazine—may allow subscribers unlimited access to its site or regularly deliver its content to subscribers by e-mail.

Many established newspapers and magazines offer both traditional print subscriptions and online subscriptions. Readers of *The Wall Street Journal* can either subscribe to the newspaper or buy a subscription to the newspaper's Web site. *The New York Times* offers most of its daily articles for free but requires subscribers to pay for special services such as access to the paper's rich archives. Some online news and information services don't charge a subscription fee but require their users to register with their sites. They then generate income by selling their subscriber lists to other Web sites that want to reach the same types of customers.

A specialized type of e-zine is the weblog, or blog. As you learned in Chapter 7, a **blog** is a public online journal kept by a writer, or blogger, whose views or activities might be of interest to readers. One of the first and most successful blogs is Matt Drudge's Drudge Report, a site that links to various news stories and occasionally features commentary. The Web site turned Drudge into a talk-radio host and celebrity. Although some blogs generate revenue through subscriptions, many of them, such as the Drudge Report, make money by selling advertising.

AFFILIATE PROGRAM Another way to generate revenue is through affiliate programs. An **affiliate program** is a partnership in which you deliver your Web site's customers to other online businesses, or *affiliates,* to buy their goods and services. In exchange for bringing business to your affiliates, you receive a *commission,* or a percentage of the sales they make on that business.

Amazon uses an affiliate program in both directions. It lets smaller stores sell on its site for a fee, and pays a fee to smaller sites that send customers to Amazon. Amazon pays affiliate Web sites a five percent commission on any sale they make for Amazon.

Autoweb is another Web site that uses an affiliate program. It currently has more than 5,000 affiliates. If a user goes to Autoweb from an affiliate's Web site and posts an advertisement for a used car, the affiliate receives a five dollar referral fee.

CONNECT

When you visit your favorite Web site, what kinds of ads do you see? Why do you think they advertise on that Web site? What does this say about you as a consumer?

E-BOOK EXPERIMENT
Stephen King used the "honor system" to sell his story *The Plant* online. Readers were asked to mail in a fee ($1–2) in exchange for each installment of the book they downloaded. Approximately 70 percent of the readers paid the fees as required; the endeavor ended prematurely after unauthorized copies of installments appeared on the Internet. *What do you think is the weakness of the "honor system" revenue model?*

QUESTION
What is intellectual property? What are some examples?

LICENSING If you are the creative type and have set up an online business to market your band, artwork, or computer animation, you need to know about licensing. **Licensing** is the granting of permission to use intellectual property such as music, photos, software programs, and inventions. As you learned in Chapter 6, no one can legally copy your material without paying you for it.

The idea behind licensing is simple. Assume you have created a product such as an e-commerce textbook or a process for running credit checks online that other companies are interested in incorporating into their Web-page design. Those companies would pay you a fee for the right to market the product, to make copies of it, or to use it in other ways specific to a *licensing agreement* signed by both parties. The purchaser of licensing rights is then able to collect whatever profit it can from the use of your product.

The main benefits to you are that you will receive your fee up front and that your product will be out on the market without you having to take any financial risks. Assuming your product works effectively, other companies may become interested in signing their own licensing deals with you.

The Internet is largely unregulated, so the stealing of images, sounds, and ideas runs rampant online. If you're offering original art, writing, or other creative work, you need to protect yourself. As discussed in Chapter 6, you can do this by copywriting your material and keeping track of anyone who visits your site.

MUSIC LICENSING RIGHTS In the case of musical compositions, licensing is a critical form of protection. Whenever a musical work is performed for the public—beyond family and friends—the performers must obtain a license.

In general, every Internet transmission of a musical work constitutes a public performance. If you transmit a song by providing it via streaming audio on your Web site, you must have a license. Radio and television stations can get international licenses to transmit original music programming online into numerous countries on the basis of a single, multiterritory license.

Don't let fear of theft keep you from selling your own intellectual property on the Web. Apple Computer sold 1,000,000 songs in the first week of its iTunes digital music store.

Section 8-1 Review

RESPOND to what you've read.

1. What is a revenue model? What is a major factor in determining the type of revenue model to use?

2. Why is it important for advertisers to know about a company's customers before advertising on a particular site?

3. What is the difference between wholesale and retail prices?

4. What is a blog? How can it generate income?

Quick TALK With a partner, brainstorm possible domain names for a handmade candy company. (Keep in mind that up to 67 characters can appear in a domain name.) *Discuss which name works best for the company and why.*

Section 8-2

E-Commerce Business Plan

AS YOU READ...

YOU WILL LEARN

- how an e-commerce business plan is defined.
- why a plan is necessary for every business.
- the major components of a business plan.

WHY IT'S IMPORTANT

Creating a company is hard work, and entrepreneurs must have a clear path, or business plan, laid out when they begin and as they continue to operate their businesses. Knowing what a business plan contains will help you to understand more about what creating and running a business entails.

KEY TERMS

- business plan
- forward-looking statements
- income statement
- balance sheet
- assets
- liabilities

PREDICT

What types of information might a business plan contain?

WHAT IS A BUSINESS PLAN?

Your business will be defined by your choices. When you have to make tough decisions, such as whether to offer a new product, you'll be able to refer to your business plan for guidance. Your **business plan** is a detailed description of a business's objectives, products and services, operations, potential customers, and financial resources. It will be used to show potential investors what you hope to accomplish and how you plan to get there. A solid business plan can be used to attract investors even if your business fails to make money at first. (See **Figure 8.3**.)

Your business plan should include not only your best ideas and loftiest goals but also a realistic and honest appraisal of the obstacles in your way. An e-commerce business plan clearly defines the business, identifies its goals, and conveys the strategies you will use to achieve them. It informs everyone, including potential investors, suppliers, the government, and future employees, about your operations and objectives. It helps you allocate your resources, anticipate and handle challenges, and make informed business decisions. Evaluating the different types of business plans used by Internet companies will help you craft the best plan for your venture.

Elements of a Business Plan

The Small Business Administration (SBA) is an excellent resource for entrepreneurs. It offers print handouts that will help you develop your business plan and gain access to an information-rich Web site (sba.gov). Its publications and services are provided free of charge. Before you begin writing your business plan, the SBA recommends that you consider four basic questions:

- What products and/or services will your business provide, and what needs do these products and services fill for consumers?
- Who are your potential customers, and why will they buy products and services from your company?
- How will you reach your potential customers to sell your products and services?
- How much capital will you need to start and operate your business, and where will you access financial resources?

To these questions, the e-commerce entrepreneur must add:

- What are the unique challenges to conducting business on the Internet, and how will I cover them in my business plan?

Figure 8.3

Business Plan

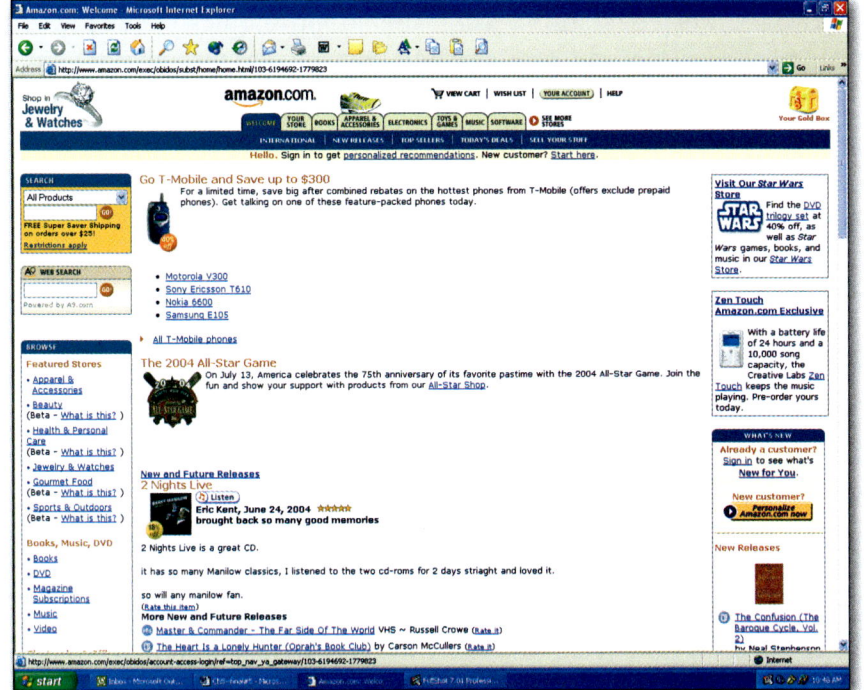

OPERATING IN THE RED After running at a loss of millions annually since its launch, Amazon reported its first quarterly profit in 2000, and, upon expanding its catalog to sell an abundance of other consumer goods, realized its second quarterly profit in 2003. *Why is Amazon able to stay afloat after losing so much money for so many years?*

© 2004 Amazon.com, Inc. All rights reserved.

Once you've answered these questions, you're ready to begin writing your e-commerce business plan. Your plan should include, but is not limited to, the following elements:

- Cover Page
- Title Page
- Table of Contents
- Introduction
- Executive Summary
- Management Plan
- Marketing Plan
- Web Site Demographics (or Market Analysis)
- Web and Retail Competition (or Competitive Analysis)
- Operations Plan
- Financial Plan

These are only the bare elements of a business plan. For the complete elements of a comprehensive business plan, see **Figure 8.5** on page 171.

THE INTRODUCTION Imagine that you're a potential investor in a company. You open the business plan and scan the introduction. If you're not immediately intrigued, you'll probably toss it in the trash and go on to the next plan submitted by an entrepreneur who's eager for your money. The introduction should move quickly toward your company's *mission statement*—a short, general description of the purpose and goals of your business. (You will read more about mission statements in Chapter 9.)

EXECUTIVE SUMMARY An executive summary is a more detailed description of your company's mission statement. This section of the plan will define the purpose of your business and highlight key facts about the company, including its goals and objectives, products and services, target market, competitive position, marketing strategies, and management and financial plans. It should convince the reader of your personal commitment to the success of the business.

The executive summary states your goals and objectives, and highlights how you will achieve them. It identifies who your customers are, how you have identified them, why there is a need for your product, and why they will buy from you. It will state your unique selling features and how your business will meet changes in the business climate. The summary should affirm the domain name for your e-commerce business and how your Web site will be designed, maintained, and updated. End the executive summary by reminding potential investors why they should support you and your idea.

Keep your executive summary brief and focused on the key points. This section is meant to convince the reader that you've planned your operation in detail and that there's a good probability of success. You should also provide evidence that there is a market for your products and services that is not already dominated by your competition. If you've identified competitors in your market, include your strategy for surpassing them. At a glance, the executive summary should give the reader an overview of everything that will be included in the total presentation.

MARKETING THE BRAND
Branding has become an increasingly popular marketing tool to raise consumer awareness not only of a new product on the market but also of the company that produced it. *Can you cite examples of product branding?*

MANAGEMENT PLAN Your management plan should give a detailed description of your business, its philosophy, and its goals. The section should also include the company's form of ownership, legal structure, and a brief history. Although you're embarking on a new venture, you can write about your relevant professional experiences and the related experiences of your partners, key employees, and board of directors. For this reason, it's important to recruit board members who have tangible e-commerce experience as either successful entrepreneurs, division heads of corporations, or investors in other Internet start-ups. If you've acquired any existing companies, you should include the history of those enterprises.

Your plan should include **forward-looking statements**, which describe the future for the company in both the short term (within a year) and the long term (generally five to ten years). Your management plan should also discuss day-to-day business operations and any long-term agreements such as building leases, equipment rentals, or loans that will have to be dealt with on a regular basis.

MARKETING PLAN The marketing plan is the first section of your business plan that explains in practical terms how you will launch and develop your business. It begins with a description of the products and services you plan to offer, including how they will be priced, packaged, and delivered.

You should describe the *terms of sale* for each of your products. Terms of sale include the methods of payment you'll accept, refund policies you'll establish, and warranties or guarantees you'll offer.

Once you've described what you're selling, explain how you're going to sell it. Outline where you plan to advertise, how often, and at what cost. Describe all partnerships you have with other companies or affiliate agreements you've made.

You'll also need to discuss the competitive environment. If you're selling a fairly new product, contact a consulting firm such as Forrester Research or an investment bank such as Lehman Brothers to learn how large the market is for your product and how large the firm thinks the market will become over the next one, three, and five years. Be sure to ask how your industry will be affected by the national and global economy, and consult a lawyer to find out if there are any impending legal or regulatory hurdles. Your marketing plan should be honest about all of the obstacles you'll face and how you intend to overcome them.

WEB SITE DEMOGRAPHICS Investors will want to know about the *demographics* of your Web site. Demographics consist of information about the characteristics of the types of customers you expect to use your Web site, such as their age, average income, level of education, and where they live. For example, you might expect your customers to be mostly teenage males who live in big cities. This kind of information, also called a Market Analysis, is important to investors because it helps them assess whether they think your business will be a success in certain markets. (For more information about demographics, see Chapters 9 and 12.)

Describe all of the site's special features, including how you'll collect information about your customers, receive feedback from them, and tailor your site to meet their needs and desires.

> **CONNECT**
> Pick a Web site you visit regularly. What methods does it use to collect information about you?

QUESTION

What are a business's competitors?

Now, draw a connection between everything you've written about your Web site and everything you know about your customers. For example, if you've started a community bulletin board like Craigslist.org, you could say that you've created a simple, text-based Web page that can be easily used by anyone with a dial-up modem connection.

If, however, you've started a high-end auction house site like Sothebys.com, you'll want to use a different approach. You could explain that you've created a site that shows every product in rich detail, both because you're selling fine works of art and because your customers are likely to have high-speed Internet connections. Describe your customers in terms of things like their age, income level, comfort with technology, and location. Finally, write about how you'll be able to change and modify your Web site as your business grows.

WEB AND RETAIL COMPETITION This section of your business plan is often referred to as Competitive Analysis. Though you've discussed your competitors in other sections, this is where you'll provide a detailed description of your most important rivals in the online and offline arenas. Remember that your business plan is meant to be read by knowledgeable investors who are interested in your business. If you downplay your competition or claim to have none, you'll look like an amateur.

Remember to list not only e-commerce competitors but also traditional bricks-and-mortar businesses that cater to your target market. If any of your rivals are public companies, you'll be able to gather information about their operations both through their filings with the

Figure 8.4

Management Plan

ORGANIZATIONAL CHART An organizational flow chart illustrates the hierarchy and interconnectedness of a company. *How does defining the management structure of a company aid in developing and selling a business plan?*

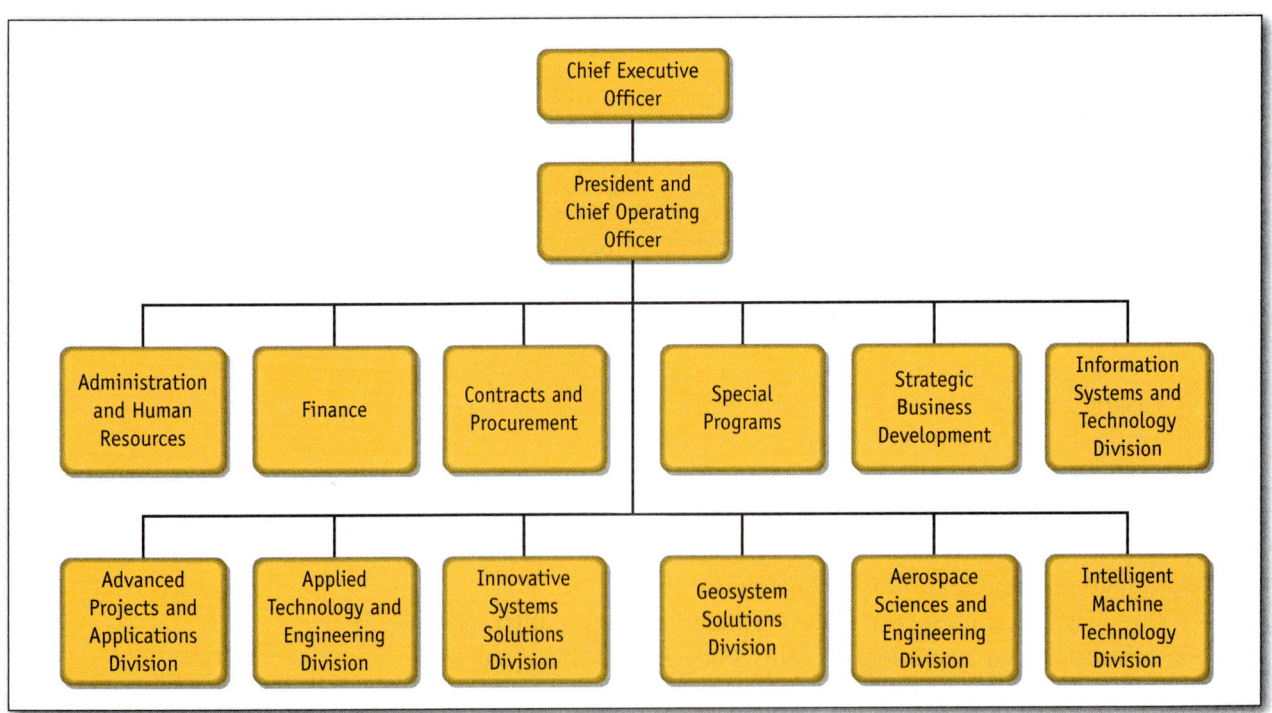

Web Site Success

LOOKING THINGS UP
Google

In 1995 Sergey Brin and Larry Page, two doctoral candidates at Stanford, collaborated on developing a search engine, which eventually became Google. Over the course of four years, Page's dorm room served as Google's data center, and Brin's dorm room became the business office. The company expanded and went international—both in office space and in searching pages written in more than 80 languages. Today, it's the Web's leading search engine, delivering more than 200 million search queries per day.

Thinking Critically
If you were Brin and Page, how would you get the money to start and grow your company?

Securities and Exchange Commission (SEC) and through the reports of various stock and industry analysts at major investment banks. In the case of large private companies, industry consultants might be able to provide you with valuable information. Also, be sure to scan the financial press for articles about public and private companies in your area. After you're done with your research, you'll be able to honestly evaluate the competition in your business plan.

OPERATIONS PLAN How will your e-commerce business get things done on a day-to-day basis? How will the home office run, what is the command structure, how will the Web site be maintained and updated, how will customer service issues be handled, and how will your product be delivered to the customer? These questions can all be answered by your business operations plan. An operations plan describes how a business is to be managed and how it produces and delivers its products and services.

A typical operations plan will include information that provides a description of the business process: how products are marketed, how customers place orders, how orders are processed, how products are delivered, and how inventories and customer service are managed. An operations plan will also include an organizational chart, such as the one shown in **Figure 8.4**, which describes the company's management structure.

If an operations plan contains forward-looking statements regarding the future of the business, then it should also contain a summary of changes being instituted at the company and an outline of the plans to achieve those goals. An operations plan is a valuable tool for understanding the business and for implementing the software necessary to bring your vision to the Web.

FINANCIAL PLAN The financial management section of your business plan should explain how much capital you'll need to start your business, how much capital you already have, and what resources you can use to obtain future capital. You should also include how long your business can operate on the money it has raised so far and how long it will take for your company to turn a profit.

For the first year, it's customary to include a monthly operating budget that highlights the company's cash flow. *Cash flow* is the difference between the amounts of money a company generates and spends each month. Cash flow should not be confused with profit. Cash flow represents money received over costs within a fixed period of time. Profits represent money made once the business's operating costs are paid.

In addition to the monthly cash-flow statement for your first year, it's traditional to include two years of projected income statements and balance sheets. An **income statement** summarizes your revenue and expenses for a period of time and shows profits or losses. A **balance sheet** shows how much a business is worth at a specific time. It lists all of the things owned by the business, which are known as **assets**, then subtracts all of its debts, which are known as **liabilities**, to show the net worth, or actual value of the business.

These projected statements all lead up to a projected *return on investment* (ROI), which indicates, over the next two to five years, how much money the company will make in relation to how much money was invested. For example, if a company starts with $100 million from investors and makes a profit of $50 million in its first year, its return on investment is 50 percent.

Finally, you must list your wages and the salaries for company directors and key executives. If, like many entrepreneurs, you've decided not to take a salary for the first year or more, list the salary you would have received, and then label it "deferred."

GROWTH AND CONTINGENCY PLAN Investors will also want to see a *contingency plan*—a plan of action for unexpected events, such as changes in economic or political conditions—in your business plan. You will also want to provide a *growth plan,* which describes how your business will expand over time. To ensure that your business plan is complete and professionally presented, it is worthwhile to hire the services of a business consultant.

THE E-BUSINESS PLAN In addition to the elements of a traditional business plan, there are elements unique to an e-commerce business plan. Just as investors want to know when a bricks-and-mortar business plans to open its doors for business, investors will want to know when you plan to launch your Web site and have it up and running. If you can't show investors samples of what your Web site looks like, you need to provide descriptions of the style of your site, how it will function, and what features it will have. Your e-business plan should explain what Web server or Web host you will use to maintain your site. In the next Unit, you will learn how to create and build your Web site and how to put it on the Internet.

Putting Your Plan into Action

When you have finished your business-plan draft, you should have it reviewed before submitting it to potential investors. Since it's important to solicit honest feedback, it might be best to contact the Service Corps of Retired Executives (SCORE), a group of former executives who

Figure 8.5

Comprehensive Business Plan Outline

MANAGEMENT PLAN
Key Management
Advisors and Service Providers

COMPANY DESCRIPTION
Business History and Description
Goals and Objectives

PRODUCT AND SERVICE PLAN
Overview of Products and Services
Product Development Status

MISSION AND VISION STATEMENTS
Mission Statement
Vision Statement

INDUSTRY OVERVIEW
Industry Trends and Growth

MARKET ANALYSIS
Target Market Demographic Profile
Target Market Projections
Market Trends and Growth
Customer Needs Analysis

COMPETITIVE ANALYSIS
Competitive Overview

MARKETING PLAN
Marketing Mix Strategies
Product Strategies
Place Strategies
Price Strategies
Promotion Strategies

OPERATIONS PLAN
Location
Property Ownership/Lease Terms
Equipment Needs
Manufacturing Processes and Costs
Suppliers and Purchasing
Storage and Inventory Control
Channels of Distribution
Quality Control Measures

ORGANIZATIONAL PLAN
Legal Structure
Record Keeping
Legal and Insurance Issues
Labor, Staffing, and Training

FINANCIAL PLAN
Investment Proposal
Projected Cash Flow Statement
Projected Three-Year Income (Profit And Loss Statements)
Projected Balance Sheet
Break-Even Analysis
Historical Financials
Financial Assumptions
Financial Ratios

GROWTH PLAN
Growth Strategies
Business Location Issues
Effects of Growth
Growth Financing
Evaluation of Growth Strategies

CONTINGENCY PLAN

EXECUTIVE SUMMARY

COVER PAGE

TITLE PAGE

TABLE OF CONTENTS

SUPPORTING DOCUMENTS

WORK-IN-PROGRESS
Developing a business plan is like outlining a strategy for turning your business idea into a reality. *Why should a business plan be a continuing work-in-progress?*

volunteer to help advise entrepreneurs. You can also find a counselor through the Small Business Development Center (SBDC).

Getting your business plan read by people willing to invest is usually one of the hardest obstacles entrepreneurs face. Some venture-capital funds accept proposals from new businesses, but since your proposal will be left in a pile with hundreds of others, every submission is a long shot.

Be sure to ask friends and family about contacts they have in the investment community. You may also want to seek out entrepreneurs who have been successful in their own businesses and might be willing to invest in another. If possible, attend conferences and seminars held by industry groups where investors and entrepreneurs can meet face to face.

If your company won't require massive amounts of capital to get started, you can take investments or loans from friends and family members. You can also acquire a loan through the SBA or a bank, though both will require collateral. Keep an eye out for government grants as well. Depending on your region, gender, or ethnicity, the government might be offering small business start-up grants that haven't been well publicized.

No matter how you choose to raise money, your business plan will serve as your primary set of credentials. If you don't need to raise money, your plan will serve as a road map for success.

Section 8-2 Review

RESPOND to what you've read.

1. What is a business plan? Why is it important? _____
   _____
   _____

2. What is an executive summary? Why is it an important component of a business plan? _____
   _____
   _____

3. What does an income statement reveal about a business? Why is it important? _____
   _____
   _____

4. What are customer demographics? Why is this information important? _____
   _____
   _____

Quick TALK With a partner, discuss the business goals and target market for your online candy company. *Use the information to create a business statement for the company.*

Name _____ Date _____

Worksheet 8-1

Creating a Road Map for Your Business

Revenue models are one of the key factors in determining the success of your company. Without a successful model, you may end up working very hard but failing. A revenue model can also be called the "engine" of the business.

1. There are many different revenue models, from simple to complex. Find an example of a site that sells:
   a. products and then ships them to buyers.
   b. services.
   c. content.
   (Do not use any examples from the chapter.)

   _____
   _____
   _____
   _____

2. There are other models used by online businesses. Find an example of a site that:
   a. provides free information but requires users to register.
   b. offers subscriptions for a fee.
   c. operates as a blog.
   (Do not use any examples from the chapter.)

   _____
   _____
   _____
   _____

3. Create a spreadsheet listing the examples you've found. Then, with a group of three or four classmates, compile the examples in one large spreadsheet.

   _____
   _____
   _____
   _____
   _____

Name _____ Date _____

Worksheet 8-2

Your E-Business Plan

Entrepreneurs see the Internet as a new and exciting opportunity. However, many entrepreneurs say they just don't have time to create business plans for their potential e-businesses. While they know business plans serve as road maps to guide them in the right direction, some people rush into new business opportunities without them. When that happens, it is very likely their businesses will fail.

1. With a partner or team of students, brainstorm an online business you would like to start. Create a business name and a domain name for your Web site. Then, using the Internet, research the Small Business Administration to learn more about writing a business plan.

   _____
   _____
   _____
   _____
   _____
   _____

2. Develop an e-commerce business plan for your business as outlined on page 165. (You do not need to develop a comprehensive business plan as outlined in Figure 8.5 on page 171.) Each section should be at least one page long. Assemble the plan in a notebook or folder.

   _____
   _____
   _____
   _____
   _____
   _____

3. Present your business plan to another group in the class or to the entire class. Use PowerPoint to create the presentation.

   _____
   _____
   _____
   _____
   _____
   _____

Chapter 8 Review and Activities

Chapter Summary

Section 8-1

- Your revenue model is determined by the type of product or service you choose to sell.
- Revenue models are developed by thinking about the product—whether consumer goods, services, or content.
- The most basic revenue model involves selling a product to customers who order from your Web site like a catalog. Other models are more complex and involve selling subscriptions, selling advertising, or setting up affiliate programs.

Section 8-2

- An e-commerce business plan is defined by your choices—your ideas, your goals, and a realistic appraisal of the obstacles in your way.
- A business plan provides a detailed description of a company's objectives, products and services, operations, potential customers, and financial resources.
- Every business needs a plan to help allocate resources, anticipate challenges, make informed decisions, and attract investors.
- The major components of a business plan include an Introduction, Executive Summary, Management Plan, Marketing Plan, Web Site Demographics, Web and Retail Competition, Operations Plan, and Financial Plan.

Chapter 8 Review and Activities

Review of Key Terms

a. revenue model
b. e-zine
c. blog
d. affiliate program
e. licensing
f. business plan
g. forward-looking statements
h. income statement
i. balance sheet
j. assets
k. liabilities

Match each term to its definition.

_____ 1. The granting of permission to use intellectual property.
_____ 2. All the things a company owns.
_____ 3. Describes how a company generates income.
_____ 4. Describes the future of a company in both the short and the long term.
_____ 5. An electronic magazine.
_____ 6. A list of all a business's debts.
_____ 7. A partnership through which an online business delivers customers to other online businesses.
_____ 8. A financial statement that shows how much a business is worth at a specific time.
_____ 9. A detailed description of a company's objectives, products and services, operations, potential customers, and financial resources.
_____ 10. A public online journal.
_____ 11. A summary of revenue and expenses for a period of time.

Applying Technology to Academics

ELA **English Language Arts—Reading**
Access the Intuit Web site. Find out what kind of information you can access for free from the site and what services you have to pay to receive. Does the site offer features that entice you to return? Explain.

ELA **English Language Arts—Writing**
Research the iTunes digital music store. Write a one-page instruction list describing how to access the store, browse for tunes by song title and by artist, make a purchase, and buy a gift certificate. Then, exchange your list with another student in the class, and see if you can follow the instructions.

ELA **English Language Arts—Speaking**
Imagine you are going to open an online business. Write a brief mission statement for your business and present it orally to the class. Does it accurately reflect your goals and abilities?

MATH **Graphing Software Program**
Track your personal cash flow for the next two weeks. Using a spreadsheet program, create a graph depicting the cash in and the cash out. Then, write a paragraph analyzing your personal cash flow and how you might increase the cash in (income) and decrease the cash out (expenses).

Chapter 8 Review and Activities

Critical Thinking
1. Explain the importance of choosing the right revenue model.
2. Which element of a business plan is the most important and why?

COMPETITIVE EVENT

You're a financial advisor who specializes in counseling entrepreneurs. Over the years you have learned that many entrepreneurs lack strong business plans. You have decided to include an explanation of business plans and their key elements in your meetings with potential clients. Your first client wants to start an online jewelry business. Explain the elements that your clients need to include in their business plans.

EVALUATION

You will be evaluated on how well you meet the following performance indicators:
1. Explain the nature of business plans.
2. Incorporate e-commerce considerations into your explanation.
3. Explain the specific elements the business owners need to include in their plans.
4. Describe the company's objectives.
5. Outline legal issues affecting businesses.

To engage in online activities, visit the *E-Commerce* Web site at **ecommerce.glencoe.com**.

BusinessWeek online

TOPIC: Developing a Marketing Plan

A company's marketing plan, whether for a single product or the company as a whole, is an outline of strategy—a plan of action intended to produce the best possible results for the company. In business those results are usually measured in dollars because profit is a firm judge of how successful a marketing plan is.

ACTIVITY Go to the Student Center at **ecommerce.glencoe.com**. Click on *BusinessWeek* Activities and open Chapter 8. There you'll learn more about what a marketing plan is and what it should include. You'll also develop a plan for a business of your own.

SELF-ASSESSMENT

Decide Your Future

Life is a series of decisions. It makes sense to develop tools to help you make them. The worksheet that follows is one such tool. It won't make decisions for you, but it will help you understand how to approach major decisions.

DECISION-PLANNING WORKSHEET

Here's how to use the worksheet on the following page. Look at the row running across the top. This is where you list three different future options from which to choose. They could consist of what to do after high school, what to do for a living, or what to study. What you choose for your options is entirely up to you. Your options could range from starting your own rock band to going to law school to working in the family business.

Now look at the left-hand column. This is where you list nine goals and values that are important to you personally. They could include learning about a particular trade, traveling, carrying on a family tradition, making money, expressing yourself, pleasing your parents, and achieving personal pride. Again, the goals and values you choose are entirely up to you.

Next to your goals and achievements, you should rate how important each one is to you on a scale of 1 to 5, 5 being the highest. If you value contributing something to society more than making a lot of money, you might give the former a 4 and the latter a 2. If you value being in a position of authority as much as doing something creative, you might give both a 3.

Now, under "Probability," list the likelihood that your goal or value can be obtained with each option. For example, if one of your options is to become a jazz musician and one of your goals is to make a lot of money, the probability for making a lot of money as a jazz musician is probably a 1 or 2. After listing the probability factor, multiply Probability (P) by Importance (I) to get a Subtotal. For example, if you ranked the importance of making a lot of money a 4 and the probability of achieving this as a jazz musician is a 2, the subtotal will be 8.

After you've completed the subtotals, add them up. Comparing the total scores for each option will help you decided which option is best for you.

If you can't think of nine goals and values, speak to friends, family members, or anyone who knows you well. They might be aware of goals and values that are important to you that you hadn't considered. Keep in mind, also, that the worksheet is a tool to help you make decisions, not a substitute for decision making.

Options:	1.		2.		3.	
Goals and Values / Importance (1 to 5)	Probability (1 to 5)	Subtotal (I x P)	Probability (1 to 5)	Subtotal (I x P)	Probability (1 to 5)	Subtotal (I x P)
1.						
2.						
3.						
4.						
5.						
6.						
7.						
8.						
9.						
Total Scores						

BusinessWeek It's Your Project

SPRINGBOARD: POP QUIZ

How old was writer S.E. Hinton when her first novel, *The Outsiders*, was published?
- **a.** 17
- **b.** 21
- **c.** 24

This lab will require some research. Go to *BusinessWeek* online for useful articles and resources.

What's Your Goal?

What do you want to do—this week, next month, in one year, or in five years? No matter what your answer, it's unlikely you'll get what you want without setting goals. This project has been designed to help you set and achieve your goals.

Think of a goal you want to reach. Choose one from the list, or create one that interests you more. Make a plan to attain it. Make sure the goal is challenging and achievable.

Here are some sample goals you could choose:
- Send 1,000 textbooks to children in Afghanistan.
- Reduce the number of teen smokers in your city.
- Write a novel or play.
- Convince the school board to fund an after-school tutoring program staffed by teens.
- Start a magazine.
- Get a new stoplight installed at a dangerous intersection.
- Run for a political office.
- Raise $5,000 to help UNICEF provide schooling for girls in developing nations.
- Open a café that caters to teens.

Five Signs of an Achievable Goal
1. **It is specific.** It describes in detail what you want to accomplish.
2. **It is realistic.** It is a goal you know you can reach.
3. **It is challenging.** It is something that requires discipline to accomplish.
4. **It is measurable.** It allows you to measure your progress toward the goal.
5. **It has a completion date.** Break long-term goals into a series of short-term goals, each one with a completion date.

Step 1 Start with *BusinessWeek*

Orient your project by conducting research in the *BusinessWeek* online archives. Use key words that relate to your topic to find articles. The articles may offer some insight as you progress through this project.

Step 2 Investigate and Engage
- Why is your goal important?
- How well attainable is your goal?
- What do you already know about it?
- How does your goal relate to technology?
- What do you want to learn from this project?

Step 3 Identify the Obstacles
- What problem will be solved when you achieve your goal?
- What difficulties have others had trying to reach a similar goal?
- What questions do you want your research to answer?

Step 4 Conduct Research and Seek Solutions
- Where could you look for answers to the questions in Step 2?
- What resources will help you better understand your goal?
- Who could you interview for information and guidance?
- Whose support must you have if you are to reach your goal?

Step 5 Select Information and Analyze Data
- How can you determine the accuracy and relevance of the information you have gathered?
- Turn your research results into statements that will persuade others to support your goal.
- Devise a step-by-step plan for reaching the goal. Include realistic completion dates for each stage of the plan.
- What future events might force you to adjust your plan?

Step 6 Connect to the Real World
Create a way to present your project—an audio-visual presentation, a talk, a Web page, an essay, or something else. It's your choice. Be sure to address these questions:
- What impact might your plan have on your friends, family, community, and beyond?
- How do the skills and knowledge you acquired from this project prepare you for the future?

Step 7 Self-Evaluation
Assess the quality of your work. Use an evaluation tool such as a checklist, a small-group discussion, a log, or some other method. Be honest with yourself. That's the only way you will have a clear idea of whether or not your plan has a chance of succeeding.

UNIT 4
WEB SITE DEVELOPMENT

In This Unit...

Chapter 9
Creating a Web Site

Chapter 10
Building a Web Site

Chapter 11
Web Site Management

E-COMMERCE Online

To learn more about Web site development, visit the *E-Commerce* Web site at **ecommerce.glencoe.com**.

BusinessWeek Byte

Rob Glaser: Real Close

Rob Glaser is out to deliver your digital content. The CEO of RealNetworks wants to do it on your computer, on your cell phone, and on your personal video-recorder. Grand aspirations, to be sure.

Too bad the Colossus of Redmond has the same ones. And Microsoft has given away server software that Real charges for and has eaten away at Glaser's lead in desktop players for consumers.

But now, signs are appearing that the strategy Glaser has crafted is turning Real into a digital-media powerhouse. His moves to lock up exclusive Web content from sources such as Major League Baseball and the National Basketball Association appear to be paying off as more broadband Web surfers and sports junkies pony up $10 per month for RealOne Super Player premium subscriptions.

Real also launched a new server product designed to take on Microsoft. Dubbed Helix, the server allows programmers to use a single piece of software to deliver a wide variety of data streams and formats over the Internet. That could save media companies big bucks by eliminating the need to encode in multiple formats.

Source: Excerpted with permission from Alex Salkever, "Rob Glaser: Real Close," *BusinessWeek*, October 1, 2002.

To read this *BusinessWeek* article in its entirety, go to **ecommerce.glencoe.com**. Click on the Student Center, find *BusinessWeek* Articles, and go to Unit 4.

THINK ABOUT IT

How many different ways has Web content evolved over the years?

BusinessWeek It's Your Project

Ever wonder why there are so many problems in the world? The Unit 4 project, "What's Good About It?" on page 240, shows you how to propose solutions.

 ecommerce.glencoe.com

183

Chapter 9
Creating a Web Site

Section 9-1
Conceiving a Web Site

Section 9-2
Planning a Web Site

PREREADING STRATEGIES

Before you read this chapter, finish these statements in your journal:

- Based on this chapter title, I predict…
- The most important thing to remember about successful e-commerce Web sites is…
- Some questions I have about planning an e-commerce Web site are:

BusinessWeek online

Go to *BusinessWeek* online. Find an article about using technology in business. Take notes on some of the tools used to build e-commerce Web sites, and summarize your findings in your journal.

Window Shopping on the Web

In some respects, e-tail Web sites are like their real-world, bricks-and-mortar counterparts. They want shoppers to stop at their stores, come inside, and browse until they buy something. To accomplish this online, Web sites must be easy to navigate and able to attract the right customers.

Drill Down

1895 X-Rays
German scientist Wilhelm Roentgen introduces the medical use of X-ray technology, which becomes a standard tool for diagnosis.

1898 Playback Time
Denmark's Valdemar Poulsen creates the first magnetic tape recorder.

1913 Around and Around
Henry Ford develops the assembly-line system for his automobile factories, and many other businesses quickly adopt his idea.

1922 Hearing Aid
The Marconi Company makes the otophone, the first viable hearing-aid device.

1937 Copy That
Chester F. Carlson patents his initial ideas for xerography and begins the process of creating photocopy machines.

1975 Videotape Era
The Sony Corporation creates the Beta VCR, while Matsushita develops VHS technology. Ultimately, Sony's project is short-lived, as VHS becomes the dominant format.

Get the Big Picture
What are some technological developments of the past—before the invention of the computer—that go into designing a Web site?

POWER READ

Be an active reader and use these reading strategies:

PREDICT what the section will be about.

CONNECT what you read with your own life.

QUESTION as you read to make sure you understand the content.

RESPOND to what you've read.

Section 9-1

Conceiving a Web Site

AS YOU READ...

YOU WILL LEARN
- the importance of having a clear plan for a Web site.
- how to identify goals and objectives when planning a site.
- how to recognize and cater to a Web site's audience.
- about different styles of presenting information to customers.

WHY IT'S IMPORTANT

Setting objectives, generating ideas, and knowing your customers are all essential to creating a successful Web site.

KEY TERMS
- goals
- objectives
- brainstorming
- feasibility assessment
- target market
- demographics

PREDICT

What is the difference between a goal and an objective?

GOALS AND OBJECTIVES

Before you begin setting up your online business, you need to consider your goals and objectives. Who is your target audience? What do you hope to accomplish? When do you want it to be up and running? Where on the Internet do you want your site to appear? Why do you want to build a Web site? These are the five Ws of any enterprise: who, what, when, where, and why. Once you've answered these, you can proceed to the next most important question: How?

Although the terms are often used interchangeably in business, goals and objectives are not the same. **Goals** are general targets you want or expect to reach sometime in the future. **Objectives** are specific targets you want to reach by a certain time, the progress toward which can be measured. In conceiving and planning your Web site, you'll want to move from the general to the specific.

Establishing Goals

Your goals may be ambitious, such as saving the environment or retiring by the age of 30. They may be as simple as promoting your band or selling your custom-made wire sculptures over the Internet. Do you want to set up a blog that your peers will find entertaining and enlightening? Do you want to break into the animation industry by enabling potential employers to see computer-animated cartoons you've created? Whatever your goals, you'll need to express them to investors, consumers, and employees.

Many Internet businesses express their goals through mission statements that are actually posted on their sites. A *mission statement* explains the purpose of a Web site and what its creators hope to accomplish with it. You can find mission statements online for businesses and organizations ranging from Eastman Kodak to the FBI. The mission statement on Ben and Jerry's Web site, for example, states three goals: to "sell the finest quality natural ice cream;" to "operate the Company on a sustainable financial basis of profitable growth" (i.e., to make money); and to "improve the quality of life locally, nationally, and internationally."

Your goals should be both short term and long term. In the long term, you might want to build a financial empire or become a recognized artist. In the short term, you might want to build a loyal market or audience. Your goals may be complex, and they may intersect with others. You might want to create a Web site that promotes not only your own artwork but also the Web sites of other artists you know, cultural events in your community, and local businesses that support the arts.

If you're planning to set up a Web site, you should write a list of your goals, with a brief outline for each. You should also look at Web sites with similar goals and read their mission statements. Once you've established your goals, you can start setting specific objectives.

Setting Objectives

Setting and sticking to specific objectives will help you accomplish your goals over the long term. What kind of Web site do you want to create? If you want to set up a blog or a newsletter, what subjects do you want to cover? Do you want your site to include links to other sites? If so, which ones? You will also need to estimate the money and resources it will take to set up and run your site; whether you'll need to raise funds from investors; how you'll organize your site; and how much time it will take to accomplish your individual objectives.

Write down your site's objectives in quantifiable terms, including how much time, information, and money they will require. This way, it will be easier to determine whether you've achieved an objective or made progress toward it. The more specific each objective is, the better. It's hard to evaluate a "soft" objective such as "Be successful by the end of the year." It's much easier to gauge your progress if you establish more concrete objectives such as "Complete site design by the end of the month."

It's important to establish objectives you think you can accomplish. An objective like "advertise Web site during Super Bowl halftime" is unrealistic (unless you have several million dollars to spare). You can work toward accomplishing multiple objectives at the same time, such as raising money from investors, setting up links to your site, and researching target markets. You can also change and adjust the priority of your objectives depending on your progress. While your goals may remain the same, your objectives might change as you build your business.

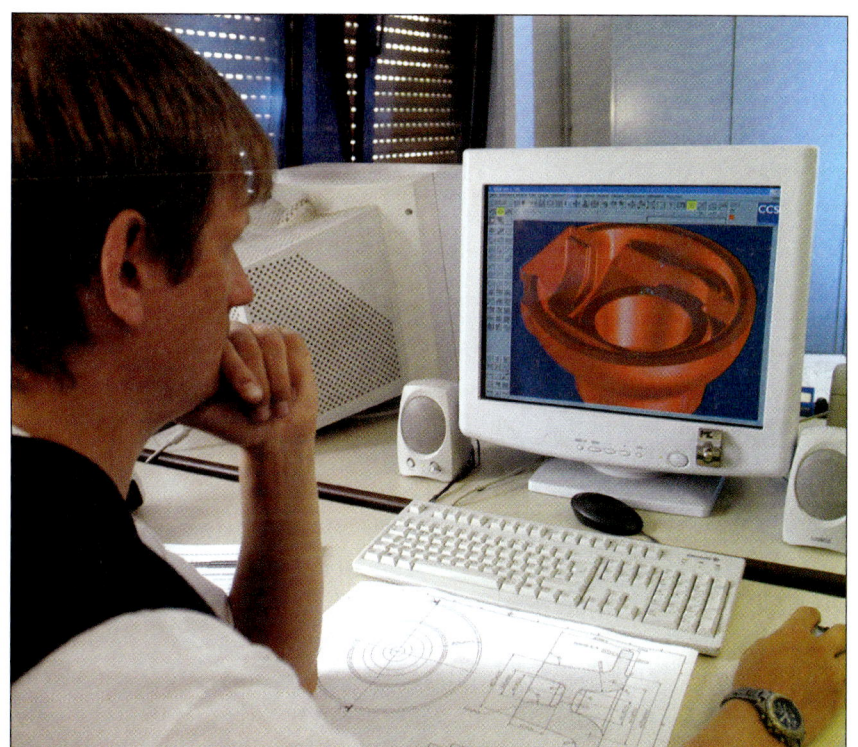

◀ **CREATE A PROTOTYPE**
Although some investors do invest in products, services, and concepts sight unseen, creating and perfecting a sample of a new product can help drive sales and investment. *What other reasons are there for working out the kinks in a concept, product, or service before opening your business?*

Chapter 9 Creating a Web Site **187**

Brainstorming

When you're planning and developing your site, it's a good idea to have a brainstorming session. **Brainstorming** is when a group of people get together to generate and share ideas. Ideally, these sessions are free from criticism or judgment. They serve as open forums where people can exchange, inspire, modify, and combine ideas. The point is to come up with as many ideas as possible.

Brainstorming is one of the most important steps in conceiving and planning a Web site. It can be used to generate ideas on various aspects of your business, from the company's mission to the features your site will have. Although you might follow up on specific ideas that are discussed, you'll want to keep your thinking general at first and not get caught up in the details. Remember that no idea is too far-fetched when you're brainstorming.

You can brainstorm on your own, but it's best to collaborate with other people. After all, two heads are better than one. By brainstorming with others, you can get different perspectives and build on one another's ideas. If you have business partners, they should be involved. If you don't, choose your participants wisely. If you can find successful Web entrepreneurs who are willing to donate their time to your effort, use them. If your business has employees, invite them all. It's especially helpful to brainstorm with people who have expertise in specific areas, such as marketing, finance, management, customer service, and advertising. Consider each person's ideas, and make sure everyone feels free to contradict you, even if you'll be making the final decisions. Write down every idea because you never know which ones will turn out to be the most useful.

Feasibility Assessment

A **feasibility assessment** takes ideas from the brainstorming process and examines how feasible, or possible, they would be to implement. Sometimes creative ideas may be too costly or technically challenging. In this stage of the planning process, the technology, marketing, and financial departments should play an active role in determining which ideas are the most practical. Generally, the most feasible ideas make it to the next step of the planning process.

Don't throw out good ideas just because they can't be executed at the moment. In the late 1980s, consultants told executives at General Motors that the continued decline of America's inner cities would make safety from crime and carjackings the primary concern of urban drivers in the United States. The consultants proposed that GM create a subscription-based satellite communications network that would allow drivers to call for help if they were being victimized.

GM's executives didn't really believe that America's cities had become too dangerous, and as cellular phones grew more and more popular, dashboard communications devices seemed unnecessary. GM chose not to pursue the idea when it was originally proposed. A decade later the company decided to launch a product called OnStar, which helps lost drivers and enables them to contact the police in an emergency. The product proved very successful. Sometimes an idea is just plain bad but often, especially in technology-based industries, good ideas are simply

---

CONNECT

When have you brainstormed with others to gather ideas? Why is it better to brainstorm with others than to come up with ideas on your own?

generated too soon. Save those rejected notions and revisit them every now and then. What's unfeasible today might be ideal a few years later.

Conceptualizing

At this step of the process, you should start sketching samples of the look and feel you want for your site. These sketches should be based on your goals and on the ideas generated during the brainstorming process. Writers, graphic artists, and programmers should all contribute ideas for the Web site's design. In keeping with your goals, think of your site as having a personality. Is it playful, elegant, classic, edgy, or practical? Your site's design and functionality should complement its personality.

Knowing Your Audience

Knowing your customers is a crucial component not only of a business plan (see Chapter 8) and a marketing strategy (see Chapter 12), but also in planning a Web site. When planning your site, consider the customers you hope to attract. Is your company selling video games and computer items? How might the appearance of a site selling these products differ from one selling furniture or household goods? Keep in mind that even within the same store, a men's clothing department will have a different atmosphere than the housewares department.

To plan and design your Web site, you will need to identify your **target market**—the specific group of customers you want to attract. To determine this, you'll need to do some basic demographic research. **Demographics** are statistics about a population, including age, income level, educational background, gender, ethnicity, and buying habits. An especially important demographic category for online businesses consists of people who are connected to the Internet and frequently buy products online. **Figure 9.1** on page 190 shows some statistics about Internet users that businesses can use to target their products and services.

Suppose you want to start a Web site to sell expensive gourmet foods, but you're not sure who will want to buy your products. You may find that gourmet foods are more likely to appeal to people over the age of 30 who are college educated, make more than $50,000 a year, and like to buy imported products on the Internet. To attract this market, you might want to design a Web site with a sophisticated, international look and feel.

Another way to gather information for your site is to look at sites with similar target markets. Suppose you want to sell designer T-shirts or an online newsletter and you already know your target market is teenagers. You can plan your Web site by looking at the sites that are the most popular with teens. **Figure 9.2** on page 191 shows the sites most frequently visited by U.S. teens according to a recent Nielsen/NetRatings survey. To appeal to teens, you might want to use original graphics, sounds, and links to other teen-oriented sites.

To get an idea of how different sites appeal to different markets, it's helpful to look at Web sites that movie studios create for their products. The Web site for *The Matrix* movies offers three versions, one for high-bandwidth connections, one for medium-speed connections, and one for low-speed connections. The site's designers expect that fans of the movies will have high-speed Internet access but don't want to prevent people with dial-up connections from going on the site. The site, like the

CONNECTION: MATH

Potential Customers

The current population of the United States is about 300 million. Suppose the target market for your online magazine is U.S. teenagers aged 15 to 19, which accounts for about 7 percent of the total population. *According to a recent survey, about one in five of all U.S. teenagers actually buy things online. What is the size of your target market?*

QUESTION

What are some types of demographic information other than those listed in the text?

Figure 9.1

Demographics of Internet Users

DEMOGRAPHIC TRENDS The Internet is used by all age groups, but predominantly by younger people. *What might explain this demographic, and why might it change in the future?*

Here is the percentage of each group that goes online. As an example, 61 percent of women go online.

Gender	
Women	61%
Men	65%
Age	
18–29	77%
30–49	74%
50–64	58%
65+	23%
Race/Ethnicity	
White, Non-Hispanic	64%
Black, Non-Hispanic	46%
Hispanic	63%
Community Type	
Urban	65%
Suburban	67%
Rural	48%
Household Income	
Less than $30,000/year	41%
$30,000–$50,000	69%
$50,000–$75,000	86%
More than $75,000	89%
Educational Attainment	
Less than High School	24%
High School	54%
Some College	78%
College +	85%

film it advertises, is designed mostly in black and has hard-to-find buttons that lead to various articles and pieces of artwork for the movie.

The Web site for the film *Seabiscuit,* a historical drama about horseracing, is quite different. The site offers trailers, biographies of the actors, and some of the history on which the film was based. The *Seabiscuit* site is designed for people who want information about the movie and might want to buy it on DVD. The site for *The Matrix,* on the other hand, is meant for tech-savvy fans of the series who might want to browse for hours.

If customers are shopping online for a particular product, it's important that the Web site appeals to them. When people feel comfortable with a site's look and feel, they're more likely to buy something and return again. Repeat business, of course, is one of the keys to e-commerce success.

Figure 9.2

Top Sites Among Teens in the United States

Top Online Destinations Most Visited by Those Aged 12–17, September 2003 (U.S., Home)		
Site	Audience Composition	Unique Audience
Originalicons.com	78%	353,000
Blunt Truth	76%	496,000
Teen People	74%	209,000
FireHotQuotes.com	73%	251,000
Buddy4u.com	73%	279,000

Source: Nielsen/NetRatings

THE TEEN AUDIENCE Teens comprise an important part of the Internet user market. *What factors should an online business consider in trying to appeal to teen users?*

Section 9-1 Review

RESPOND to what you've read.

1. What is a feasibility assessment? Why might some ideas be unfeasible? _____

2. What is a target market? Why is it important to know who your target market is when you're creating a Web site? _____

3. What are demographics? How can they benefit an online business? _____

Quick TALK With a partner, list some of the demographic characteristics of your class. Break the class down into as many different demographic groups as possible. *Think of Web sites that might appeal to these different groups.*

Section 9-2

Planning a Web Site

AS YOU READ...

YOU WILL LEARN
- why it's important to organize your Web site.
- about the function of a home page.
- how to create site maps, storyboards, and navigation schemes.
- ways for potential customers to find, access, and contact your site.

WHY IT'S IMPORTANT

A well-designed and well-organized Web site that's easy to navigate is more likely to attract customers who will visit, browse, buy, and return.

KEY TERMS
- navigate
- home page
- layout
- site map
- storyboards
- navigation scheme

PREDICT

How can storyboards be helpful in the Web-site design process?

ORGANIZING THE SITE

Once you've established your goals and objectives, you're ready to plan your Web site. The first thing you'll want to do is design a home page. What do you want it to look like? Do you want it to have just text and graphics, or do you want it to have sounds and video, too?

You'll want to make it easy for visitors to find their way around, or **navigate**, your site. That includes getting from page to page within your site, as well as getting to and back from other sites that your site links to. Just as important, you'll want to make it easy for customers to access your site and contact you with any questions or problems they might have. Remember, every aspect of your site should be designed and organized with your customers in mind.

The Home Page

A **home page** is a site's main page and usually the first one that appears when you log on to a site. If you have an online business, a home page is like your storefront—it's the first thing your customers see. For this reason, you'll want it to make a good impression.

Your home page should attract visitors' attention, communicate what your site offers, and reflect the business's personality. You could welcome visitors with bright colors and strong graphics or with dark images and stark graphics, depending on your audience and the image you want to project. You could include a short video clip and music, but some potential customers might be put off by loud music or overwhelming effects. If your site opens with a long video montage, it might take too long to load. Be creative, but also be sure to consider the point of view of your audience.

Your home page needs to contain certain elements, such as the name of your business, your logo, and basic information about the site. It should also have a simple and logical layout. **Layout** refers to the arrangement of the elements on the page, such as text, graphics, and headlines. If your home page is cluttered or disorganized, potential customers might become impatient and leave.

Site Maps and Storyboards

You don't want to cram everything on your site on to the home page. Instead, your home page should be a springboard to other pages where customers can learn more, shop for products, and contact you. Each category should have its own Web page that links to the home page like a branch of a tree. Each category might be broken down into smaller categories that require separate Web pages. For example, your home page might link to a page that displays the products and services you're offering, which, in turn, links to a page for subcategories or services.

To organize your site logically, you should draw a **site map**—a diagram of the site's overall structure. A site map is like an outline or table of contents. **Figure 9.3** shows a typical site map. To organize and lay out each page, you should use **storyboards**—pencil sketches or computer-generated images that show what individual Web pages will look like. They should illustrate everything that will appear on the Web site, including text, graphics, color schemes, links, and menus. The storyboards will act as blueprints for programmers and graphic designers to use in building the site.

Navigation Schemes

While site maps and storyboards show the visual layout of your Web site, your **navigation scheme** shows how different Web pages *relate* to one another and link to other sites. Suppose your home page has links to pages for News, Products and Services, Mission Statement, Contact Information, and Ordering. Each Web page might have a link to another page, or at least a link back to the home page. Your Products and Services pages should link to the Ordering page but may not link directly to the Mission Statement page. A navigation scheme can be very complex, like a web, but should be designed so visitors can easily navigate and interact with your site.

Figure 9.3
Sample Site Map

WEB SITE STRUCTURE Companies spend a great deal of time, effort, and money to develop and maintain their Web sites. *Why is a well-structured Web site important?*

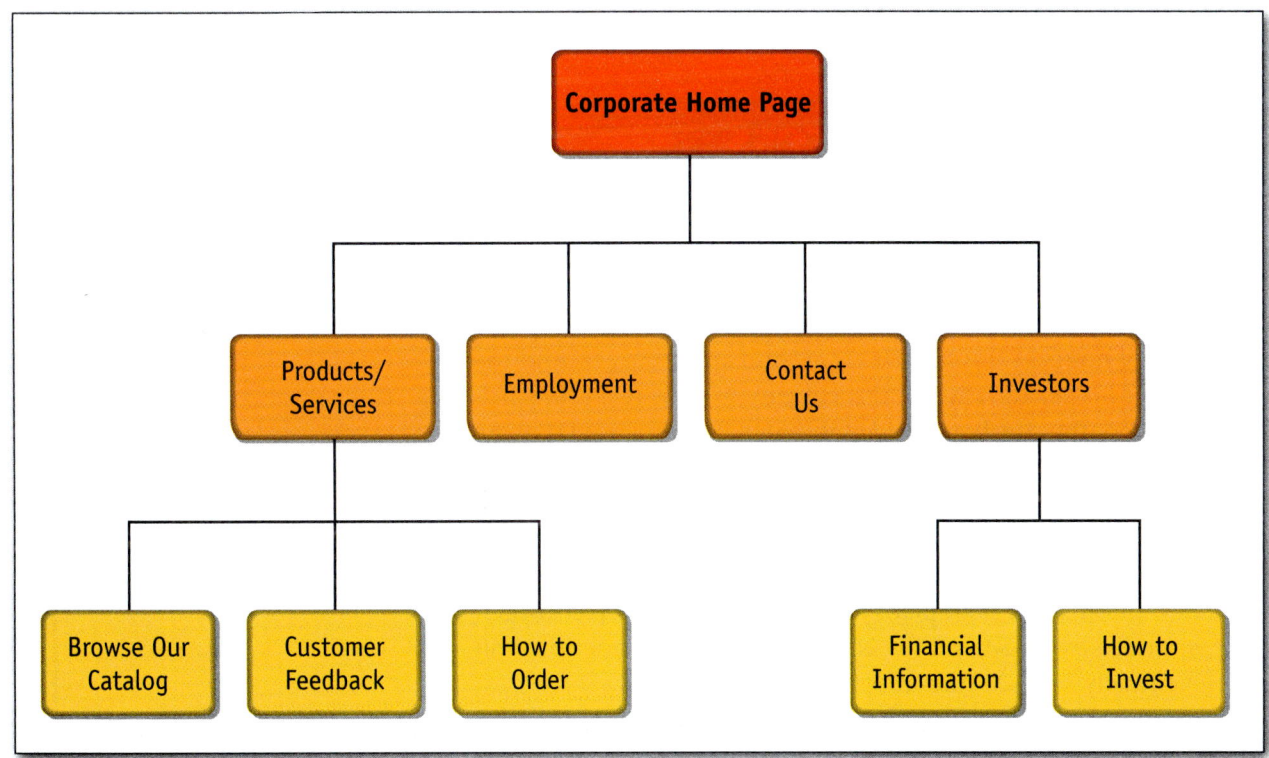

Chapter 9 Creating a Web Site 193

CONNECT

Log on to some of your favorite Web sites. What kind of navigation schemes do they use, and how do they differ? Do they use any creative arrangements or interesting visuals?

You'll also want to create navigation links that are easy for your visitors to locate and use. Almost all e-commerce sites have a *navigational bar* that is clearly located on the home page and often appears on every page. You can be creative in setting up your links. You can arrange them in a vertical column on the side of the page; use color-coded tabs, photos, or icons as links; or use a combination of elements. Again, you should be creative, but also keep your audience in mind.

Reaching Your Audience

You not only need to design your site so it's easy for customers to navigate; you must also make sure it's easy for them to locate, access, and contact your site. How will potential customers find your site? Once they do, how will they access it? How can your customers reach you if they have questions or concerns?

ACCESSING YOUR SITE Customers visiting your site for the first time will probably find it using a browser. For this reason, it's wise to support—or make your site readable by—all of the major Internet browsers, such as Microsoft Internet Explorer, Netscape, and Mosaic. Customers are easily deterred when they're told that a Web site doesn't support their particular browser. Most browsers are available for free, so you could provide a link that enables your customers to download them if they don't already have them.

Even if your site supports all the major browsers, it might be hard for potential customers to find it among the millions of sites on the Internet. To locate particular products or information, Internet surfers often use search words. One way you can attract the right customers is to create an HTML file that includes search words for your site. For example, if your online business is selling designer skateboards, you can create an HTML file with words like "skateboards," "designer," "custom-made," "decks," and "boards." That way, whenever someone

Web Site Success

A VIRTUAL HELICOPTER RIDE
Terrafly.com

Your family is shopping for a new house. Instead of spending hours in the car driving around and trying to find the perfect home, Terrafly has made it easy to view real estate listings. Terrafly—the brainchild of the High Performance Database Research Center in association with the computer science school at Florida International University—offers a vast database of aerial maps and satellite imagery. Do you want to see what houses are for sale in Miami? Scan beachfront properties and view photos and floor plans for it? This online project, which is funded by NASA and the National Science Foundation, will eventually hold 20 trillion bytes of data.

Thinking Critically
What other industries might utilize the functions of Terrafly?

REACHING THE CUSTOMER Customers can now use the Internet or the phone to place orders for 30-minute home delivery of groceries, household staples, and prepared food. *Why is a reliable system for product delivery important to online companies?*

types those words into a search engine, they will be led to your site, along with similar sites. Users often type in specific search words to suit their interests and needs, so the more specific your search words are, the better. (You will learn more about registering with a search engine and using HTML files in Chapter 10.)

If you're selling a product to a tech-savvy demographic, you might want to design a site that uses video streaming, fancy graphics, and sound effects. If your customers need special software to browse your site that is available for free on the Internet, provide a downloadable link to it. For example, many sites offer information in *portable document format* (PDF) files, which can be read by Adobe's Acrobat Reader. Adobe allows users to download its PDF reader for free, so it's customary for sites to provide a link to it. Major Web sites, however, like Yahoo!, eBay, and Amazon, use basic designs that don't require any special software to access and navigate. Remember, if your site has too many special features that take too long to download, it may deter customers.

COMMUNICATING WITH CUSTOMERS Once your customers have found you, you need to set up a means for them to contact you to establish good customer relations. According to a Forrester Research study, 71 percent of online users preferred e-mail as a means to contact customer-service representatives. It's important to include a Contact Us link on your home page that provides customers with an e-mail address. You can also provide a Guest List for customers to make comments and leave their own contact information.

Some customers don't like to wait for e-mail responses and prefer to speak with customer-service representatives directly. Also, if problems arise with your e-mail account, customers will need other ways to contact you. For these reasons, it's a good idea to provide your customers with phone and fax numbers. Toll-free numbers are available to homes and small businesses for as little as $3 per month, plus the cost of calls received.

QUESTION

How can you set up your site so customers can contact you?

Editing Your Site

As you develop your site, you might want to change its style or design. To store all your information and keep track of changes, you need to create a folder on your local hard drive that contains all of your Web files. You can create and edit your changes there.

Once your pages are designed and coded, you can test your Web site. It's best to test your Web files on various computer platforms, such as Windows, Macintosh, and Linux, before copying them onto the Web server. When you feel you're ready to publish your site, you can copy the folder to a Web server, where it will be published for all the world to see. If you want to make changes once the site is live, it's best to temporarily disable the site by taking down the content, editing it as you like, and then reuploading it for the public.

You might want to change your site as you learn more about Web design or your customers. Be careful not to change it too much or you might confuse your customers. The design and functionality of major Web sites haven't changed much over the years. Google is still just a logo over a search window on a plain white screen. It offers more services now than it did when the company started, but the site doesn't look much different. Although you shouldn't be afraid to improve your site, don't change it so radically that customers who are familiar with it can no longer find or recognize it.

Section 9-2 Review

RESPOND to what you've read.

1. How is a home page like the front of a store? _____

2. What is the difference between a site map and a navigation scheme? _____

3. What can you do for customers who need special software to use your site? _____

4. How can you keep track of your Web files so you can change them? What should you do before you copy them onto a Web server? _____

Quick TALK With a partner, create a storyboard for one Web page of an imaginary sporting-goods, vacation-planner, or office-supply business. *Include as many actual parts of the page as possible.*

Name _____ Date _____

Worksheet 9-1

Planning Is Key

Planning is essential to any successful e-commerce Web site. Usually the easy parts are the design and layout, selecting "bells and whistles" to attract customers, and choosing the features you'll incorporate into it. Before you begin, you must take the time to prepare and plan your site. When you plan, you have a much better chance of success.

1. Choose two Web sites. One must be the online site for a big-box store. The other may be an e-commerce site of your choice. Describe the layout of each site.

   _____
   _____
   _____
   _____

2. As you analyze the sites, try to determine what objectives the companies had in mind when planning their sites.

   _____
   _____
   _____
   _____

3. Who is the target audience for each Web site?

   _____
   _____
   _____
   _____

4. Describe the navigational bars and buttons for each.

   _____
   _____
   _____
   _____

5. Choose one of the sites. How would you improve it?

   _____
   _____
   _____
   _____

Name _____ Date _____

Worksheet 9-2

Customer Care

Customer service, often called customer care, is essential to the success of any business. Customer service can mean the difference between happy customers who will return to your site and customers who you will never see again. Most e-tailers agree that their customer service must be excellent.

1. Refer to the two Web sites you chose for Worksheet 9-1. Do you think the companies provide excellent customer service? Explain.

   _____
   _____
   _____
   _____

2. As you analyze each site, describe how you would contact a customer-service representative.

   _____
   _____
   _____
   _____

3. Explain how you could determine the status of your order and how you could return an item you do not want.

   _____
   _____
   _____
   _____

4. Imagine you are a consultant working with a company that is planning to launch an e-commerce site. Write a memo to the owner of the company describing the customer-service options you would recommend for the business.

   _____
   _____
   _____
   _____

Chapter 9 Review and Activities

Chapter Summary

Section 9-1

- Before you create your e-commerce Web site, it's important to establish your goals and express them in a mission statement.

- Setting and sticking to specific objectives will help you to accomplish your goals.

- Brainstorming and feasibility assessment are useful methods for generating ideas and determining how practical they are.

- When planning your Web site, consider the demographics of your target audience—the customers you want to attract.

Section 9-2

- A site should be organized so it's easy for customers to read, navigate, access, and contact it.

- A home page is like a storefront. It's the first thing customers see, so it should make a good impression.

- Site maps, storyboards, and navigation schemes are useful tools for organizing a Web site.

- Customers can locate and access sites using browsers and search words.

- Web sites should provide contacts via e-mail, phone, and fax.

Chapter 9 Review and Activities

Review of Key Terms

a. site map
b. objectives
c. brainstorming
d. demographics
e. home page
f. navigate
g. feasibility assessment
h. goals
i. storyboards
j. layout
k. target market
l. navigation scheme

Match each term to its definition.

_____ 1. Statistics about a population's age, income level, and educational background.

_____ 2. Drawings of what individual Web pages will look like.

_____ 3. A diagram of the overall structure of a site, resembling an outline or table of contents.

_____ 4. General targets you want or expect to reach sometime in the future.

_____ 5. An examination of how practical ideas are to implement.

_____ 6. A group session of people generating ideas.

_____ 7. Specific targets you want to reach by a certain time.

_____ 8. The arrangement of the elements on a Web page, such as text, graphics, and headlines.

_____ 9. To find one's way around a site from page to page or to and from other sites.

_____ 10. Shows how different Web pages relate to one another and link to other sites.

_____ 11. The specific group of people to whom a company wants to sell its products.

_____ 12. The main page of a Web site.

Applying Technology to Academics

English Language Arts—Reading
Choose a Web site you enjoy visiting. Analyze the site, and determine what goals you think the company used to define and shape its vision. Explain why you think the company chose those goals.

English Language Arts—Writing
Select a Web site that sells a product you might like to purchase. Write a paragraph describing the various ways in which the product is displayed on the site (e.g., text, pictures, sounds, video). Explain which feature you find most attractive.

English Language Arts—Speaking
With a group of four or five classmates, imagine you are going to create a Web site for your new e-commerce business. Brainstorm ideas that will help shape the site's look, feel, and features.

Calculation
Research the Web sites of UPS, Federal Express, and the U.S. Postal Service. Determine the various costs of shipping a package that:

- will travel from Orlando, Florida 32801 to Los Angeles, California 90189;
- weighs 5 pounds;
- is 10 inches long, 15 inches wide, and 5 inches tall;
- is valued at $100; and
- needs to arrive the next day.

Which shipper is least expensive?

Chapter 9 Review and Activities

Critical Thinking

1. How do you feel about technology that can scan your e-mail messages and then deliver advertisements to you based on the content of your messages? Is this kind of technology an invasion of your privacy? Explain.
2. Define customer service and what it means when dealing with an e-commerce site.

COMPETITIVE EVENT

Assume the role of assistant manager of an 18-hole golf course. The course is located near a small Southern town 15 miles from a popular tourist area that enjoys golfing weather year-round. Your course does a fair amount of business during late spring, summer, and early fall, but business drops off considerably in winter. You have determined that many tourists do not know about your golf course and that a Web site would help to generate more business for it. Create a presentation for the golf-course manager (judge) about building a Web site for the course. The golf-course manager (judge) is not familiar with creating Web sites, so you must explain the goals and objectives.

EVALUATION

You will be evaluated on how well you meet the following performance indicators:
1. Identify ways in which technology impacts business.
2. Describe tools used in Web-site creation.
3. Demonstrate basic search skills on the Web.
4. Identify strategies for protecting a business's Web site.
5. Pinpoint capabilities of Internet/Web programming.

INTERNET ACTIVITY

To engage in online activities, visit the *E-Commerce* Web site at **ecommerce.glencoe.com**.

BusinessWeek online

TOPIC: Careers in Web Design

In the world of e-commerce, the company Web site is often the only representation of the business the consumer will ever see. Not surprisingly, Web designers and developers are important members of an e-business team.

ACTIVITY Go to the Student Center at **ecommerce.glencoe.com**. Click on *BusinessWeek* Activities and open Chapter 9. There you'll learn more about the different aspects of setting up an e-business Web site and what it takes to be a Web designer.

ecommerce.glencoe.com

Chapter 10
Building a Web Site

Section 10-1
Fundamentals of Web Design

Section 10-2
Creating an Attractive Site

PREREADING STRATEGIES

Before you read this chapter, finish these statements in your journal:

- Based on this chapter title, I predict…
- The most important thing to remember about creating an e-commerce Web site is…
- Some questions I have about how Web sites are created for e-commerce are:

BusinessWeek online

Go to *BusinessWeek* online. Look for information about companies that offer Web-design and Web-support services. Write a short article describing how these services can benefit someone setting up an online business and how to contact these companies.

Technological Tools of the Trade

All the resources you need to start an online business can be found on the Internet. With a little technological know-how, you can set up shop on your own.

Drill Down

▸ **1805 Keeping It Cool**
American inventor Oliver Evans designed the first refrigerator, revolutionizing the ability to obtain and keep perishable food.

▸ **1843 Get the Fax**
Alexander Bain created the communications technology for the fax machine, which became a standard office fixture in the late 1980s.

▸ **1899 Paper Clips**
Norway's Johan Vaaler patented the paper clip, a simple design that has proven essential for binding papers and maintaining documents.

▸ **1945 Tupperware**
DuPont chemist Earl Tupper used polyethylene plastic to design home-storage supplies that ultimately became Tupperware, an enormously popular line of household items.

▸ **1968 Microprocessor**
Intel's Ted Hoff designed the first microprocessor. The invention would ultimately lead the way for significant developments in computer technology.

Get the Big Picture
List the various ways in which a Web site could help promote your new e-commerce business. What features in particular do you think would generate customer interest and increase company revenue?

POWER READ

Be an active reader and use these reading strategies:

PREDICT what the section will be about.
CONNECT what you read with your own life.
QUESTION as you read to make sure you understand the content.
RESPOND to what you've read.

Section 10-1

Fundamentals of Web Design

AS YOU READ...

YOU WILL LEARN

- how Web pages are designed and created.
- about coding languages used by Web designers.
- how hyperlinks can be used effectively.
- how style and enhancements can be added to Web pages.

WHY IT'S IMPORTANT

Dynamic Web-page design is no accident. Although Web-page authors must learn the intricacies of design software and coding methods to produce quality sites, Web-site creation is possible today even for those with limited computer expertise.

KEY TERMS

- tags
- frames
- tables
- hyperlinks
- image map
- Cascading Style Sheets
- JavaScript
- Dynamic HTML
- Java

PREDICT

What alternatives does a business have when seeking to create a Web site?

WEB DESIGN BASICS

To do business on the Internet, you need to build a Web site. Although you can pay someone to create your site for you, this can be expensive. The cost of designing and setting up an online store can easily run thousands of dollars. If you're just starting out or you're on a tight budget, you may need to design and build your own site. While this task might seem daunting at first, it can actually be a lot of fun.

Specifics of Web Design

To build your own Web site, you either need to learn HTML or use software that writes it for you. Most beginning Web authors prefer to use WYSIWYG (pronounced "wizzy wig") software. WYSIWYG stands for "What You See Is What You Get." This kind of software enables you to design your page visually, without having to learn HTML.

WYSIWYG programs such as Adobe PageMaker, Microsoft FrontPage, and Macromedia Dreamweaver allow you to see your Web page as it will appear on the Internet. These programs also have added features that make it easier to manage Web sites. For example, they come with link checkers that enable you to make sure all your links work. Often these programs include simple graphics software so you can edit the pictures you're posting online.

One popular alternative to WYSIWYG software is Macromedia Flash. Instead of using HTML, Flash creates Web pages with SVG (Scalable Vector Graphics) technology. An advantage to Flash is that it enables you to create animation and make your site more interactive.

If you enjoy writing your own code, you might want to try using an HTML editor such as HTML-Kit, Cute HTML, CoffeeCup HTML Editor, or AceHTML. These programs allow you to write HTML but make the coding process easier by keeping the HTML tags and attributes right before your eyes.

If you're really adventurous, you can write Web pages with a simple text editor like Windows Notepad or SimpleText for Macintosh. After all, a Web page is nothing more than a text document with special markup added.

For a beginner in Web authoring, WYSIWYG programs are generally the way to go. Even if you're using a WYSIWYG program, however, it's good to have at least a basic knowledge of HTML.

HTML As you learned in Chapter 1, HTML stands for "hypertext markup language." HTML markup tells the browser how to structure a Web page, which fonts and colors to use in the display, which images to include, and so on.

HTML markup consists of *elements,* each of which usually has two tags: an opening tag and a closing one. **Tags** are formatting codes that define what the parts of a document will look like. An opening tag looks something like this: <html>. The *tag name* consists of the letters inside the brackets. In the previous example, the tag name would be "html." A closing tag looks very similar, but with a slash added before the tag name: </html>. If you want to see what HTML code looks like, right click on any Web page, and click your browser's "View Source" option.

Anything contained between the element's two tags takes on the characteristics (shape) of that element. For example, to add a headline to your page, just put your text inside the headline element: <h1>Name of headline</h1>. When your page displays, everything inside the <h1> tags appears in large letters, like a headline. Sometimes an element does not contain text. Instead, the element may be an instruction to insert a line break or an image. These *empty elements* are written this way: <img />.

Every HTML document or Web page must include the following required elements: <html>, <head>, <title>, and <body>. These elements must be placed inside one another, as illustrated in **Figure 10.1**.

Figure 10.1

Sample HTML Document (Required elements in blue text)

WEB LANGUAGES HTML uses elements made up of tags to control the appearance of Web-site pages. *If HTML seems too challenging, what other options do you have for designing your own Web site?*

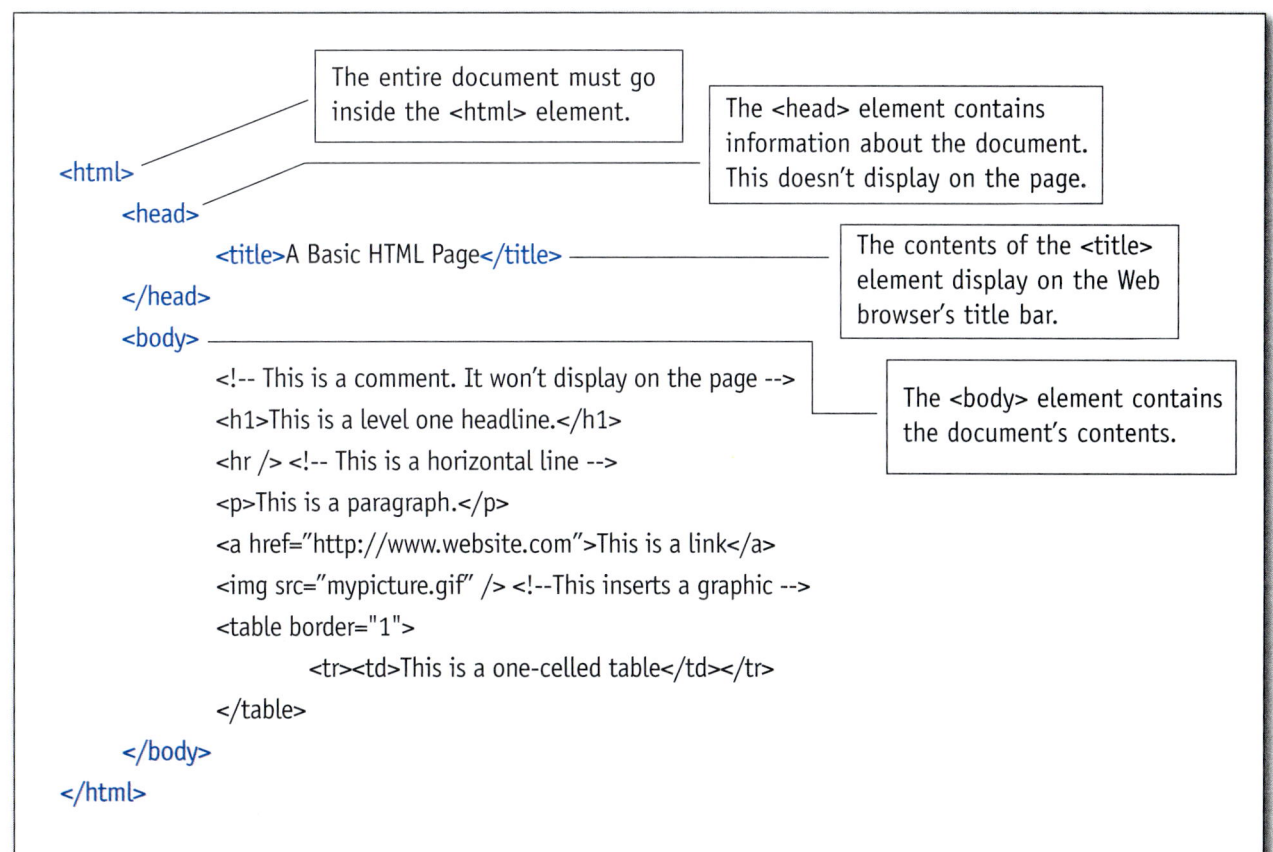

CONNECTION: SCIENCE

The Language Behind the Page

There are many ways to create a Web page, but most rely on HTML. How do you know if your coding is correct? An HTML validator is a software program that allows you to review your work before you launch your site on the Internet and is available for free on many Web sites. *Why might it be important to run your finished Web pages through an HTML validator?*

Once you understand how elements work, you're almost ready to work with HTML. However, before you can create a Web page, you need to learn about *attributes* and *values*. Attributes describe an element. For example, if you were an HTML element, one of your attributes would be hair color; others might be eye color, height, and weight. However, for an attribute to be of any use, it must have a value. For instance, you might have hairColor="blond," eyeColor="blue," height="5 feet," weight="125lb." Now we have a good description of you. Attributes and values work the same way in HTML. For example, perhaps you want to create a headline that will display in a blue font. You would write your headline tags like this:

<h1 style="font-color: blue">Your Headline</h1>.

In this case you used the style attribute and added a value of "font-color: blue." HTML has many different attributes and values that can be used to modify the various elements of your page.

FRAMES Frames enable you to display more than one Web page in a single browser window. In other words, when you look at a frames-based page, you are actually seeing several Web pages displayed at the same time. One use for frames might be in placing a Web site's navigation links at the top or left-hand side of the page. These links would always be available to the user in the same location on each page of the Web site but would be much easier to update. For example, an online clothing store might have a section for new fashions, another for sale items, and another for customer questions and comments. As you add new items and delete old ones, you would normally have to update the links on every page in the site. However, with frames you only need to update the navigation page.

Using frames on your Web site can lighten your workload. However, frames have some drawbacks. The main page (frameset) of a frames-based site has no content. It functions like a blueprint, telling a browser how to build the frames. If the main page of your site is a frameset, a search engine will have difficulty indexing it, and your ranking may be affected. Also, too many frames on a single page tend to make a Web site look choppy, cluttered, and confusing. If you want to use frames, keep them simple, and don't make your home page a frame page.

TABLES Tables were originally developed to display lists of information that have multiple rows and columns of data. For example, you might use a table to display product inventory and pricing information. In the following illustration, one column of data cells gives product ID numbers, another column gives product names, and another gives the price for each item.

Product ID	Product Name	Price
22334554	CD Storage Tower	$24.95
22334555	CD Replacement Cases	$9.95
22334556	CD-R 50 Pack	$15.95

Figure 10.2

Table Based Formatting

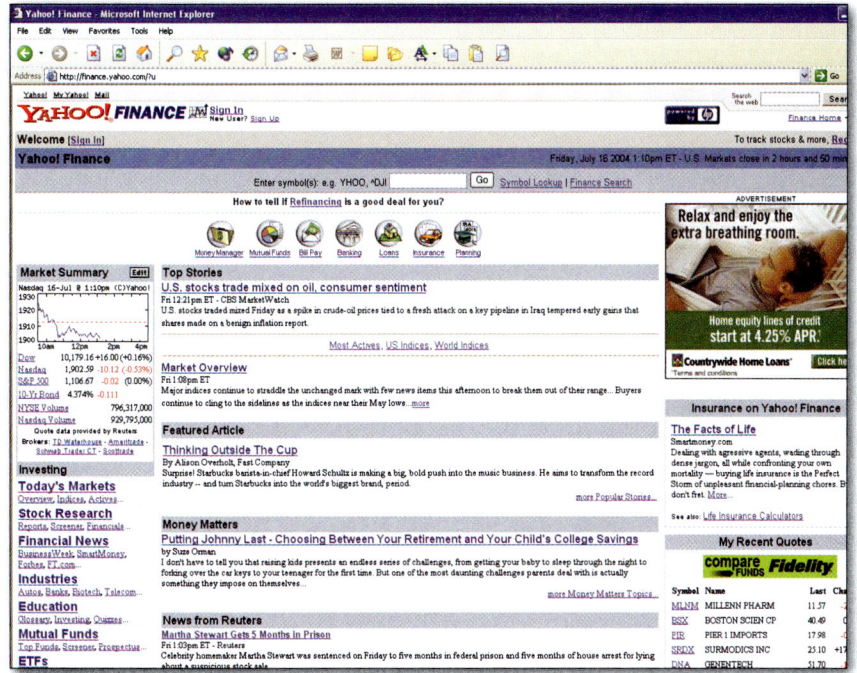

Reproduced with permission of Yahoo! Inc. © 2004 by Yahoo! Inc. YAHOO! and the YAHOO! logo are trademarks of Yahoo! Inc.

TABLE DATA Tables were the first tools used to design Web pages, though in recent years CSS-based design has gained popularity. *What questions should you consider before deciding which approach to use when developing a Web site?*

However, although tables were designed for displaying data, they're most commonly used to create Web-page layouts. **Figure 10.2** shows what can be done with table-based formatting.

Why are tables such a popular design tool? This question can be answered with one word—control. By putting graphics and text inside of table cells and controlling each cell's background color separately, you can create some very attractive page layouts—just with tables.

Even though tables can be used to create layouts, the developers of HTML try to discourage Web authors from doing this. Table-based page layouts make it difficult for special browsers to properly understand your page. For example, a Braille Web browser understands tables only in terms of spreadsheet-type data. If you use a table for your layout, your content will make no sense to the browser, and it won't be able to properly translate the page into Braille. Nevertheless, because of the control they offer, tables remain the most popular layout tool for Web designers.

HYPERLINKS AND IMAGE MAPS **Hyperlinks**, also called *hypertext links* or simply *links,* connect the document you're currently viewing on the Internet with another location in the same document, another document on the same Web site, or another document somewhere else on the Web. A blue, underlined font usually identifies hypertext links.

When a visitor to your site clicks on a link, they instantly move to another location. This interconnectedness is a major reason for the success of the Internet. By means of simple hyperlinks, you can access the Louvre Museum in Paris as quickly as your hometown newspaper. Whether the information comes from around the corner or around the world, it's all accessible because of hyperlinks.

CONNECT

New languages and technology are constantly being developed for the Internet. Since you started using the Internet, has it become easier or more complicated to use?

QUESTION

Why should you take the time to make sure all the links on your Web site are working?

Hyperlinks can be especially useful on an e-commerce Web site. They can connect your customers with information about products, ordering, shipping, and customer service. Your links can make it easy for your customers to find needed information, to select products, and to complete a purchase. Well-designed links make a Web site user-friendly and leave your visitors with a favorable impression of your business. However, poorly designed or difficult-to-understand links can just as easily drive customers away.

You may wish to create links for things other than Web pages. A *mailto:* link launches your customers' e-mail programs and allows them to send e-mail directly to your company. You can also create links that enable visitors to download a music file or product brochure created in a word-processing program. These types of links are known as *file links*. For example, an online music store might provide links that download sample clips from CDs they have for sale. A car dealer might include a link to informational brochures about its newest models.

Once you've created your links, you'll need to continually review and update them. When links are no longer accurate, they need to be changed or removed. Nothing is more frustrating to a Web-site user than clicking on links that don't work. These are referred to as *broken links*. A broken link is a link to a Web page or file that is no longer available or cannot be found. A user quickly leaves a Web site that has too many broken links, so take the time to make sure all your links work.

Text links are easy to spot on a Web page because of the formatting features of color and underlining used to make the text stand out. If you want to use a picture or other image as a link, it also needs to stand out so users recognize it as a link. You can do this by creating graphical *buttons* or by using *icons* that symbolize what the link will do. For example, many Web authors use a small image of a mailbox to identify e-mail links.

Another way to create graphical links on Web pages is with an image map. An **image map** is a single graphic that has several different *hot spots*. The hot spots serve as links to more than one location.

▶ **DIRECTIONAL IMAGES** With an image map, users select the area of the image that represents the topics in which they are interested. *What are the benefits of using an image map in the design of a Web site?*

For example, you might use a map of the United States to assist your customers with shipping information. When they click on the picture of a particular state, a link takes them to a page that describes the shipping costs for a package sent to that state.

To create an image map, a Web designer starts with a picture, chart, map, or virtually any graphic. Then the designer must map out the exact location of the hot spots and identify them as links. This can be quite complicated to do by hand. Most Web authors use inexpensive image-mapping software such as CuteMap or CoffeeCup Image Mapper to perform this task.

Advanced Web Design Tools

When you start creating Web pages, it won't take you very long to discover that HTML is not a good design tool. In fact, HTML isn't even supposed to be used for creating designs and layouts. At its best, HTML can create a static, brochure-like Web page.

To add style and interactivity to your online venture, you need to give HTML some help. Technologies such as CSS, JavaScript, Java, DHTML, Flash, and others can help you bridge the gap between a dull, lifeless Web site and one filled with action and interactivity.

Technologies for Adding Design and Interactivity

Once you've decided your Web site needs some help, you need to be careful. Adding bells and whistles to an e-commerce site is like adding salt to food: A little bit helps, but too much ruins it. A good rule of thumb for adding design and interactivity to your Web site is to use only what you *need*. Spend some time thinking about what will appeal to your customers and improve their shopping experience.

CASCADING STYLE SHEETS **Cascading Style Sheets** (CSS) was developed to address HTML's design weaknesses. It is not part of HTML but instead allows Web designers to use HTML for page content, while using CSS to apply style, layout, and design. You can add style to more than one page at a time with an *external style sheet*. An external style sheet is a separate document that can be linked to a page or to an entire Web site. You can also control the style of a single page by means of an *embedded style sheet*. It is called "embedded" because it is placed in the page itself. **Figure 10.3** on page 211 provides an example of an HTML document with an embedded style sheet. A third way to apply style to a Web page is by adding style rules on a line-by-line basis. This is known as an *inline style sheet*. You can add an inline style sheet using the "style" attribute inside any element on a page.

It is possible for more than one style sheet to apply to the same Web page. Which style wins when there's a conflict? That's where the term *cascading* comes in. The *cascade* describes the order in which styles are applied on a Web page. Figuring out the cascade can become quite complicated. However, the basic rule is that an inline style sheet comes first, an embedded style sheet second, and a linked style sheet third.

CSS enables Web authors to define colors, fonts, link colors, layout, and much more. Because a single style sheet can apply a cohesive design to every Web page, you can change the look of your entire site by editing only one document. Without CSS, you would have to change the style of each individual page, which can be time-consuming and tedious. However, the usefulness of CSS is not just in its ability to control the look of an entire Web site. CSS gives you the power to create precise layouts for individual pages, something not possible with HTML alone. It gives you the power to design special effects such as fonts that change color when someone moves a cursor over them.

With CSS, you can even create special instructions for printers, making your pages "printer-friendly." You can also use CSS to include audio instructions that tell an aural (audio) browser how to read your page out loud. These are only a few of the options available to the Web author who knows how to use CSS.

In spite of the positives of CSS, it also has some drawbacks. The biggest problem with CSS is that Web browsers such as Internet Explorer and Netscape support it differently and sometimes not at all. Often, a page designed with CSS looks completely different on two different browsers. On some older browsers, your layout might completely break down. For this reason, many Web authors prefer not to use CSS. However, despite its weaknesses, CSS is a technology every Web author needs to learn. With it, you can create a unique look for your e-commerce site.

JAVASCRIPT As helpful as CSS can be, it lacks the power to do much more than add style to a page. The fact is, more and more Web-site visitors are looking for *interactivity*. In other words, they don't want a Web site that simply sits there like an electronic brochure. They want one that responds to them as they explore it. One tool Web designers can use to add interactivity is JavaScript.

JavaScript is a scripting language. It is not a programming language in the truest sense of the word because its scripts will not work as standalone applications. They must run in connection with a Web page.

With JavaScript, you can write short bits of code that add functionality to a Web page. For example, you can create a script that checks a customer's order for mistakes or omitted information. If you've ever filled out a form online and had it "kicked back" by the computer because you left out a zip code or phone number, you have seen JavaScript at work. You can also use JavaScript to animate buttons, change images and page colors automatically, and much more. One popular e-commerce application for JavaScript is in processing order forms.

You can write a script that totals the items in an online customer's shopping cart, figures in tax and shipping, and checks the order form to make sure the customer hasn't forgotten to give you important information. Because this script is embedded in the Web page, all of the processing work is done on your customer's computer. This speeds up the ordering process and ensures you give your customer the best service possible.

How can you make use of JavaScript on your Web pages, especially if you're not a programmer? WYSIWYG programs such as FrontPage and Dreamweaver can write the scripts for you automatically. HTML editors such as CoffeeCup HTML Editor and AceHTML have libraries of prewritten scripts built in. All you have to do is select a script you need and tell the editor where to insert it.

Figure 10.3

Sample HTML Document with Embedded Style Sheet

CASCADING CONFLICTS Web-page designers use CSS to add style, color, layout, and other design elements to HTML formatting. *What is the main drawback of using CSS formatting?*

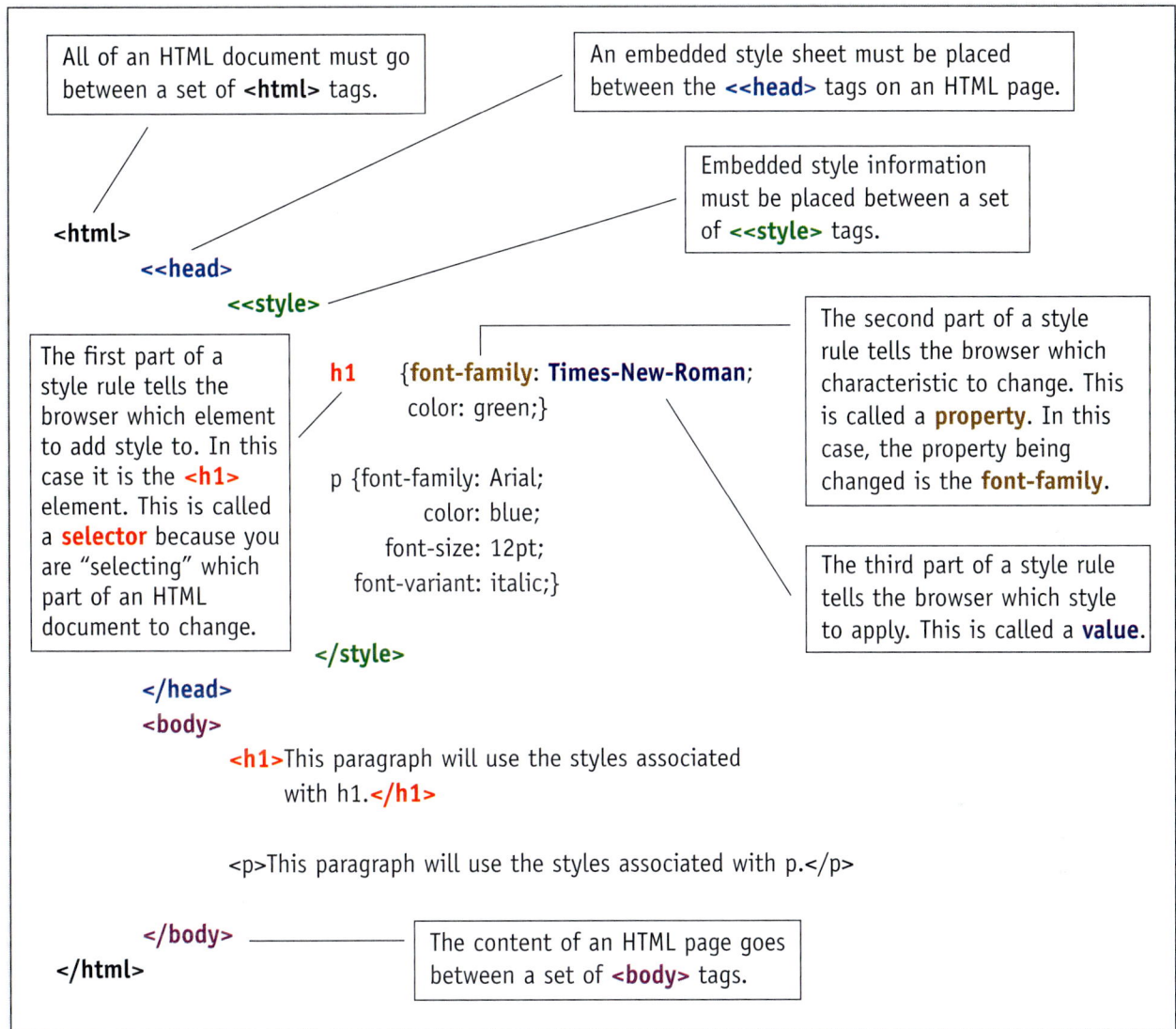

If you don't have access to WYSIWYG or HTML editor software, you can visit a *script library* Web site such as the JavaScript Source. Sites such as this offer thousands of free "cut-and-paste" scripts you can use on your Web page. They also offer tutorials if you're interested in learning how to write your own scripts.

DHTML Dynamic HTML (DHTML) is not a separate programming language or a new version of HTML. DHTML is a hybrid technology that combines HTML, CSS, and JavaScript to add dramatic effects and animation to Web pages. For example, if you've ever visited a Web site where one page "dissolves" into another when you click on a link, you have seen DHTML in action. The name, Dynamic HTML, describes what this technology is intended to do. Standard HTML is static; it can only be used to structure and display documents. DHTML is dynamic, allowing Web authors to create stunning visual effects for their Web sites.

You can use DHTML to create drop-down menus that offer customers a list of links when they move the cursor over a certain category. For example, suppose you created an online bookstore with navigation tabs for fiction, nonfiction, magazines, gift books, and children's books. When a customer "mouses over" the fiction tab, a drop-down menu appears with options for historical fiction, mysteries, romance, inspirational fiction, and so on. A menu system created with DHTML can help you design a clean, uncluttered site, even if it features a lot of links.

As with any Web technology, DHTML has its downside. People who are not programmers may have a difficult time using DHTML because it can be quite complicated to write. To write your own DHTML, you need to have a solid working knowledge of HTML, CSS, and JavaScript. Although WYSIWYG software such as FrontPage and Dreamweaver can create DHTML effects for you, and HTML editors sometimes include cut-and-paste DHTML, there are other problems that make this technology less effective than it could be.

The Microsoft and Netscape corporations took completely different approaches in developing DHTML. As a result, the two versions are incompatible. This means some older Netscape browsers may not be able to display DHTML effects created for Internet Explorer and vice versa.

All things considered, DHTML is interesting, but there are better ways to add interactivity to your site.

JAVA Another technology available to Webmasters is **Java**. Not to be confused with the JavaScript scripting language, Java is a programming language in its own right. Java programs for use on Web pages are called *applets*. An applet is a mini program that can carry out a specific function on a Web page. For example, you can include applets that create text animations, menus, buttons, MP3 files, slide shows, and even arcade games.

A popular use of Java nowadays is in creating dynamic navigation buttons. Software such as CoffeeCup Button Factory enables you to create animated Java buttons without needing to learn Java programming. Also, Web sites such as the Java Boutique offer free downloadable applets you can use on your Web pages.

An applet can be inserted on a Web page by means of HTML elements. However, your end users must also have Java installed or enabled on their browsers for the applet to work. Java works on a user's computer by means of a virtual machine. When someone loads your Web page and applet onto his or her computer, the Web browser starts the Java virtual machine. The Java virtual machine then runs the applet. If your end user doesn't have the virtual machine or has turned it off, the applet will not run. This can be a serious problem if you have created your navigation buttons with Java. If your buttons don't display, your customer has no means for navigating your site. Never forget that a well-designed Web site should always have a dual navigation system. It is fine to use Java to create fancy or dynamic link buttons. However, be sure to also include a set of text-based links so your customer can still navigate your site easily.

Section 10-1 Review

RESPOND to what you've read.

1. What are frames? Why can they be an asset to a Web page's appearance? _____

2. Why is WYSIWYG software so helpful to novice Web page creators? _____

3. How are tables useful in Web-page design? _____

4. What is an image map? _____

5. What are the three types of cascading style sheets? What are the major differences among them?

Quick TALK With a partner, discuss whether the idea of creating a Web site for an e-commerce business sounds daunting or like something that could be fun. *What methods might you use to create a site, and why?*

Section 10-2

Creating an Attractive Site

AS YOU READ...

YOU WILL LEARN

- which common files are used to create Web sites.
- how graphics, audio, and video can affect a Web page.
- why font size and type are important considerations in Web design.
- how visual considerations affect Web-page planning.

WHY IT'S IMPORTANT

When creating a Web page, every component must be considered. Particular attention should be paid to seemingly ordinary elements, such as fonts and graphics, so each part of the page flows together and is user-friendly.

KEY TERMS

- JPEG
- GIF
- button
- MP3
- streaming video

PREDICT

How can graphics enhance a Web page's appearance?

VISUAL ELEMENTS IN WEB DESIGN

As you begin to plan your e-commerce site, you need to give some thought to how you want it to look. Part of this process involves thinking about what your customers are seeking and what type of Web site they find attractive. Following basic Web-design principles will help you plan a pleasing site.

Color Consistency

A consistent color theme is an important part of Web-site design. Drastic changes in page color and imagery cause users to become confused, making it harder for them to find your products and services.

Use colors that are pleasing to the eye and reflect the company's image. From a technical perspective, you have up to 16.7 million colors to work with. However, very few computer monitors display those colors the same way. What does this mean for a Web-page designer? Don't count on your selected colors looking the same when displayed on another person's monitor. A red might be an orange, an orange might be a brown, a blue might be a violet, and so on. Your best bet is to stick to a palette of 216 colors, sometimes called the "browser-safe" palette. One of the best reference sites for learning about the browser-safe palette is VisiBone.com.

BACKGROUND COLOR Background is perhaps the most important use of color on your Web page. Most designers recommend white or off-white due to its high readability levels. A white background contrasts with black text without being harsh on a reader's eyes. *Contrast* is the difference in brightness between two related elements (such as the text and the background). Balanced contrast increases the readability of a Web site. Too much or too little contrast makes the page difficult to read and can cause eyestrain. **Figure 10.4** shows how color contrast can affect the readability of your site.

Visual Balance

Web-site designers must make sure there is a proper balance of text, graphics, and blank space (called *white space*) on their Web pages. A page with too many graphics is cluttered and slow to download. On the other hand, too much text makes a site difficult to read. Minimize your graphics and text, and use white space as areas for the eye to rest. Much like pleasing background colors, a balanced use of text, graphics, and white space heightens readability.

Figure 10.4

Color Contrast

| Text is easier to read when it is used with an appropriate background. The color and size of the text as well as the color and pattern of the background influence readability. Is this text easy to read? | Text is easier to read when it is used with an appropriate background. The color and size of the text as well as the color and pattern of the background influence readability. Is this text easy to read? |

READABILITY Make sure there is enough contrast between your text color and background color. *Which example do you find easier to read?*

Graphics

While text may be the most common way information is sent over the Internet, graphics are the most powerful. Web-page graphics are any images used in Web-site design. Graphics help designers create Web sites that most effectively present their products and services. Proper use of graphics can make a Web site look and feel like a magazine, offering colorful illustrations and powerful images instead of just plain text.

DESCRIPTIVE FILE TYPES The most important thing to consider in creating Web graphics is file size. Large graphics files download slowly, so you need to save your graphics in a file type that can be *compressed,* or compacted to save space. The two most commonly used formats for graphics on the Web are **JPEG** (or JPG) files and GIF. JPEG (pronounced jay-peg) stands for Joint Photographic Experts Group. This format is most often used for photographs. The JPEG file format provides designers with high-quality images that can be saved in very small files. Because JPEG files allow images to be saved with millions of colors, this format is best suited for photographs.

GIF (pronounced "jif") stands for Graphic Interchange Format. GIF files are commonly used for animations, cartoons, and logos. The GIF format only allows images to be saved with 216 colors, but the image can be compressed with no loss of quality. This is because GIF images are compressed with a special technique that inventories the colors and locations of the different pixels in the image. The Web browser reads the inventory of pixels and colors and re-creates the image perfectly. This makes GIF images perfect for buttons, logos, icons, and any other graphics that don't suffer from a limited color palette.

Another consideration in creating graphics for Web pages is the *resolution* you use when saving the picture. Resolution refers to the number of pixels per inch in an image. Think of a Web graphic as being made up of thousands of tiny dots. The more dots in the picture, the sharper it will be. Each dot is called a *pixel* (short for "picture element"). If you are creating images for the Web, you should save them at a resolution of 72 pixels per inch.

Web Site Success

STRIKING SITE SELLS FOR THE SMALL SET
BabyBunz.com

After deciding in 2000 to grow its online sales presence, baby-product catalog company Baby Bunz & Co. hired Toolhouse Design Company to create an effective, attractive Web site. Designer Greg deVeer knew he had to avoid using too many fancy graphics, which can undermine a site's ultimate function. To sell products, he had to achieve a balance between efficiently outlining Baby Bunz's inventory and leveraging the emotional impact of cute babies. deVeer decided to use large, compelling baby photos in the background and let users do their shopping right over the top of them. The large graphics provide the emotional punch without compromising the shopping experience. This makes for a fun, visually appealing site that's still easy to navigate.

Thinking Critically
Baby Bunz hired a multimedia design company to create its Web site, but many companies choose to create their own sites. *What are the pros and cons of hiring a designer versus doing the work in-house?*

CONNECT
Take a look at two or three of your favorite Web sites. How many types of fonts are used on each site's home page? Why do you think they were designed this way?

FONTS Believe it or not, Web-page designers think a lot about fonts. *Font* is a typography term that categorizes a typeface or family of typefaces, such as Times New Roman, Arial, Helvetica, and so on. Different fonts convey different moods to the reader. Fonts come in six categories: old style, transitional, modern, slab, sans serif, cursive, text letter, and decorative. Something as simple as choosing an appropriate font may help to create the right image for your Web site. Not all computers can read all fonts, so it's better to use a more common font such as Arial, Times, or Courier than a more exotic one.

Once you select a font that conveys your business's message, decide on the font's *point size*. If the point size is too small, customers will have difficulty reading the product information. If the font is too large, you won't be able to fit the copy on the page. You need to balance the font style and size with the amount of information presented on the Web page. The three elements work in unison.

BUTTONS Web designers use buttons to help visitors navigate quickly and easily from one area to another. A **button** is usually a GIF file a visitor can click on, although it is possible to use Cascading Style Sheets to create a button without using a graphic. Buttons on a Web site serve the same purpose as large signs hanging from the ceiling of a supermarket; they help you find what you're seeking. Buttons are usually placed at the top and/or sides of a Web page to attract your attention. A well-designed Web site also makes use of text links at the bottom of the page—just in case the buttons don't display.

There are many types of buttons you can use on a Web site. Sometimes buttons are created with a programming language such as Java or with a vector graphics program like Macromedia Flash.

Multimedia

Multimedia can transform a boring Web site into a vibrant Internet destination. Reflect about your favorite places to visit on the Internet. What makes them stand out? Most likely these sites are interactive. In other words, there are things for you to *do:* audio to listen to, video to watch, maybe even games you can play. The same multimedia techniques that make Web sites fun for you can make your site truly extraordinary.

USING AUDIO ON THE WEB You can use audio in various ways on a Web site. A song may play softly in the background as potential customers browse the site. You can include a welcome message, thanking visitors for shopping at your online store. You can add audio descriptions of the products, services, and features offered on your site. Soft "blips" and "bleeps" can be used to let visitors know when they are moving over links or buttons. These audio cues are effective in helping people with visual impairments navigate your site. Take care, though. For many Net surfers, Web-site background music and narration can be irritating—particularly if the audio loops endlessly. If you plan to use audio, be sure to provide an option for turning the sound off.

Often, audio used on Web sites is taken from traditional sources such as CDs, cassette tapes, or record albums. If you anticipate using music such as this on your site, be sure to consider copyright issues. If you don't get permission to use someone else's music on your site, you may be in violation of the law.

To use audio on your Web site, you must convert it to a format that can be stored on the Internet. *Converting* is the act of changing the format of a file. There are many different audio formats that work on the Web. However, because of their small size and high quality, **MP3** files have quickly become the most popular format. Free or low-cost programs such as Windows Media Player 9, Musicmatch Jukebox, and GoldWave make creating digital audio a simple task. All you need to do is plug your CD player (or other sound source) into the input jack on your computer's sound card. Then the software records the audio

AUDIO ONLINE Streaming audio has contributed to the creation of Internet "radio" stations, which can broadcast live or recorded music over the Web on a continuous basis. *What scripting languages can handle audio?*

Chapter 10 Building a Web Site **217**

onto your hard drive as the music plays. Once you convert the audio into MP3 format, adding it to your Web site is as simple as creating a link to the sound file.

Although MP3 files are small when compared to other formats, a long clip can still be 10 to 20 megabytes. If you plan to use long audio clips, consider using a program like Apple's QuickTime or Windows Media Encoder to format the files for **streaming audio**. Unlike downloaded audio, which must download completely before you can listen, streaming audio enables visitors to listen to the files in real time rather than waiting for them to download.

USING VIDEO ON THE WEB It is now possible for video broadcasts to be transmitted over Web sites because of faster Internet connections and improved technology. Video can be a powerful tool, offering rich multimedia experiences for your site visitors. In fact, there are some good reasons to consider using video as well as audio on an e-commerce site.

With video *and* audio, a salesperson can give a hands-on demonstration of a product while delivering a persuasive sales presentation to visitors. Other Web sites, such as news-related sites, might make video clips of news reports available. Motion-picture companies often put trailers for their latest films on their Web sites. The potential uses of video on the Web are countless. However, because video files can be quite large, using streaming video is almost a must.

QUESTION
Why do you have to be careful about using music on your Web site?

Section 10-2 Review

RESPOND to what you've read.

1. Why are font size and type significant in Web-page design? _____
   _____
   _____

2. Are Web pages with lots of graphics generally preferable to those with very few? Why or why not?
   _____
   _____

3. What are MP3 files? What makes them so popular? _____
   _____
   _____

Quick TALK Choose ten Web sites that represent a variety of industries. *Discuss with your partner which sites might benefit from the addition of video and which would not.*

Name _____ Date _____

Worksheet 10-1

Design It!

Creating Web pages takes a little practice, but once you get the hang of it, it's a lot of fun. The first step is to decide what you want to publish on your Web site. Creativity, imagination, and a little skill are all you need.

1. Develop a concept for an e-commerce Web site. You can use one you created for a previous worksheet or create a new one.

   _____
   _____
   _____
   _____
   _____

2. Will you use tables and frames? Explain how. If you don't plan to, explain why.

   _____
   _____
   _____
   _____
   _____

3. Describe the graphics, audio, and video you will use on your site. If possible, locate some examples.

   _____
   _____
   _____
   _____
   _____

4. Describe the fonts, colors, and other basic graphic elements you will use.

   _____
   _____
   _____
   _____
   _____

5. If you have the software available, create your Web site, implementing the design elements described above.

Name _____ Date _____

Worksheet 10-2

Decode it!

Java technology is computer language that helps Web designers to create exciting and fancy buttons, applets, and other animations that are fun and creative. Java was invented by Sun Microsystems in 1995 and is embedded in cell phones, PDAs, pagers, video games, TVs, and personal computers.

1. Research Sun Microsystems. Find out what types of products and services it produces, where the headquarters is located, and the name of the CEO.

   _____
   _____
   _____

2. Find a picture of the Java logo. Draw it below. Why do you think Sun chose this logo to represent Java?

   [   ]
   [   ]
   [   ]
   [   ]
   [   ]

   _____

3. Go to the official Java Web site. Search for the educational tools. Choose one of the tools, and explore it. Which one did you chose? What did you like about it? Would you use it again?

   _____
   _____
   _____

4. Find out where you could take a course in Java in your community. Check your own high school, technical college, or computer store. What is the cost of the course? What is the length of time for the class? Do you think you would like to learn Java? Explain why or why not.

   _____
   _____
   _____

Chapter 10 Review and Activities

Chapter Summary

Section 10-1

- Web sites can be built using HTML code, WYSIWYG software, or SVG code.
- Some of the components of a Web page include frames, tables, hyperlinks, and image maps.
- HTML markup tells the browser how to structure a Web page, which fonts and colors to use, and which images to include.
- Cascading Style Sheets, JavaScript, Dynamic HTML, and Java are some of the popular coding languages used by Web-page designers.

Section 10-2

- The two most commonly used format files for graphics on the Web are JPEG and GIF.
- Graphics, audio, and video can make a boring Web site much more appealing.
- Different fonts convey different moods to the viewer. Fonts help to create a Web site's image.
- Follow basic design principles when planning the visual aspects of your Web site. Consider color consistency, background color, visual balance, graphics, fonts, and multimedia.

Chapter 10 Review and Activities

Review of Key Terms

a. tags
b. frames
c. tables
d. hyperlinks
e. image map
f. Cascading Style Sheets
g. JavaScript
h. Dynamic HTML
i. Java
j. JPEG
k. GIF
l. button
m. MP3
n. streaming audio

Match each term to its definition.

_____ 1. The most popular audio format.
_____ 2. A file format most often used for photographs.
_____ 3. Most commonly used to create Web-page layouts.
_____ 4. Used to connect documents you are currently viewing on the Internet with another location.
_____ 5. Enable you to display more than one Web page in a single browser window.
_____ 6. A file format most commonly used for animations, cartoons, and logos.
_____ 7. A GIF file that a Web-site visitor can click on.
_____ 8. Program used to create applets, or mini programs.
_____ 9. A single graphic that has several different hot spots.
_____ 10. You can use it to write short bits of code that add functionality to a Web page.
_____ 11. Hybrid technology that combines several languages to add dramatic effects to a Web page.
_____ 12. By using this, you enable your visitors to listen to audio files in real time.
_____ 13. Developed to address HTML's design weaknesses.
_____ 14. Formatting codes that define what the parts of a document will look like.

Applying Technology to Academics

English Language Arts—Reading
Using the Internet, find some Web sites that would be helpful for someone who is learning to write HTML code. Choose two, and compare the information they provide. Which one would be more helpful? Explain why.

English Language Arts—Writing
Access your favorite Web site. Using word-processing software, write a memo to your instructor evaluating the visual design elements used on the Web site. Include all the elements discussed in the chapter.

English Language Arts—Speaking
Research the process you would need to follow to create streaming audio. Produce a one-page flyer explaining the process. Then, teach someone else in your class how to format the files needed to create streaming audio.

Calculation
Using your favorite search engine, search for "Web design principles." How many results are reported? How many results appear per page? How many pages would you have to scroll through to see all the results?

Chapter 10 Review and Activities

Critical Thinking

1. Discuss the advantages and disadvantages of designing and creating your business's Web site yourself.
2. Do you think it's possible to use too many design elements on a Web site? Explain.

COMPETITIVE EVENT

Assume the role of e-commerce manager for a manufacturer of holiday ornaments and related products. Your company does a good business from its online catalog, but you feel that updating the catalog would make your products even more attractive to new and existing customers. Your idea is to add video and sound both to the company Web site and to the online catalog. Explain to the company's owner (judge) how these two elements—video and sound—can be added to the Web site and how they will help to increase sales.

EVALUATION

You will be evaluated on how well you meet the following performance indicators:

1. Identify capabilities of Internet/Web programming.
2. Identify ways in which technology impacts business.
3. Discuss trends in e-commerce.
4. Persuade others.
5. Make oral presentations.

INTERNET ACTIVITY

To engage in online activities, visit the *E-Commerce* Web site at **ecommerce.glencoe.com**.

BusinessWeek online

TOPIC: Personalized Web Sites

Online shoppers are being greeted more and more by Web sites that recognize their repeat customers. Sites not only remember customers' previous purchases, but also make recommendations based on those purchases. Even in cyberspace, shopping is no longer anonymous.

ACTIVITY Go to the Student Center at **ecommerce.glencoe.com**. Click on *BusinessWeek* Activities and open Chapter 10. There you'll learn more about what personalizing Web sites entails and how it can help businesses and customers.

Chapter 11
Web Site Management

Section 11-1
Positioning a Web Site

Section 11-2
Back End Management Tools

PREREADING STRATEGIES

Before you read this chapter, finish these statements in your journal:
- Based on this chapter title, I predict...
- The most important thing to remember about positioning a Web site is...
- Some questions I have about managing an e-commerce Web site are:

BusinessWeek online

Go to *BusinessWeek* online. Check out the site's Search Tips. In class, discuss how you could use these tips to make your own Web site easier to find in a search engine.

224 ecommerce.glencoe.com

Web Sites with Appeal

Where can you turn for ways to improve the service your e-commerce site provides? The Internet is full of inspiring examples. It's up to you to decide what works best for your business.

Drill Down

1887 Seeing Clearly
Switzerland's Dr. Eugen Frick develops the first practical contact lenses.

1913 It's a Steel!
Harry Brearley of the Brown Firth research laboratory invents stainless steel.

1936 Here Comes the Sun
L'Oreal becomes the first company to mass market suntan lotion, a product that soon generates billions of dollars annually.

1938 Penultimate
Hungarian brothers Laszlo and George Biro patent the ballpoint pen, which remains the primary writing device throughout the world.

1959 Buckle Up
Volvo engineer Nils Bohlin creates the first three-point seatbelt for cars. The company allows other manufacturers to use the idea for safety reasons.

1975 At the Drive-Thru
A McDonald's in Sierra Vista, Arizona, becomes the first location to use a drive-thru window, which establishes a new national standard for convenience.

Get the Big Picture
Since the first IBM PC was introduced to the market in 1981, computer technology has advanced with accelerating speed. Companies need to keep up technologically to compete in e-commerce. What developments in business technology do you predict for the future?

POWER READ

Be an active reader and use these reading strategies:

PREDICT what the section will be about.
CONNECT what you read with your own life.
QUESTION as you read to make sure you understand the content.
RESPOND to what you've read.

Section 11-1

Positioning a Web Site

AS YOU READ...

YOU WILL LEARN
- how Web sites are positioned through search engines.
- why positioning is important to generating site traffic.
- how keywords are integral to positioning.

WHY IT'S IMPORTANT

When a user conducts a Web search, site arrangement is significant. A number of factors are taken into consideration in search-engine lists. Since most viewers give their primary attention to the first few Web pages listed, businesses strive to be among those at the top.

KEY TERMS
- position
- title element
- meta data
- robots
- pay-per-click

PREDICT

How do you think search engines "know" in what order to put search results?

MAXIMIZING YOUR WEB SITE IN A SEARCH ENGINE

Radio, television, and print advertisements are all about visibility. Cereals, shaving creams, and pharmaceutical companies fight to purchase airtime during the most popular shows, all with the goal of reaching the most people at one time. The same is true online. To reach as many Internet users as possible, you can register your Web site with a search engine.

To register with a site, open your Web browser and go to a search engine. Some of the major search engines are Google InfoSeek, AltaVista, Excite, Lycos, and DogPile. Locate the search engine's site submission page. (See **Figure 11.1**.) Type in your site's address, or URL (Uniform Resource Locator) into the search engine's form and submit it.

There is a main disadvantage to registering a site with a search engine. Users receive many "hits" when conducting a search. If there are too many hits before your URL appears, users may not visit your site. For this reason, it's important to secure a good position on search engines.

Placement of a URL

Position refers to the placement of a URL in search-engine results. For example, a search for books on Google shows Barnes&Noble.com first, Amazon.com second, and several hundred Web sites below. Since most search tools only display 10 or 12 results per page, and since most customers will not look beyond the first page, position is crucial for success. Think about it. Reading a brightly lit screen is exhausting. Most customers do not have the patience or time to dig through hundreds of pages of URLs.

If you want to get your site closer to the top of the results pages, pay attention to the position of the keywords on your site. Poorly ranked pages are often the result of faulty placement. For example, search engines don't read keywords located in frames or graphics. Instead, place your keywords in HTML text somewhere on the upper portion of your page. Also, make sure your pertinent information is placed on the home page, rather than buried inside your site's pages. If you follow these rules, there's a better chance that search engines will find your site before millions of others.

Because position is so important, Web designers and copywriters create pages with crawler-based search engines in mind. Crawler-based search engines create listings automatically, using programs that move through links and pages. There are several critical elements your page must contain to boost its position with these search engines.

226 Unit 4 Web Site Development

Figure 11.1

Registering with a Search Engine

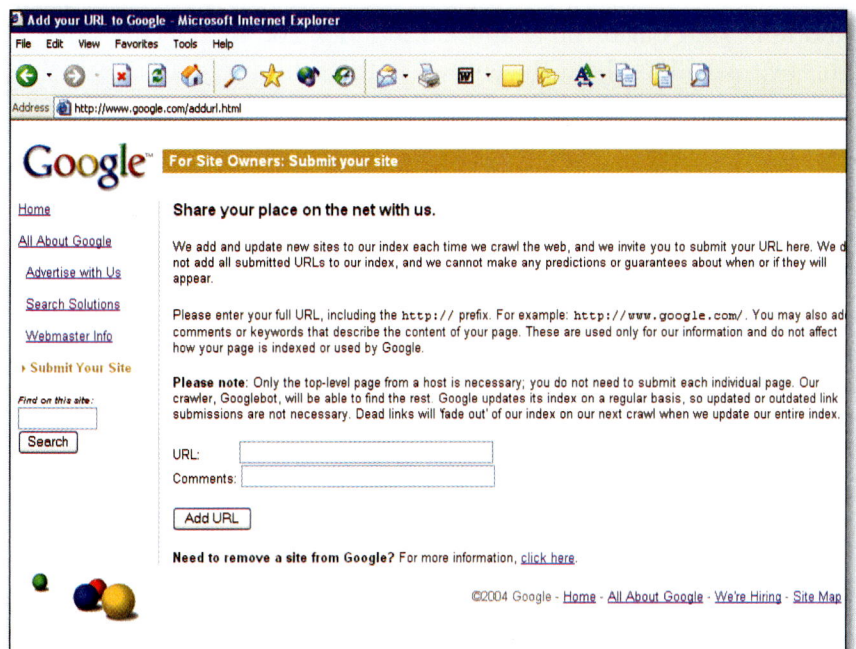

REGISTERING A WEB SITE One of the best ways an online company can reach consumers is by registering its Web site with an Internet search engine or directory service. *What is the main commercial benefit of registering a Web site?*

First, every HTML document must have a **title element** in the head section of the page code. As with every HTML element, the title is contained within a tag. Tags are the bits of codes that define the elements. As you learned in Chapter 10, when you visit any Web page and select the "View Source" option on your browser, the HTML code is revealed.

It is important to create titles that clearly identify what your page is about for the browser. "Page One" and "Home" are not particularly descriptive. "History of Jazz" is much more specific and meaningful.

META DATA One way to boost your position with crawler-based search engines is to use meta data. **Meta data** is information about a document contained within a tag called a *meta tag*. Like an HTML tag, a meta tag provides information about a Web page. Unlike a standard HTML tag, a meta tag does not alter a Web-page display. Instead, these tags specify meta data properties such as keywords, author, or content. Include meta tags in the "head" area of your Web page. **Figure 11.2** on page 229 shows how meta tags and meta data are used.

Internet surfers do not see the meta tags, but robots do. **Robots** are automated programs that follow links to visit Web sites on behalf of search engines or directories. They process and index the code and content of each page and store it in the search engine's database. The right meta data will land you in the right part of the database.

Although meta data and content-rich title elements will not guarantee top listings with the search engines, they will definitely help. For example, a student searching for a recently published book may not know that an online bookstore stocks several used and new copies. Fortunately, the bookstore designs individual pages for each book, with

CONNECT When did you last use a Web search engine? Did you click on the first few listings or scroll through all of the results? Why do viewers often click only on sites near the top of the list?

QUESTION What is meta data? What do search engines do with meta tags?

Chapter 11 Web Site Management **227**

CONNECTION: MATH

Finding an Affordable Web Server

These days, Web-hosting offers are a dime a dozen. Look up "Web hosting" in any browser, and you'll be bombarded by countless options. *Research the different prices you can find for a year's worth of hosting for a 100 MB Web site. Determine which Web host is offering the best deal.*

each title element identical to the book's title. The student will find the page among the search results.

Put some thought into the keywords you want to include in your page's meta tags. Crawler-based search engines use the keywords to locate and rank your page. You're allowed to index 200 to 250 characters in this text (which is about the entire length of this paragraph, including spaces). Try to duplicate words written on your home page in the text of your meta tags. This way, the crawler will be more likely to locate your page.

Some Web designers obsess over writing the "right" text for meta tags, but don't get too carried away. Not all crawlers support it. Generally, Web-site administrators, or Webmasters, care the most about crawler-based search engines and how those engines index pages.

PAY-PER-CLICK As with traditional advertising, Internet search tools also sell companies a better position in their database. Search services such as Overture.com list Web sites in search-engine results on a **pay-per-click** basis. With pay-per-click agreements, companies only pay for qualifying clicks to the destination site based on a prearranged rate.

For example, Overture maintains a list of popular search terms. It also partners with affiliate search engines, such as MSN, Yahoo!, AltaVista, InfoSpace, AlltheWeb, and NetZero. Companies bid on specific terms by naming the price they are willing to pay for each click. The higher the price a company is willing to pay, the closer to the top it will appear in the search results.

PAY-FOR-PERFORMANCE Some search engines, such as Yahoo! and Google, display *pay-for-performance* search results separately from crawler-based listings. Yahoo! calls them "Sponsor Results" and displays them prominently at the top of the search list. Google displays them as text advertisements to the right of the search results.

To see how pay-per-performance works, go to Yahoo! and search for "cheap tickets." Cheap Tickets owns the maximum bid on this phrase, followed by Expedia and Orbitz. A search for "cheap airfare" shows that Expedia owns the maximum bid on that phrase.

PRODUCT PLACEMENT
As with e-commerce, positioning plays a major role in the success of bricks-and-mortar businesses such as grocery stores, where manufacturers sometimes pay a "restocking" fee to keep products in prominent positions on the shelves. *How do such fees affect a company's ability to compete in any given market?*

Figure 11.2
Meta Data/Meta Tags

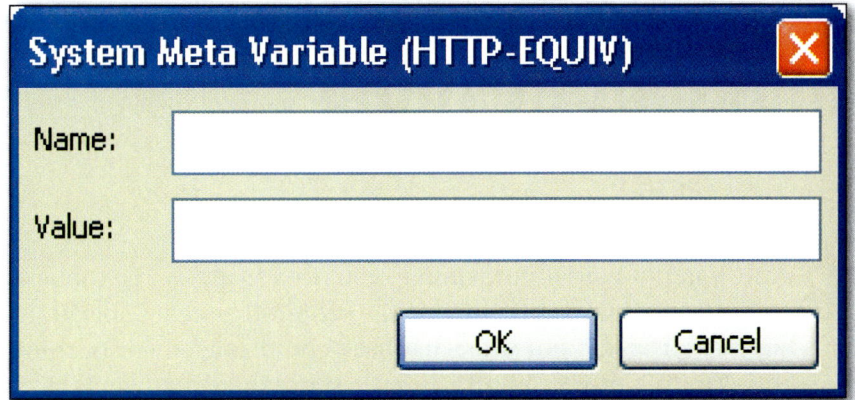

META DATA A meta tag specifies properties about a Web-page display, such as keywords, author's name, or content. *What is the purpose of including this information in a Web page?*

Pay-for-performance is an effective way to reach customers, especially since these customers are already searching for your products and services. Unlike with traditional advertising, these customers actually come to you.

Section 11-1 Review

RESPOND to what you've read.

1. What is a pay-for-performance search result?

2. What is a crawler-based search engine?

3. What is pay-per-click? How does it affect Web-site placement on a listing?

4. What are robots on a Web page?

Quick TALK Brainstorm words that Amazon might use as keywords in its meta tags. (Keep in mind that this allows for only 200 to 250 characters). *Compare your keywords with a partner, and discuss why some words seem to work better than others.*

Chapter 11 Web Site Management **229**

Section 11-2

Back End Management Tools

AS YOU READ...

YOU WILL LEARN

- how effective back-end management can lead to a successful e-commerce business.
- which factors to consider when choosing a server.
- how database-driven sites customize a visitor's experience.
- what components run online shopping carts.

WHY IT'S IMPORTANT

In e-commerce what goes on behind the scenes can in many ways be more important than what the customer sees on a site. Carefully planning the operational components is a key step toward ensuring a successful online business.

KEY TERMS

- back-end management
- server
- bandwidth
- Structured Query Language
- PHP

PREDICT

Do you think good product selection and competitive prices are enough to make a Web site successful?

BACK END MANAGEMENT

Before you put your e-store online, you need to give some thought to **back-end management**, or the behind-the-scenes operations of a business. If you're not sure what back-end management is, think of it this way: When you go into a music shop, there are two parts of the store. There's the front end, which consists of all that you as a customer see—the products, the prices, the displays, the clerks. Suppose you pick out a CD you want to buy, take it to the checkout counter, and discover that there isn't any cash register? You ask the sales personnel to help you, but they don't have any way of taking your money or giving you change. Maybe you want a CD by a particular artist, but you can't find it on the shelves. You ask some clerks to help you, but they tell you they have no idea what the store has in stock. Even worse, they don't have any way to find it. Do you think you'd go back to that store or recommend it to someone else? Probably not. The store might have had attractive, well-designed displays, but its poor back-end management will probably drive it out of business.

Back-end management is the infrastructure of your online company—the part that helps people find, buy, and receive what they're seeking. (See **Figure 11.3**.) If your back-end management is organized and efficient, your customers will have a good experience and probably come back. More importantly, they'll tell their friends about your store. If your customers don't have a good experience, have to hunt for the product they want, or wait for a slow site to download, they will leave—and, chances are, they won't come back. They certainly won't recommend you to their friends. In other words, good back-end management makes for a successful e-store. Bad back-end management means you won't be in business very long.

Good back-end management will at least include a secure Web server to host your site, a way to take credit-card payments, shopping cart software, and an efficient way to maintain your online catalog. You'll also want to consider shipping, taxes, order tracking, inventory tracking, and more. Thinking about all these issues might seem a bit overwhelming. However, if you're wondering where to start, start with a server.

Servers

Your first step in setting up an e-commerce Web site should be choosing a Web host or **server**, a computer system that gives you access to the Internet. With hundreds, if not thousands, of possible servers from which to choose, this can be a difficult decision. As you look for a server

Figure 11.3

Back-End Management

WEB SITE INFRASTRUCTURE
If your back-end management is organized and efficient, customers will be able to find, buy, and receive what they're looking for online. *What sort of back-end management concerns might be involved in selling a product online?*

to host your e-business, you should seek out certain features that may not normally be offered by a standard Web host. For example, your server should be able to provide a secure connection for credit-card transactions. It should have a "CGI Bin" for form processing. If you plan to use Microsoft FrontPage, the Web-site publishing software, to build your site, your server will need to install FrontPage Server Extensions. You'll also need to consider storage space and bandwidth.

When it comes to storage space for your Web site, the more you can get the better. Generally, e-commerce sites require 800 megabytes to one gigabyte of space, particularly if they're using audio and video. **Bandwidth** determines the amount of traffic that the server will allow on your site. Bandwidth could be compared to lanes on a freeway. The more lanes on the freeway—the wider it is—the more traffic it can carry. If you expect to have a lot of traffic on your Web site, be sure to find a host that can provide you with maximum, and preferably unlimited, bandwidth. This is important because if you run out of bandwidth, your site will go offline temporarily, or you will be billed extra for hosting too much traffic.

Another consideration when looking for a host is reliability. How much downtime does the server have? If your server is down, you can't make any sales.

Once you've found a suitable server, you're almost ready to set up your Web site. However, as an online store your Web site will have certain requirements that other Web sites don't have. You will need to manage your stock, and you will need to manage sales.

CONNECT

Have you told a friend or family member about a good online shopping experience? How about a bad one? How is this type of discussion crucial to a business's success?

Chapter 11 Web Site Management

AVOIDING GRIDLOCK
Like the number of lanes on a freeway, bandwidth determines the amount of traffic your Web site can handle. *What happens when there's more traffic than a Web site's bandwidth can support?*

 **QUESTION**
What is back-end management, and what does it include?

Databases

Have you ever wondered how Amazon or eBay manages to create so many different Web pages? If you observe carefully when you're on these sites, you'll notice that they appear to be creating Web pages "on the fly." These are examples of *database-driven* Web sites.

If your online business sells large numbers of different products, and your stock is constantly changing, you might want to consider developing a database-driven site. You can build a site that automatically customizes your visitors' pages using **Structured Query Language** (SQL) and **PHP**. SQL (pronounced "sequel") is a standardized language for requesting information from a database. PHP is a special scripting language that can be embedded in HTML.

Suppose you had an online music store and a customer visited your site looking for Mozart CDs. With a standard HTML- or XHTML-based site, your visitors might have to search through dozens of Web pages for all the different varieties of the product. You would have to anticipate their search and create in advance a special Mozart Page listing all the available Mozart CDs.

However, a database-driven site would take the search information, pull together all the products in the database fitting the description, and generate a Web page with all the relevant products. Your customers can't buy what they can't find, so a database-driven site is a great option for a large online store.

Once you've set up your database, you're almost ready to go. You also need to give your customers a way to check out of your store.

E-Stores and Shopping Cart Software

As your customers move through your store, they will need a way to keep track of the products they want to buy, and you'll need a way to accept their payments. If you're launching a small e-business on a tight budget, you might want to set up a merchant account with a service such

Figure 11.4

Online Shopping Cart

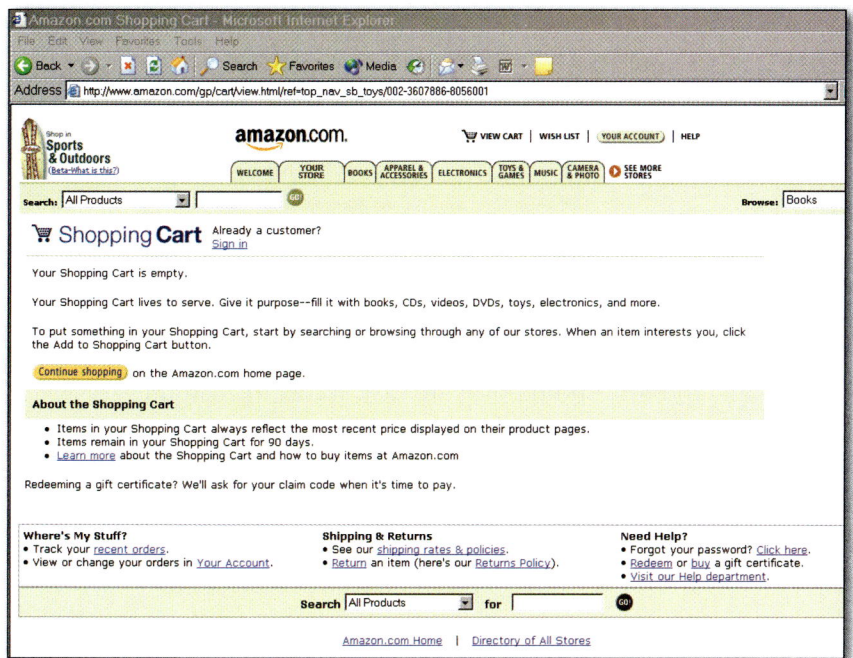

© 2004 Amazon.com, Inc. All rights reserved.

SHOPPING TOOLS You can pay an annual fee to an e-store and receive a shopping-cart tool for your Web page, or you can learn how to create a shopping cart on your own. *Why do you think the shopping-cart icon is so popular?*

as PayPal, an eBay company that enables any business or individual with an e-mail address to send and receive payments online. PayPal now provides online shopping carts and even enables you to accept credit cards. **Figure 11.4** shows how an online shopping cart works.

Web Site Success

SEARCH ENGINE SURGE
Yahoo!

In recent years, the online search market has been receiving increasing attention. Advertising revenue on such sites has been growing at a rate that has outpaced the rest of the online market. In addition, competition for that revenue among search services has been heating up, and companies have been investing in new technologies to improve search-engine capabilities. Yahoo! has been making great strides in a market where three-quarters of all search requests were once processed by one company, Google. Yahoo! was one of Google's largest customers until 2004, when it started to use its own in-house database to capture its own share of the Internet search market.

Thinking Critically
Yahoo! is anticipating drawing an even larger search audience with a paid in-house database, featuring only those Web pages that pay a fee for inclusion. *What are ways a search-engine Web site makes money?*

Chapter 11 Web Site Management **233**

Another, slightly more expensive option is to work through an *e-store* such as Yahoo! Stores. For an annual fee, an e-store provides all the back-end management tools you need, including a shopping cart. However, if you want to do it yourself, you can buy shopping-cart software that will help you set up a shopping cart on your Web site. Some of the software, such as SalesCart's FrontPage E-commerce, is designed to work with your Web-design software, while other stand-alone software, such as QuikStore, sets up the cart on your Web site. Some software, such as AAcart, works in connection with an online service that processes your orders for you.

As you plan your e-store, you need to structure everything to make the customer experience as pleasant as possible. They should be able to find the items they want without difficulty. The checkout process should be easy to follow *and* secure. Your shipping and product delivery should be smooth, and you should provide easy access to customer service. If you can do this, you have a good chance of building a successful online business.

Section 11-2 Review

RESPOND to what you've read.

1. What is bandwidth? What considerations must a Web-site author be aware of when adding bandwidth to a site? _____
_____
_____

2. What factors should be considered when choosing a Web-site server? _____
_____
_____

3. How can a database-driven site make customers' searches very efficient? _____
_____
_____

4. What are some ways in which e-commerce businesses can collect payment for customers' purchases?
_____
_____

Quick TALK Have you ever tried to purchase an item from a Web site but had problems completing a product search or checking out? *With a partner, discuss how consistent Web-site problems can drive a customer away from a particular site.*

Name _____ Date _____

Worksheet 11-1

Type It!

Typography can be defined as the style, appearance, or arrangement of type on a page. Good typography is just as important on a Web page as it is in a book. It should be pleasing to look at and easy to read. People who are interested in lettering and art can pursue careers as type designers.

1. Choose three crawler-based search engines, and conduct a search for typography. How many sites did you find for each engine?

   _____
   _____
   _____
   _____
   _____
   _____

2. Research the history of typography. Write a one-page paper highlighting what you found.

   _____
   _____
   _____
   _____
   _____
   _____

3. Find examples of at least four of the font categories described in the chapter. Which one do you like the best? Explain why.

   _____
   _____
   _____
   _____
   _____
   _____

4. Using one of the Web-crawler search engines, locate examples of GIF buttons. Print two or three that you like, and explain to the class why you like them.

Name _____ Date _____

Worksheet 11-2

Behind the Scenes

Back-end management is a key to the success of any online business. If your back-end management is efficient, is organized, and appears seamless to your customers, they will have a good experience and will return to do business with you again and again.

1. Think about a bricks-and-mortar bookstore or CD store you frequent. What components make up its back-end management, or the business side of the business?

   _____
   _____
   _____
   _____
   _____
   _____

2. Using the Internet, access a Web site that sells books or CDs. Describe the components that you think make up the back-end management.

   _____
   _____
   _____
   _____
   _____
   _____
   _____

3. Are there any similarities between the business side of a bricks-and-mortar business and an online business? Any differences? Explain.

   _____
   _____
   _____
   _____
   _____
   _____

4. Using the Internet, research shopping-cart software. Select one, and write a one-page article describing the software's features and price, as well as the customer support offered once the product is purchased and installed.

Chapter 11 Review and Activities

Chapter Summary

Section 11-1

- To position your Web site closer to the top of the search-engine results, pay attention to the position of the keywords on your Web page.

- Position is crucial for the success of an e-commerce Web site. Most search tools display only 10 to 12 results per page, and most customers do not look beyond the first page.

- Place your keywords in HTML text somewhere on the upper portion of your Web page to increase the chance that search engines will find your business.

- A Web site can secure a prominent position in search results by using pay-per-click and pay-for-performance.

Section 11-2

- Back-end management is the business part of your online business and, when organized efficiently, helps to ensure a good customer experience.

- When choosing a server, consider security, form processing, storage space, bandwidth, and any application software you will be using.

- Database-driven sites customize a visitor's experience by creating Web pages "on the fly."

- Online shopping carts allow customers to keep track of the products they want to buy and offer them a way to make their payments.

Chapter 11 Review and Activities

Review of Key Terms

a. position
b. title element
c. meta data
d. robots
e. pay-per-click
f. back-end management
g. server
h. bandwidth
i. Structured Query Language
j. PHP

Match each term to its definition.

_____ 1. Information about a document located inside the meta tags.
_____ 2. Automated programs that follow links to visit Web sites on behalf of search engines.
_____ 3. What every HTML document must have in the head section of the page code.
_____ 4. A standardized language for requesting information from a database.
_____ 5. With these agreements, companies only pay for qualifying clicks to the destination site.
_____ 6. The placement of a URL in search-engine results.
_____ 7. The amount of traffic a server will allow on your site.
_____ 8. A computer system that gives you access to the Internet.
_____ 9. The business part of your online business.
_____ 10. A special scripting language embedded in HTML.

Applying Technology to Academics

ELA — English Language Arts—Reading
Choose one of the advanced authoring tools discussed in the book. Using the Internet, locate five facts about the language used not mentioned in the chapter. Then, determine whether you would be interested in learning more about it. Explain why or why not.

ELA — English Language Arts—Writing
Using the Internet, research current trends in wireless technology. Write a one-page paper summarizing your findings. Include at least one paragraph describing how you think wireless technology will affect your life during the next three years.

ELA — English Language Arts—Speaking
Imagine you work for an e-commerce business and your boss has asked you to purchase a new server to host the business site. Using the Internet, research at least two servers you think will meet your company's needs. Make an oral presentation of your findings.

MATH — Calculation
After researching shopping-cart software, you have settled on two products that would serve your needs. Product A costs $79.95 per month, while Product B costs $989 per year. Which is less expensive?

238   Unit 4   Web Site Development

Chapter 11 Review and Activities

Critical Thinking

1. Do you think services like pay-per-click and pay-for-performance give an unfair advantage to companies that can afford them?
2. How can a database-driven Web site help you to increase sales?

COMPETITIVE EVENT

Assume the role of manager of a company that specializes in creating Web sites for small businesses. You have approached the owner of a small shoe store (judge) that specializes in hard-to-find sizes. The shoe store owner (judge) has been reluctant to add a Web site for the business because of a lack of knowledge about how to design and manage a Web site. The shoe store's owner (judge) is able to understand the potential a well-designed Web site could offer the business and is open to your suggestions. You are to meet with the shoe store's owner (judge) and explain the advantages of going online, including how to strategically position the site on a search engine to reach a maximum number of customers and how to streamline customer service using back-end management tools.

EVALUATION

You will be evaluated on how well you meet the following performance indicators:

1. Identify ways in which technology impacts business.
2. Demonstrate how a search engine works.
3. Explain strategies for positioning a Web site in a search engine.
4. Identify back-end management tools needed to operate an online business.
5. Describe the types of software available.

INTERNET ACTIVITY

To engage in online activities, visit the *E-Commerce* Web site at **ecommerce.glencoe.com**.

BusinessWeek online

TOPIC: International Commerce: Amazon.com

Amazon.com is one of the most successful retailers online. Amazon now "powers" other businesses' Web sites (like Borders and CDNOW), and its influence is being felt well outside the United States.

ACTIVITY Go to the Student Center at **ecommerce.glencoe.com**. Click on *BusinessWeek* Activities and open Chapter 11. There you'll learn more about Amazon's international expansion and how Amazon caters to different markets.

ecommerce.glencoe.com

BusinessWeek It's Your Project

SPRINGBOARD: POP QUIZ

Approximately how many acres of the Amazon rain forest are destroyed each hour?
- **a.** 50
- **b.** 500
- **c.** 5,000

This lab will require some research. Go to *BusinessWeek* online for useful articles and resources.

What's Good About It?

Problem: The school buses are jammed with students. **Solution:** Remove the seats! **What's good about the idea?** More riders could fit into each bus. The school district wouldn't have to buy a new bus or repair seats. **What's bad about the idea?** Standing up, riders wouldn't be able to do much homework. They could fall and get hurt. The school district could get sued. There's something good about every proposal. Wacky ones often suggest useful solutions. This project will show you how to propose and assess solutions to problems.

Choose a problem you want to solve. Pick one from the list, or use another problem that interests you. Investigate the problem and suggest a solution, using the seven steps as your guide.

Here are some problems for you to consider:
- acid rain
- child labor in developing countries
- global warming
- high insurance rates for teenage drivers
- lack of schooling for girls in developing countries
- low voter turnout among young Americans
- overfishing of swordfish
- pollution of major lakes and rivers
- protecting America's freedoms against the threat of terrorism
- teen smoking
- the high cost of college
- the "technology gap" between groups with computers and groups without them
- the obesity crisis in the United States

Step 1 Start with *BusinessWeek*

Orient your project by conducting research in the *BusinessWeek* online archives. Use key words that relate to your topic to find articles. The articles may offer some insight as you progress through this project.

Step 2 Investigate and Engage

Write down the problem to which you are seeking a solution. Then ask yourself these questions:
- What do you already know about this problem?
- What must you know before you can propose a solution?
- What do you want to learn from this project?

Step 3 Identify the Obstacles

- What solutions have others proposed and tried?
- Which proposals have had the best results?
- Why hasn't the problem been solved before?
- What are the sticking points—minor issues that need solutions?

Step 4 Conduct Research and Seek Solutions

- Where will you look for answers to the questions in Step 2?
- How might books, computers, and other technologies help?
- What problems has your research been unable to solve?

Step 5 Select Information and Analyze Data

- Turn your research results into statements that relate to your problem.
- Brainstorm and list potential solutions. To the right of the list, make two columns: "What's Good About It" and "What's Bad About It." Jot down at least two positives and two negatives next to each proposed solution.
- Decide which solution or set of solutions shows the most promise. Write down your reasons why they are promising.

Step 6 Connect to the Real World

Create a way to present your project—an audio-visual presentation, a talk, a Web page, an essay, or something else. It's your choice. Be sure to address these questions:

- What impact might your solution have on the world in which you live?
- How do the skills and knowledge you acquired from this project prepare you for the future?

Step 7 Self-Evaluation

Assess your work. Use an evaluation tool like a checklist, a small-group discussion, a log, or some other method. Be honest with yourself. That way you'll know where you should work harder the next time you undertake a major research project.

UNIT 5
MARKETING IN THE DIGITAL WORLD

In This Unit...

Chapter 12
Fundamentals of Internet Marketing

Chapter 13
Distribution in E-Commerce

Chapter 14
Customer Service and Web Site Personalization

Chapter 15
Advertising for E-Commerce

E-Commerce Online
To learn more about marketing in the digital world, visit the *E-Commerce* Web site at **ecommerce.glencoe.com**.

BusinessWeek Byte

These Sites Are a Shopper's Dream

Comparison sites such as Shopping.com and BizRate are growing fast—and making money bringing buyers and merchants together.

Nirav Tolia loves online shoppers. The chief operating officer of price-comparison site Shopping.com, Tolia waxes poetic on the virtues of being able to hunt down bargains and product ratings of 149 kinds of goat cheese from the seven online merchants that feed their price and product info to his site. Every goat-cheese lover who clicks through from the listing to a seller's site earns 5 cents for Tolia's company.

Gourmet foods are just one of many categories that surfers can browse at Shopping.com. Higher-value categories, such as computers, earn it 30 cents per click.

Traffic to Tolia's site actually exceeds visitor numbers at MSN Shops and AOL's shopping section, he claims. By the measurements of some Web-analysis firms, Tolia says, it even tops Yahoo!'s formidable shopping site.

Source: Excerpted with permission from Alex Salkever, "These Sites Are a Shopper's Dream," *BusinessWeek,* November 25, 2003.

To read this *BusinessWeek* article in its entirety, go to **ecommerce.glencoe.com**. Click on the Student Center, find *BusinessWeek* Articles, and go to Unit 5.

THINK ABOUT IT

The types of products available for online purchase have grown astronomically in recent years. Are there any manufactured goods or services left that *can't* be purchased via the Web?

BusinessWeek It's Your Project

Are you a teenager, a son or daughter, a teammate, a student, a citizen of the world, or all of these things at the same time? The Unit 5 project, "What's Your World?" on page 324, asks you to consider the different worlds you live in and how they shape your life.

 ecommerce.glencoe.com

243

Chapter 12
Fundamentals of Internet Marketing

Section 12-1
Marketing Basics

Section 12-2
Market Research

PREREADING STRATEGIES

Before you read this chapter, finish these statements in your journal:

- Based on this chapter title, I predict…
- The most important thing to remember about e-commerce marketing is…
- Some questions I have about marketing on the Web are:

BusinessWeek online

Go to *BusinessWeek* online. Search the site for articles about or links to companies that offer Web-site marketing research services. Make a chart comparing the types of marketing research they offer and the prices they charge.

Making It Big

Developing a product, service, or idea to sell is the first step in building a business. Getting the attention of potential customers is necessary to succeed in business.

Drill Down

▶ **1795 Keeping It Fresh**
Frenchman François Appert invents the preserving jar for food storage.

▶ **1873 Jean Pool**
Levi Strauss begins manufacturing the first blue jeans in San Francisco. The product gains enormous popularity and ultimately becomes a worldwide fashion staple.

▶ **1898 Flash Flood**
Russian emigrant Conrad Hubert invents the flashlight, which he markets as a novelty item.

▶ **1903 True Colors**
Edward Binney and Harold Smith create crayons—a phenomenon that appeals to children of all ages.

▶ **1924 Stuck on Band-Aids**
To bandage his wife's cuts, Johnson & Johnson's Earle Dickson inserts pieces of gauze into strips of tape. The corporation learns of the invention and begins mass-producing the items as BAND-AIDS®.

Get the Big Picture
Make a list of products and services you might buy on the Internet. Identify how you learned about each product or service. What strategies or techniques did the sellers use to generate your interest?

POWER READ

Be an active reader and use these reading strategies:

PREDICT what the section will be about.
CONNECT what you read with your own life.
QUESTION as you read to make sure you understand the content.
RESPOND to what you've read.

245

Section 12-1

Marketing Basics

AS YOU READ...

YOU WILL LEARN
- what marketing means in e-commerce.
- about marketing challenges faced by e-commerce businesses.
- the four Ps that make up the marketing mix.

WHY IT'S IMPORTANT

Relationships between buyers and sellers are more personalized and dynamic than ever before. To reap the benefits—and avoid the risks—of these advancements, both buyers and sellers must recognize marketing fundamentals.

KEY TERMS
- marketing
- marketing mix
- market segments
- cross selling
- promotion
- virtual marketing
- personalization
- cookies
- e-mail marketing
- permission marketing

PREDICT

How do you think Internet marketing differs from other forms of marketing?

MARKETING

Marketing is the process of planning and carrying out the production, distribution, promotion, and pricing of products and services. As such, it encompasses countless business decisions, from Web-site designs to delivery options.

The famous cow-themed Gateway computer boxes and monthly membership rates at Netflix.com are both examples of marketing. When Zappos.com e-mails tracking numbers to its customers, it is striving to maintain customer satisfaction and loyalty—the main goals of any marketing effort.

The Marketing Mix

How can you attract new customers and keep old ones coming back while navigating the complexities of the global marketplace? To start, you need to create a marketing mix. The **marketing mix** consists of four critical elements, collectively known as the four Ps: product, price, place, and promotion. (See **Figure 12.1**.) Each one must be tailored to meet the needs of your target market. If any one of the four Ps fails to meet those needs, the whole strategy could fail. The following section describes each element of the mix in detail.

PRODUCT A *product* is any item manufactured for sale. Tennis shoes, computers, coffee mugs, clothing, and pencils are all products. Products are not the only things businesses sell, however. Many businesses provide services. *Services* are not material goods, but rather work done in exchange for payment, such as printing brochures, tailoring clothing, and editing video.

The line between products and services is not always clear. Netflix offers DVD rentals through the mail, with no late fees or time limits. Its customers get both a product *and* a service. On AbleStock.com, Web designers can find digital images in three file sizes, instead of hiring photographers to take the pictures. When deciding what to sell, you should consider the ways in which products and services intersect. Sometimes a creative combination—like that offered by Netflix—fills a need no one else has even noticed.

In marketing a product, you need to consider the level of *product quality* you intend to provide. Product quality includes the durability, reliability, and aesthetic features of the construction materials, as well as the service customers receive. Warranties, guarantees, and secure shipping all contribute to product quality.

Decisions regarding product quality and service levels will depend largely on your target market, and some target markets include more

Figure 12.1

Elements in the Marketing Mix

MARKETING PROGRAM Consider some successfully marketed products that are targeted at teens. *How do the marketing programs for these products satisfy the elements in the marketing mix?*

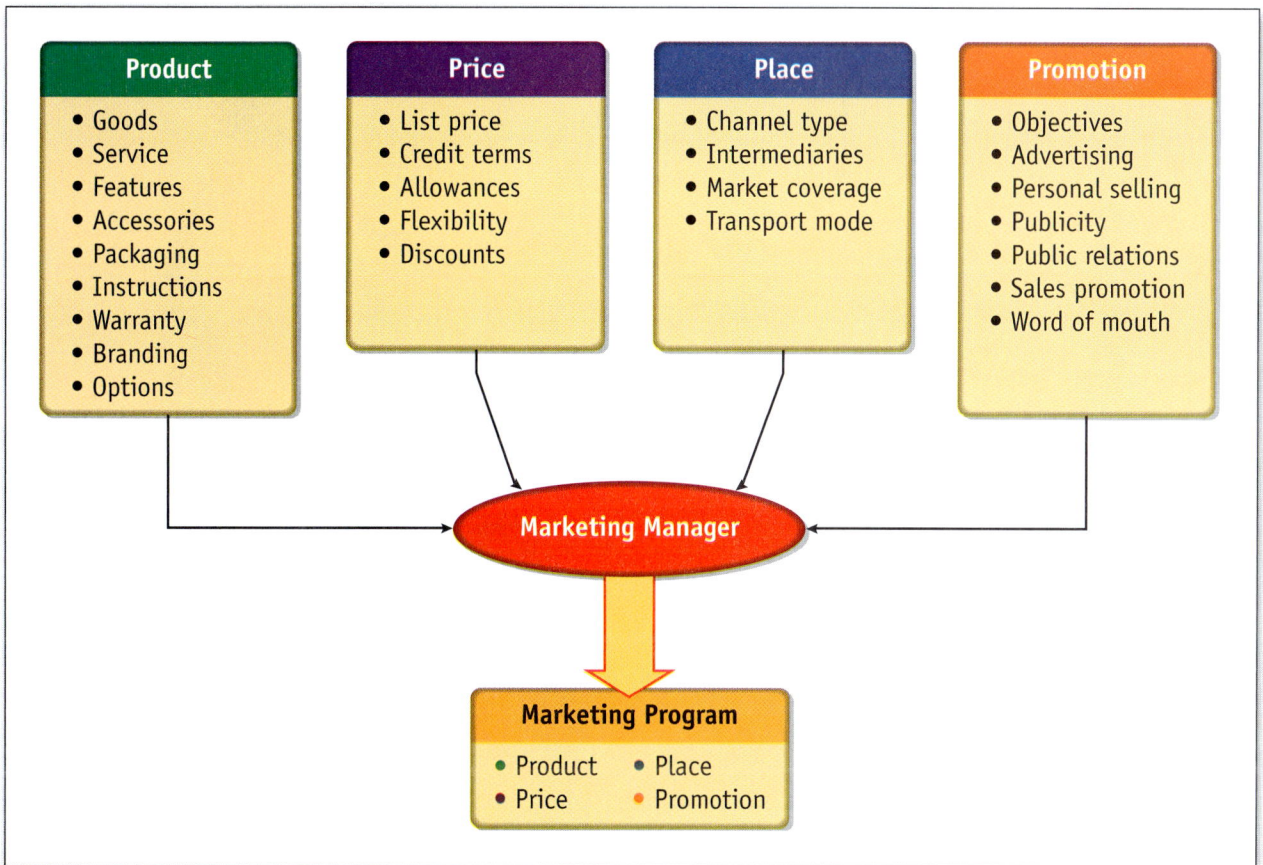

Source: McGraw-Hill Ryerson, *Understanding Canadian Business,* 4/e, William Nickels, University of Maryland; James McHugh, St Louis Community College; Susan McHugh, St Louis Community College; Paul Berman, McGill University and John Abbott College; Rita Cossa, McMaster University

than one market segment. The U.S. Small Business Alliance (USSBA) suggests several **market segments**, or distinct groups of people who share certain characteristics. Market segments are identified in the following ways:

- *demographically,* which groups people by measurable statistics such as income, age, or gender;

- *psychographically,* which defines people by lifestyle preferences;

- *geographically,* which categorizes them by region, neighborhood, or district; and by

- *product benefits,* which identifies people according to the satisfaction they get from products, such as saving money, convenience, or luxury.

Figure 12.2

Market Segment

MARKET DIVISIONS Nike and Nuala pursue different target markets within the same market segment. *What specific segment of the market does each site target?*

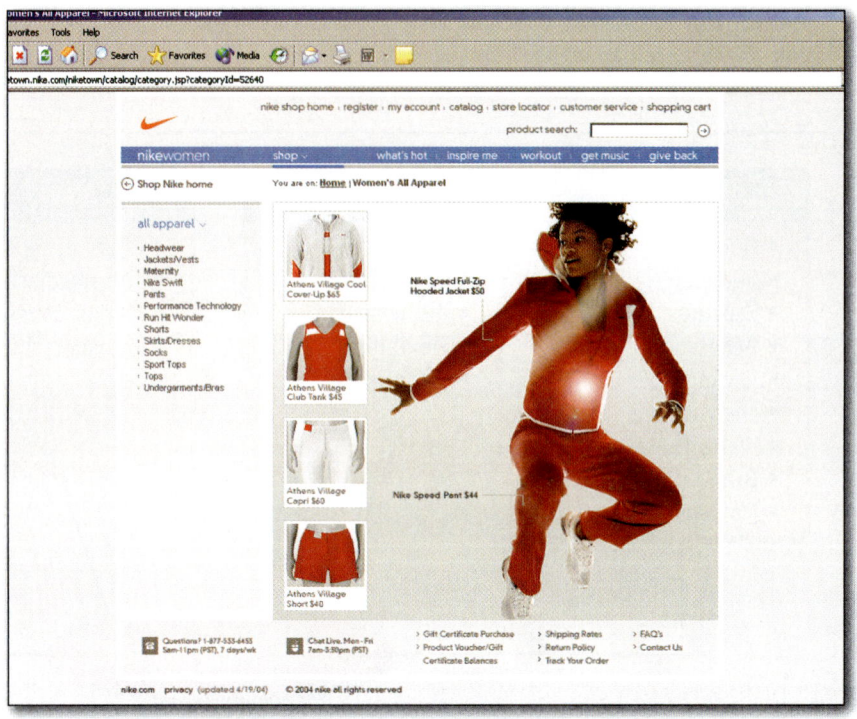

Figure 12.2 illustrates how different sites appeal to different market segments.

Your *product line*—a group of related products that complement one another—is also important. Pacific Sunwear (PacSun) has a product line designed to anticipate and meet the needs of teenage customers, with each product fitting into a young lifestyle. Purchase a pair of jeans, and you can easily find shoes to match. Product lines can include several brands, styles, and sizes.

The full set of products offered for sale, including all product lines and categories, is your *product mix*. A good product mix offers more than just variety; it represents a strategic marketing opportunity. Log on to the J. Crew Web site to buy a new suit and you'll find related items—or "other popular picks"—at the bottom of the screen. The idea is to generate another sale by offering related items, or **cross selling**. Cross selling helps customers find all the items they want in one product line and deters them from visiting the competition for that perfect belt or tie.

PRICE Price is the amount of money a business receives in exchange for a product or service. A product's price should cover all manufacturing, marketing, and related costs, as well as generate a profit. These are not, however, the only factors in deciding how much to charge. You must also consider your target market. What are your customers *willing* to pay and what do they *want* to pay?

Price also contributes to the image of your products, services, and business as a whole. Stores with "Everyday Low Price" formats such as Wal-Mart and Target attract price-conscious customers who purchase many items per visit. On the other hand, stores that use a HILO format—selling at higher everyday prices, but also offering deep discount sales—attract an entirely different type of consumer. These customers shop more frequently but purchase less each time. They are more interested in quality, service, and convenience than in saving money.

The times and conditions of payment, price increments, and quantity or promotional discounts make up your pricing structure. AbleStock.com offers a choice between yearly memberships or per-image payments. America Online (AOL) provides many different membership levels and prices. Some companies even offer items at different quality or performance levels to reach more markets. Apple Computer not only sells high-end professional computers but also consumer-oriented all-in-one desktops, with smaller hard drives, less memory, and lower prices. Additional discounts are provided to educational institutions.

E-commerce creates even more pricing options. When you purchase a plane ticket from Priceline.com, you name your own price. This is known as *variable pricing*. When properly carried out, it can result in higher profits for a business. Why did it work for Priceline? According to the Wharton School of Business, Priceline created a situation in which variable pricing would not offend those customers who paid full price. Airlines were already using *dynamic pricing*, in which seat prices fluctuated, or changed, based on flight time and capacity. Priceline added a new ingredient to the mix: inconvenience. Customers who were willing to give up their choice of airline and exact travel time and date could buy tickets on their own pricing terms. A little inconvenience resulted in significant savings.

Virtual auctions, like eBay, allow buyers to name their own price as well. With this model, they also compete against other bidders. The site charges sellers a nominal listing fee and also takes a small percentage of the sale.

> **CONNECT**
> When have you visited more than one Web site to compare prices for a particular product? Is paying the lowest possible price always your main goal when shopping online, or do other factors play a role?

> **CONNECTION**
> **LANGUAGE ARTS**
>
> **Where Do You Find Rich Media?**
> They overlap, leap, and intrude on your space. Glitz, movement, and flash draw your attention. The days of banner ads have been replaced with "rich media," a term supposedly coined by an Intel Corporation insider. Rich media is online advertising with television ethos. Macromedia Flash or Unicast and Eyeblaster formats have been used to create these pop-up wonders. *If you were going to invest in producing a rich-media advertisement, what sites would you put it on to get more "bang for your buck"?*

PLACE When it comes to business, place means more than just the physical location of a store. It's where customers shop—whether in a bricks-and-mortar store, online, on the phone, or through the mail—the various purchasing options they have, and how the product or service will be delivered. For example, Netflix has numerous distribution centers throughout the United States so that no customer will have to wait too long for delivery.

As an online entrepreneur, you need to decide where your customers will most likely buy your products and services. Will they purchase directly through your site, or will your site direct them to local resellers? Birkenstock.com directs online shoppers to local Web-enabled stores instead of selling through the site. This allows Birkenstock to provide product information and strengthen its brand identity while keeping its resellers happy. The company does not want to compete with its own retailers.

You must also think strategically about the impact place has on delivery and price. When Jeff Bezos founded Amazon.com, he chose to locate the company in Seattle for several reasons. First, Washington state has a low sales tax, combined with a relatively small population. Since sales tax is charged only in an online store's home state, this meant that very few customers would be forced to pay sales tax on their transactions. Second, the state had a large pool of technical talent. Third, one of the biggest wholesale book distributors was located in nearby Roseburg, Oregon, so Amazon wouldn't need to invest in a large warehouse right away. How many issues will affect your choice of place?

PROMOTION Beyond providing great products at fair prices and having a good location, you must inspire potential customers to visit your site and purchase your products or services. **Promotion** educates potential customers about your products and services. E-commerce makes it possible to target specific groups as never before, with personalized e-mail, e-zines, newsletters, chat rooms, and pay-per-click search-term placements (discussed in Chapter 11).

▼ **SUSAN B. REVIVAL** After discontinuing minting the Susan B. Anthony dollar coin in 1981, the U.S. was left with a stock of 550 million coins, which they managed by selling them for use in Postal Service stamp machines, vending machines, and city subway systems. By 1999, demand for new Susan B. Anthony dollars had grown so great that the coin was minted again. *Is effective marketing the key to successfully launching a new product, or can you count on word-of-mouth alone to sell a product?*

One of the best ways to promote your business is also one of the most traditional: word-of-mouth. Research by Taylor Nelson Sofres (TNS) found that 98 percent of customers recommend e-commerce sites that provide a positive experience. Only 1 percent of unhappy customers will do the same. If your site is well designed, is easy to navigate, and provides prompt customer service, the majority of your customers will tell a friend.

You can enhance your word-of-mouth promotion with **virtual marketing**, a clever and cost-effective strategy that uses customers to help promote your business. When you send electronic Hallmark cards, you are actually sending a link to the original site.

When recipients click the links to view their cards, they also pay a visit to Hallmark.com, where they can view cards, locate stores, and send more cards.

You can also encourage word-of-mouth promotion by designing a tell-your-friends or e-mail-this-page dialog box for each product or service. CNN.com allows you to e-mail stories to friends, but since the idea is to lure new readers to the site, these messages do not contain actual articles. Your friends receive a link that takes them to the story page, where they may click links to other stories. Similarly, Amazon provides several methods, including wish lists and discounts, to share items with friends.

User reviews are another effective promotion strategy. Zappos.com, Amazon, and Drugstore.com all encourage users to review their products—whether good or bad. This not only increases trust among the sites' customer bases; it also allows the companies to gauge their customers' reactions so they can change their product mixes accordingly. Chat rooms take this a step further by allowing people to talk in real time.

Online promotional campaigns often include newsletters or e-zines (electronic magazines). PacSun sends a newsletter called *PacScene,* which invites customers to check out all the products and happenings at PacSun and its stores. *Affiliate programs* (discussed in Chapter 8) divide revenues between online advertisers or merchants and online publishers or salespeople. Amazon uses a popular model, in which referring Web sites are paid commissions for each click that results in a sale.

Another way to reach your target market is via pay-per-click advertising. In this model, you bid for specific search terms, with the highest bidder showing up first on the search-results list. You only pay when someone clicks to your page. This is a highly targeted promotional strategy, reaching people who are already interested in the products or services you provide—otherwise known as qualified customers. However, this strategy must be used judiciously, since you could end up paying more for total clicks than you make back in sales. Overture is an example of a pay-per-click search engine, while Google offers AdWords, a pay-per-click model that displays small text advertisements at the side of a search-results page. (See **Figure 2.3** on page 252.)

PERSONALIZATION To the four Ps of the marketing mix, a fifth could be added: personalization. **Personalization**, which means establishing one-to-one relationships with your customers, is potentially the most powerful aspect of e-commerce. When Amazon greets you by name, offering book and movie recommendations, it is using a personalization strategy.

Many online businesses use cookies to personalize their marketing. **Cookies** are small text files written directly onto visitors' hard drives, containing information such as log-in names, shopping-cart contents, or unique identification codes. Cookies also store user preferences, facilitating targeted advertising and customized content, as the Web site "remembers" each customer's patterns. Using cookies, businesses can collect data and analyze the behavioral patterns of their customers to market personally to them.

E-mail marketing is another way in which you can establish personal relationships with your customers. E-mail allows you to send customized messages, tell customers about new products and sales, or remind

Figure 12.3

Pay-per-Click

TAKING BIDS Google's pay-per-click model displays small text advertisements at the side of a search-results page. *What kinds of search terms might you be willing to bid on?*

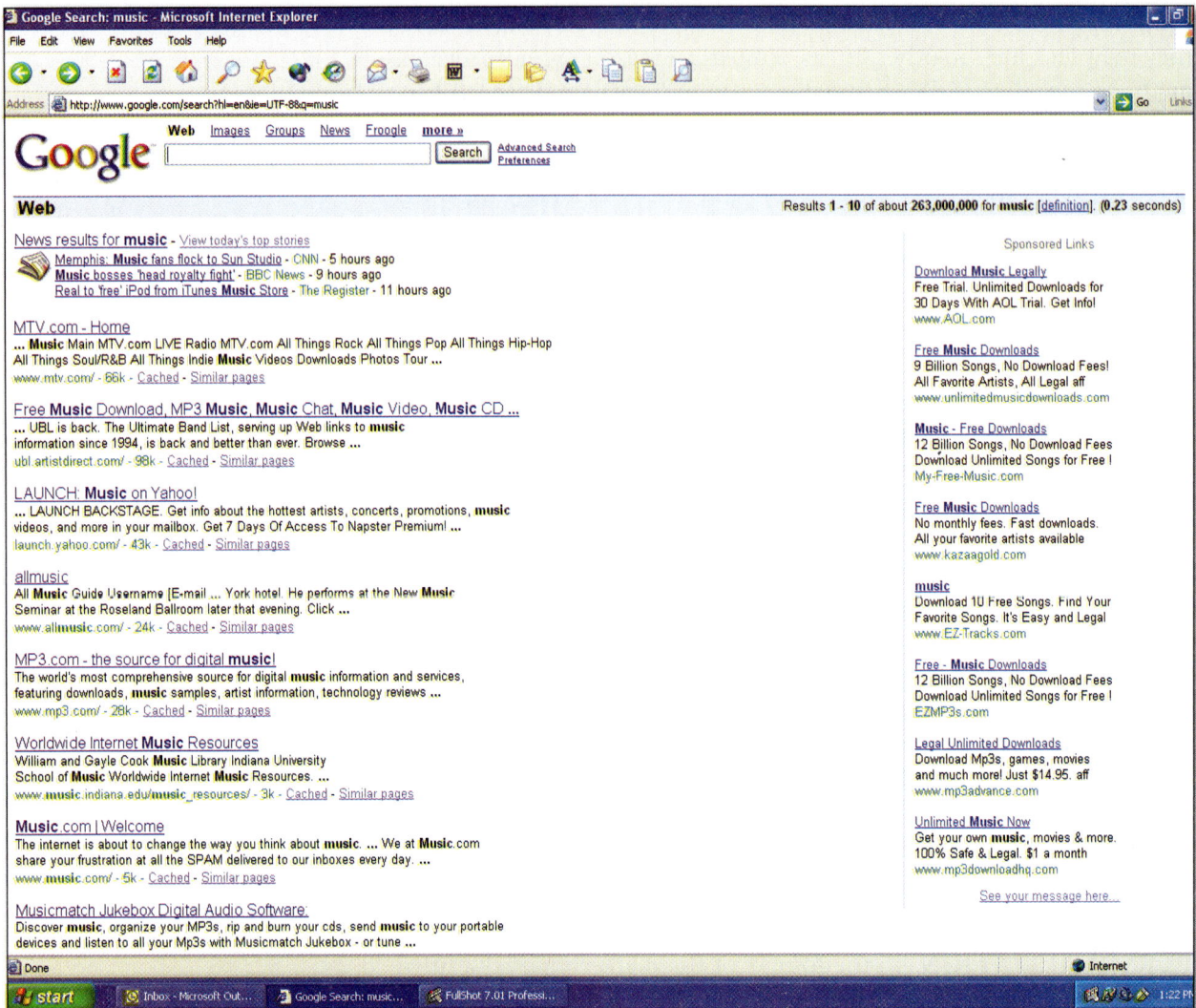

QUESTION

Why isn't e-mail marketing as effective as some other e-commerce marketing schemes?

them to order gifts for holidays or special events in their profiles. According to The Direct Marketing Association, e-mail marketing provides the highest return on investment (ROI) of any direct-marketing medium. It is the least expensive way to reach customers, with one of the highest response rates. There is, however, a major disadvantage to e-mail marketing. Many consumers consider all forms of e-mail marketing spam and will delete it without reading it or use a filter to block it.

Think about how companies traditionally try to get your attention. They interrupt your favorite television programs with commercials, print slick ads in your favorite magazines, and slip Sunday specials into the newspaper. They rent billboard spaces and pay radio stations to play commercial jingles between songs. Blimps pass overhead with corporate marketing messages.

Fast Company reports that Americans are exposed to more than one million marketing messages per year. This is nearly 3,000 messages a day. As a consumer, you have little choice whether or not you have to see and hear these messages.

Recently, many companies have started using a new marketing strategy called **permission marketing**. Permission marketing is based on one simple principle: The potential customer has *consented* to receive marketing messages and will only receive advertisements about products and services they have requested.

When deodorant company Carter-Wallace wanted people to know about its newest deodorant, Arrid XX, it used permission marketing. It offered consumers the opportunity to enter a contest in which the winner received a luxurious Caribbean trip. To enter the contest, respondents provided their e-mail addresses and consented to receive daily e-mail with product messages and clues to help them win the contest.

During the contest period, Carter-Wallace sent more than 20 e-mail messages to each participant. A follow-up survey showed that nearly 25 percent of the people who participated in the permission marketing had purchased the new deodorant and an additional 50 percent of the respondents said they were likely to purchase it in the future.

Permission marketing has proven extremely successful for e-commerce sites. When companies identify people who are interested in their products, they are able to deliver specific sales messages that are well received by the viewer.

Section 12-1 Review

RESPOND to what you've read.

1. What is marketing, and how can it impact an online business?

2. How can a good product mix create successful marketing opportunities in e-commerce?

3. Of the four Ps of the marketing mix, how can the component *place* be best described?

4. What is permission marketing?

Quick TALK With a partner, pick a product to market online. *Brainstorm ways to market the product using the tools in your e-marketing mix.*

Section 12-2

Market Research

AS YOU READ...
YOU WILL LEARN
- what customers generally want from their online shopping experience.
- how marketing research provides information about customers and their wants.
- the varied types of research conducted by e-commerce businesses.

WHY IT'S IMPORTANT
To develop a strong marketing mix, a company must understand its customers' needs, wants, and buying patterns. Researching the market enables businesses to cater more accurately to their customers.

KEY TERMS
- market research
- marketing research
- clickstream data
- metrics

PREDICT
Why do you think businesses do market research?

IDENTIFYING MARKET OPPORTUNITIES

Traditional bricks-and-mortar retailers have always collected and analyzed market data. When you walk into your local sporting-goods store, employees may immediately determine your approximate age, income bracket, gender, and tastes. If you carry shopping bags from a competing store, they'll probably notice this as well. Managers may also watch as you walk through the store, taking note of which sections generate the most interest or highest sales.

Market Research

In the world of e-commerce, companies have their own ways of collecting market data. **Market research** encompasses all of a business's efforts to collect data regarding the economic climate, competitor prices or strategies, governmental regulations and policies, consumer demographics, and emerging technologies.

There are two main types of research. One type is called *secondary research,* which is gathered from literature, publications, broadcast media, and other nonhuman sources. Popular secondary-research sources include:

- U.S. Small Business Alliance (USSBA)
- U.S. Census Bureau
- American Marketing Association
- University publications and libraries

You can also pay for up-to-the-minute research from firms such as JupiterResearch, Forrester Research, or WebTrends. Trade publications and magazines are another option.

The other type of market research is *primary research*—data gathered directly through one-to-one interviews, focus groups, and surveys. Primary research consists of collecting original information for a specific marketing purpose, such as to gauge how a new product will sell or find out why consumers prefer a particular brand. Have you ever received an e-mail or telephone call from someone asking questions about a recent customer-service experience? This is primary market research. Secondary research is usually conducted before primary research because much of the information a business wants to gather might already exist. For example, if a business wants to determine what types of people would most likely be interested in buying folding bicycles, there might be statistics available in trade publications or from marketing research companies.

254   Unit 5   Marketing in the Digital World

Marketing Research

Marketing research, as distinct from *market* research, focuses specifically on consumer behaviors and the reasons behind buying habits. It collects information about consumers, especially a company's target market, and analyzes demographics, buying habits, product preferences, and concerns about technology. Marketing research is conducted by groups ranging from the entertainment industry to the U.S. government. (See **Figure 12.4**.)

Marketing research efforts at PacSun and dELiAs Corporation include regular customer surveys. Many sites create contests or registration incentives to learn more about their customers.

Clickstream data, or the information recorded in most server logs, is another rich source of information. These are a visitor's electronic footprints, including their unique IPs, page views, downloaded images, and URLs. For maximum benefit, this data must be analyzed in terms of patterns, not just as isolated bits of information. The number of page "hits" means nothing if marketers cannot determine how many hits resulted in sales. Patterns found in clickstream and other data are called **metrics**. Metrics can be used to improve site navigation, create better privacy policies, and determine which pages generate the most interest.

Marketing research for a consumer-orientated business can be conducted in one of two ways: on an ad-hoc basis or on a continuous basis. Companies use *ad-hoc research* when they need to focus on specific questions. In most ad-hoc research, the question or questions are asked only once, helping businesses to identify flaws in products or services before they hit the market. Mail, telephone, and online surveys may be used, along with product samples.

Continuous research is more time-consuming and complex. In continuous research, consumer panels are asked questions over a period of

Figure 12.4

Marketing Research Revenues by Industry Groups

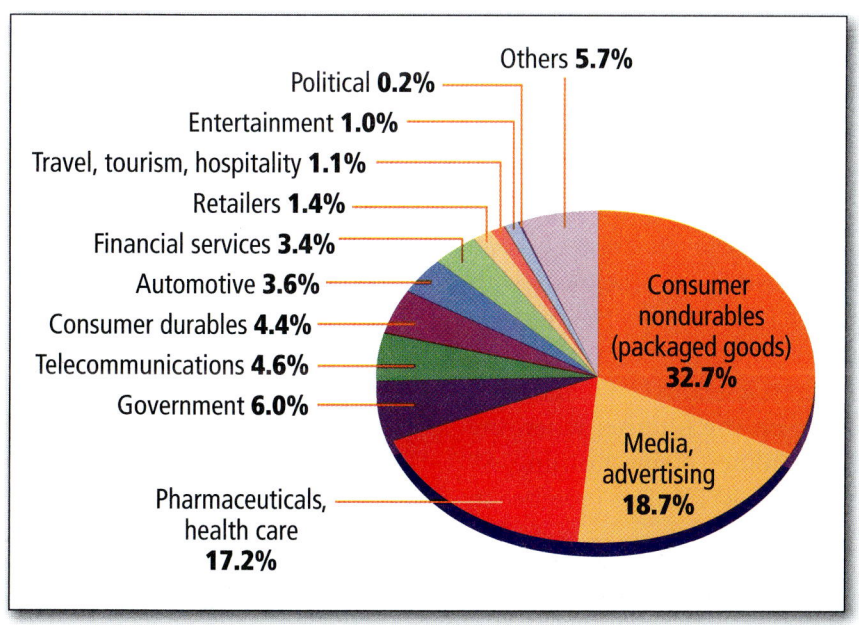

MARKETING RESEARCH REVENUES Over $3 billion is spent annually by client industry groups for marketing research. *Why does the government spend money on marketing research?*

Chapter 12 Fundamentals of Internet Marketing

time. *Consumer panels* are paid to answer questions or provide feedback at crucial intervals. In response to this feedback, companies change products or services in time for the next panel meeting. A major advantage of continuous research is that it helps companies gauge the effectiveness of each change, since consumers respond more than once. Microsoft and Apple often perform *beta testing*—sending out an early version of software to selected users—to catch disastrous coding errors or system crashes prior to the public release of new operating systems.

When companies need to test an advertising campaign, they turn to *advertising research*. This measures consumer response to both the message and the medium, as consumers evaluate everything from fonts and image quality to the feelings they evoke. Some companies have even invited customers to vote on their favorite commercial.

Product research focuses on the best way to produce and distribute products or services in the current technological or environmental context. When Ben and Mena Trott decided to launch a personal publishing Web site called TypePad, they started with a beta test. People who chose to receive prelaunch updates were invited at random to try the new site for free. They reported software bugs, problems with memory limitations, and other glitches. They also suggested design and interface improvements. By the time TypePad was launched, it had already been tweaked several times.

Sales research studies sales and market data to determine how well a product or service should do in the marketplace. For example, a grocery delivery service might start in a limited market to test its appeal.

Research in business operations can help a company improve the effectiveness and efficiency of distribution. Are the majority of customer complaints about slow shipping? What is happening in the warehouse to slow down order processing?

As you can see, e-commerce companies rely heavily on research to develop a successful marketing mix for their target market.

CONNECT

How might you feel about being part of the marketing campaign for a new product and receiving a daily e-mail containing contest clues? Why might this type of permission marketing be unappealing to some people?

Web Site Success

FINDING THE URBAN VIBE ONLINE
BET

Like many good businesses, Black Entertainment Television (BET) decided to diversify, from a cable television channel into an online portal. BET was launched in 1980, before the age of 24/7 CNN and MTV, and eventually became a leading portal for African-American consumers. Before the portal went live, however, the company had to address a serious challenge: How could it reach an urban audience on the Internet when the majority of that audience wasn't online? A smart offline, $9 million marketing campaign lifted the BET online presence to the forefront of its community.

Thinking Critically
According to Forrester Research, income is the strongest indicator of whether people have Internet access. *What options exist for the person without a home computer who wants to go online?*

NET GROWTH As the Internet grows and more users join the online community, the demographics that describe that community—age, gender, ethnicity, etc.—increasingly resemble those of the population as a whole. *What is the most effective approach to selling a product in this increasingly diverse online market?*

ONLINE CUSTOMERS Internet users are a diverse group of people. Children, senior citizens, professionals, and college students all use the Internet. In the United States, both high- and low-income households are online. According to Nielsen/NetRatings:

- Senior citizens are using the Internet in increasing numbers. They currently represent 5.9 percent of online users in the United States.

- Two in ten Internet users were between the ages of two and 17 in September 2003.

- Fifty-two percent of Internet users in the United States are women.

The market research firm InsightExpress reports that Internet users look more and more like the population as a whole. According to JupiterResearch, African-Americans make up 8 percent of the U.S. online population, and AOL found that this group uses broadband at a higher rate than the general online population. English-speaking Asian-Americans are the most wired, according to the Pew Internet & American Life Project.

What do shoppers want from the online experience? While personalization has often been played up as the key, a JupiterResearch study found that only 14 percent of consumers buy more often because of personalized offers. Fifty-four percent of the same respondents wanted faster-loading pages and easier navigation. One Genex survey found that 65 percent of respondents will not do business on poorly designed Web sites, and almost 30 percent will reject a bricks-and-mortar store if its Web-site design is not appealing.

Privacy is also a big concern. According to the Annenberg Public Policy Center of the University of Pennsylvania, when adults were shown ways in which Web sites monitor and use information, 85 percent disapproved. They did not want their personal behaviors tracked, even on their favorite sites.

What does this say about Internet users? Despite their geographic diversity and potentially long distance from your store, they are not

QUESTION

Why is the senior citizen demographic important for businesses to consider when marketing over the Internet?

Chapter 12 Fundamentals of Internet Marketing

very different from traditional shoppers. Both groups want similar things: easy navigation, whether through aisles or pages; quick and simple checkout; privacy and security; fair prices; and great products. In fact, many online shoppers use the Internet to research prices and features before heading out to the store. According to DoubleClick, 45 percent of online shoppers visited retail stores to make their final purchase, while 17 percent of bricks-and-mortar window shoppers ended up purchasing online. As it turns out, the two groups are not really two groups at all.

Section 12-2 Review

RESPOND to what you've read.

1. How are Internet and traditional shoppers similar?

2. Why will a business conduct research into their distribution operations?

3. What is the difference between primary and secondary research? Provide examples of each.

Quick TALK With your partner, brainstorm ten questions you could ask customers about your newly test-marketed soft drink LimeTime Smoothie. *Discuss how the responses to these ad-hoc questions might affect the way you market the drink.*

Name _____ Date _____

Worksheet 12-1

Shop Till You Drop—No More!

According to *Newsweek*, in fall 2003 an estimated 85 million shoppers spent $1.5 billion dollars per week online. Many of these shoppers chose to compare prices to find the best bargains available. There are several Web sites consumers can use to make this process easier.

1. Access three different comparison-shopping Web sites. Describe how they are different and how they are similar.

   _____
   _____
   _____
   _____
   _____

2. Why do you think these types of sites are available to consumers, and who do you think sponsors them?

   _____
   _____
   _____
   _____
   _____

3. Reviewing the sites you selected above, see if you can find out whether there are any hidden fees or special charges.

   _____
   _____
   _____
   _____
   _____

4. Imagine you own a business. How would you go about getting your company's products and prices posted on the sites you selected for review?

   _____
   _____
   _____
   _____
   _____

Name _____ Date _____

Worksheet 12-2

School Spirit

E-zines, or electronic magazines, are created for people who are interested in particular content. Businesses create e-zines to advertise their products or services. Organizations create e-zines to provide information to their members and potential members. No one really knows how many e-zines there are because new ones keep popping up each day. Today is your chance to add one more.

1. Research several e-zines, some that businesses have created and some that organizations have created. Which features are similar, and which are different?

   _____

   _____

   _____

   _____

2. With two or three classmates, design an e-zine for your school. Determine your target market, then create a sketch with a logo, a name, and a list of features you think students would find interesting.

3. Write the articles for your e-zine's inaugural issue.
4. If possible, launch the e-zine on the Web.

Chapter 12 Review and Activities

Chapter Summary

Section 12-1

- Marketing is the process of planning and carrying out the production, distribution, promotion, and pricing of products and services.

- The four Ps of the marketing mix are product, price, place, promotion.

- To the four Ps of the marketing mix, a fifth could be added: personalization.

- Methods online businesses use to establish personal relationships with customers include cookies, e-mail marketing, and permission marketing.

Section 12-2

- Research has found that online consumers want faster-loading pages, easier navigation, and security and privacy when shopping online.

- Market research touches on almost every aspect of a business, from economic climate and competitor prices to governmental regulations, consumer demographics, and emerging technologies.

- E-commerce businesses analyze patterns found in clickstream and other data to learn about their customers.

- Marketing research, as distinct from market research, includes ad-hoc research and continuous research, product research, sales research, and beta testing.

Chapter 12 Review and Activities

Review of Key Terms

a. cross selling
b. permission marketing
c. cookies
d. marketing mix
e. market research
f. marketing
g. clickstream data
h. market segments
i. e-mail marketing
j. virtual marketing
k. marketing research
l. personalization
m. metrics
n. promotion

Match each term to its definition.

_____ 1. A marketing strategy in which potential customers have consented to receive marketing messages.
_____ 2. A cost-effective strategy that enlists customers to help promote a business.
_____ 3. The process of planning and carrying out the production, distribution, promotion, and pricing of products.
_____ 4. Includes product, price, place, and promotion.
_____ 5. The concept of generating another sale by offering related items.
_____ 6. Small text files written directly onto visitors' hard drives.
_____ 7. Collecting information about consumers, especially those in a business's target market.
_____ 8. Information recorded in most server logs.
_____ 9. Distinct groups of consumers who share certain characteristics.
_____ 10. The broadest form of research, encompassing almost every aspect of business.
_____ 11. Patterns found in clickstream and other data.
_____ 12. Provides the highest return on investment of any direct-marketing medium.
_____ 13. Educates potential customers about products and services.
_____ 14. Establishing one-to-one relationships with customers.

Applying Technology to Academics

English Language Arts—Reading
Pick any three of the key terms in this chapter. Go online or use business reference books to find three differently worded definitions for each of the terms. Read the definitions in class and pick which ones you think are the best in terms of how easy they are to understand.

English Language Arts—Writing
Imagine you own an online jewelry business and have worked very hard to establish personal relationships with your customers via e-mail. Write an e-mail message to a prospective customer introducing your company. Write another to a customer who has not made a purchase in several months to find out why.

English Language Arts—Speaking
Pick a type of product you purchase often, such as CDs, soft drinks, or magazines. Create a chart, with illustrations, showing how the product gets from the manufacturer to you using the five elements of the marketing mix. Present your chart to the class.

Graphing Software Program
Conduct a marketing survey of your class. Pick a popular movie. Survey the students in your class to find out who's seen it, who wants to see it, or who has no interest in seeing it. Break down the results into demographic categories like gender and hair color. Then, using a computer program, graph the results.

Chapter 12 Review and Activities

Critical Thinking

1. How is traditional marketing similar to or different from Internet marketing?
2. Do you think businesses should be allowed to place cookies on your hard drive without your permission? Explain your answer.

COMPETITIVE EVENT

Imagine you're the assistant manager of a company that grows and sells organic vegetables. The business began as a small farm and has expanded to become the premier regional seller of organic vegetables. The business manager (judge) has asked you to develop a marketing plan to begin selling the company's organic vegetables online. Include the four Ps of marketing plus personalization as they relate to Internet marketing in your marketing plan. Discuss an outline of your Internet marketing plan with the company manager (judge).

EVALUATION

You'll be evaluated on how well you meet the following performance indicators:

1. Explain the nature of marketing plans.
2. Develop a marketing plan.
3. Explain the concept of marketing strategies.
4. Identify your online target market.
5. Make oral presentations.

To engage in online activities, visit the *E-Commerce* Web site at **ecommerce.glencoe.com**.

BusinessWeek online

TOPIC: Internet Marketing

The worldwide reach of the Internet has made it a place rich with marketing opportunities. At the same time, the sheer popularity of the Internet has made it possible for an e-commerce site to drop into obscurity if it is not marketed carefully.

ACTIVITY Go to the Student Center at **ecommerce.glencoe.com**. Click on *BusinessWeek* Activities and open Chapter 12. There you'll learn more about what makes a successful strategy for marketing online.

Chapter 13
Distribution in E-Commerce

Section 13-1
Channels of Distribution

Section 13-2
Physical Distribution

PREREADING STRATEGIES

Before you read this chapter, finish these statements in your journal:

- Based on this chapter title, I predict…
- The most important thing to remember about distribution and e-commerce is…
- Some questions I have about how distribution works with e-commerce business are:

BusinessWeek online

Go to *BusinessWeek* online. Find a story about or a sponsored link to a company that offers online E-CRM services. Make a list of the features it provides and explain how they can benefit an online business.

264 ecommerce.glencoe.com

Bringing It Home

E-commerce companies rely on many of the same delivery methods as their bricks-and-mortar counterparts. Finding a fast, reliable way to deliver product to the customer can raise a company's profile.

Drill Down

1883 Counting Cash
James Ritty and John Birch receive a patent for the first mechanical cash register. The device helped make business transactions quicker and more accurate.

1927 Television
American Philo Farnsworth transmits the first television image.

1950 On Credit
Diners Club founder Frank McNamara creates the credit card.

1952 Bar Codes
Joseph Woodland and Bernard Silver patent a system of bar codes that allows computer scanners to read product information.

1968 ATM
Don Wetzel introduces the first successful ATM (Automatic Teller Machine), which greatly improved upon the original ideas of Luther George Simjian.

1970 RAM
The Intel Corporation releases the first RAM semiconductor chip to the public. RAM technology continues to evolve and develop with tremendous growth.

Get the Big Picture

In 490 B.C. a Greek messenger ran 26 miles to deliver news of the Athenian victory over the Persians in the Battle of Marathon. How has the distribution of news changed since then in terms of who delivers information and how they deliver it?

POWER READ

Be an active reader and use these reading strategies:

PREDICT what the section will be about.
CONNECT what you read with your own life.
QUESTION as you read to make sure you understand the content.
RESPOND to what you've read.

265

Section 13-1

Channels of Distribution

AS YOU READ...

YOU WILL LEARN
- what distribution means in e-commerce.
- common methods of distribution used by e-commerce businesses.
- the roles of cybermediaries.

WHY IT'S IMPORTANT

Businesses can sell all they are able, but if their distribution methods are not well planned and executed, the business will quickly lose credibility with its customers.

KEY TERMS
- distribution
- channel of distribution
- intermediary
- cybermediary
- Web distributor
- Web affiliate
- aggregator

PREDICT

Why do you think an efficient distribution system is important to an e-commerce business's success?

FOUNDATIONS OF DISTRIBUTION

Feeling a sweater's fibers isn't possible when buying it online. You can only imagine how soft it feels. Imagination, in part, is an active participant in online shopping, as well as your expectation that the e-tailer follows through with the delivery of your sweater. You expect your purchase to arrive within days after the transaction. What you might not have considered is the amount of work and details an e-tailer has to consider before that purchase is in your hands.

In business, one of the important functions is delivering the right product or service in the right quantities in the right place at the right time. The plan for how a product will get from one place to another is often referred to as *logistics*. Logistics range from managing the inbound movement of raw materials and supplies to managing the outbound movement of finished products. Logistic activities include receiving goods, warehousing, inventory control, vehicle scheduling and control, and distribution of goods.

In an e-marketplace, one-stop shopping should be as error free as shopping at a bricks-and-mortar store. The online shopping location is called an *e-marketplace*. When a product is purchased for business use, the end user is called an *industrial user*. When a product is purchased for personal use, the end user is called a *consumer*. In this section you will learn how e-commerce companies build a distribution system to deliver their products to you in a timely manner.

How Distribution Works

Distribution refers to the way the producer or manufacturer delivers products, services, or information to the consumer. Trucks, trains, airplanes, and ships are possible transportation methods. The products and services you purchase through an e-commerce transaction can be tangible, such as sporting goods or designer clothing, or intangible, such as computer software and online newspaper subscriptions. If the online purchase is delivered using a conventional method, delivery charges are determined by the size and weight of the product. With a large or heavy product, the delivery costs may be high and wipe out any money you saved by not paying sales tax for the online purchase. If delivery is too expensive, it may discourage you from making the online purchase.

Changes in the methods for conducting business and delivering goods can lead to confusion. Until all parties understand and agree on the distribution process, there will be small hurdles to overcome. For example, in the gift-giving season of 1999, some bricks-and-mortar businesses that had recently expanded their online operations offered more than they could deliver in time for the holidays. Toys "R" Us included information about its new online site in its 1999 catalog, which is mailed to 60 million households. The unexpectedly enthusiastic response overwhelmed the Web-site server and caused bottlenecks at the company's Web site. They also ran out of many popular items, leaving thousands of orders unfilled and missing its holiday-shopping guarantee to its customers.

Toysrus.com was not the only e-tailer to miss its deadlines due to errors in distribution during this period. Other e-commerce sites, including Macys.com and CDNow.com, also missed delivery deadlines. In response to consumer outrage, the Federal Trade Commission (FTC) launched Project Toolate.com, an investigation into whether online businesses delivered goods as promised during the holiday shopping period. At the conclusion of the inquiry in July 2000, the FTC fined seven stores— Toysrus.com, Macys.com, CDNow.com, KBKids.com, HoneyBaked Ham, Minidiscnow.com, and Patriot Computer Corp.—a total of $1.5 million.

Channels of Distribution

A **channel of distribution** is the path a product takes from producer or manufacturer to consumer. It is a particular way to direct products to consumers. Producers use a specific channel to move goods to the end user. Some online products and services, such as newsletters, software, and sales discount coupons, can be distributed directly to customers over the Internet. Delivery over the Internet cuts down the producer's costs, which can lower costs for consumers, and allows consumers to receive information instantly rather than waiting several days for it to be delivered through the mail.

CONNECT

How would you react if you ordered goods from an online business for holiday delivery and they arrived late or not at all? What could the company do to make up for it?

Web Site Success

FLYING OFF THE BOOKSHELVES
Powells.com

It's been heralded as "the greatest bookstore in the world, bar none," but Powell's Books in Portland, Oregon, has grown into one of the most successful online bookstores, as well. Stocking miles of bookshelves since 1971, the independent bookstore added a modest online store in 1994, which has grown to 40 percent of its total sales. Much like the store's early, innovative idea of mixing used, new, hardcover, and paperback books caught on with other booksellers, so has a popular facet of its online distribution practice: Long before other e-tailers offered discounted delivery charges at a certain price point, Powells.com included free shipping on all orders of $50 or more.

Thinking Critically
By offering free shipping to customers, Powells.com had to pay the shipping charges itself. *Why might Powells.com have chosen $50 as its price point for free shipping? How might changing this price affect sales?*

Chapter 13 Distribution in E-Commerce **267**

Figure 13.1

Channel of Distribution

CHAIN LINKS This flow chart illustrates a channel of distribution, which is the way online products move from manufacturer to consumer. Have you purchased something online recently? *Can you trace the steps undertaken by the manufacturer to get the product to you?*

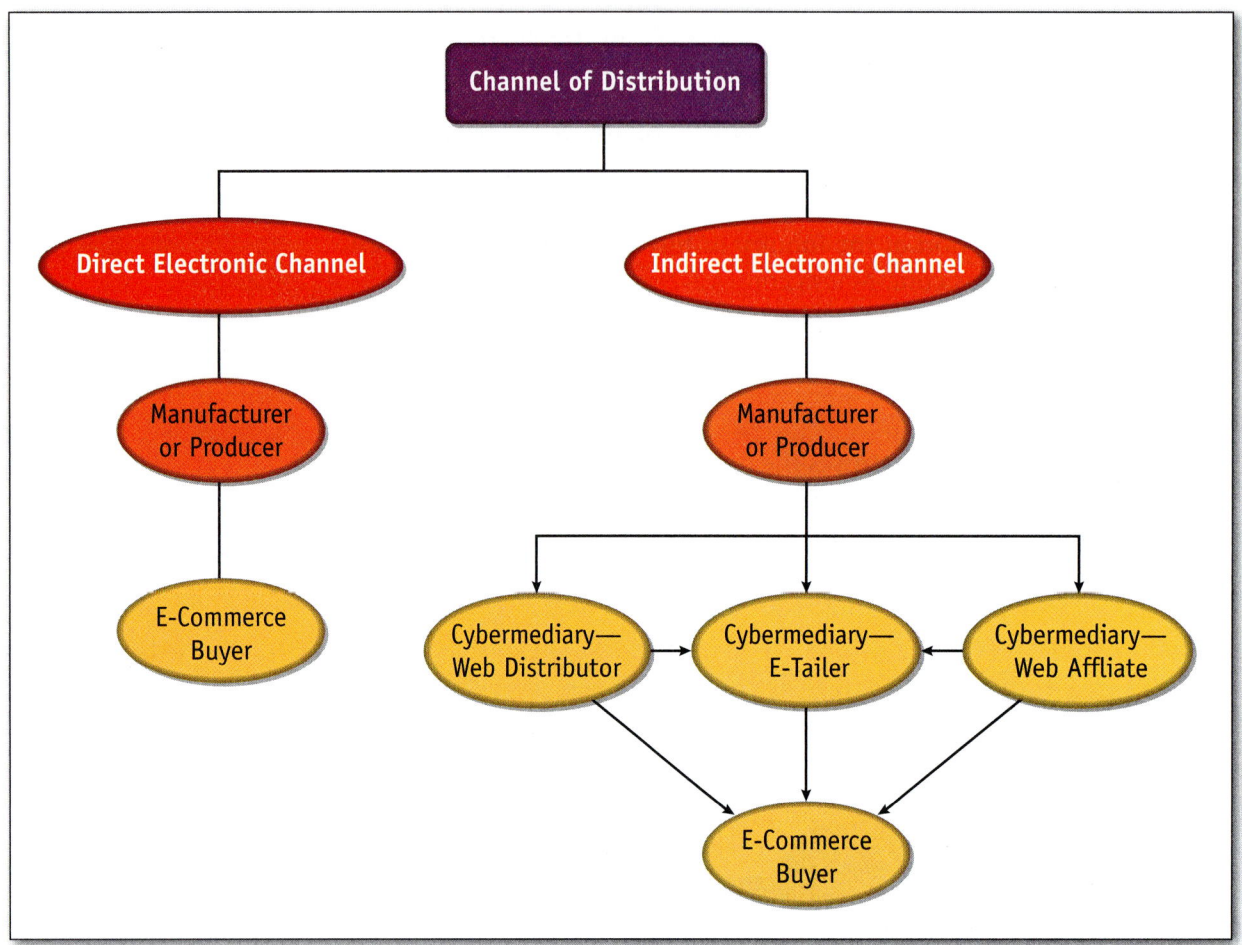

QUESTION

Is Amazon.com an example of direct or indirect distribution? Explain your answer.

DIRECT AND INDIRECT CHANNELS OF DISTRIBUTION

With many online purchases, you receive the product directly from the manufacturer or producer. This is called *direct distribution*. Direct distribution occurs when the goods and services are sold from the producer directly to the customer. Dell, the computer company, uses direct distribution.

A business that doesn't produce the goods or services it sells, but gets them from producers and then sells them to consumers, uses *indirect distribution*. Indirect distribution involves one or more intermediaries. An **intermediary** is a business that acts as a third party or go-between in moving products from the manufacturer to the end user. For example, if you start a magazine about pop culture, you might want to find a company to distribute if for you that specializes in independent publications. This company distributor would in turn distribute to an independent bookstore near you. **Figure 13.1** shows how a channel of distribution in e-commerce works using both direct and indirect distribution.

CHANNEL MEMBERS E-commerce uses online intermediaries called cybermediaries. A **cybermediary** is an Internet channel of distribution that helps move products from the manufacturer to the consumer or industrial user. Often a cybermediary does not have a bricks-and-mortar component and does not manufacture the products it sells. While a manufacturer focuses its attention on the product's success, a cybermediary has expertise in displaying, merchandising, and selling items. A cybermediary can come in many different forms, including:

- an e-tailer, which sells to the end user. Popular e-tailers include Amazon and eBay. To review e-tailers, refer to Chapter 3.

- a **Web distributor**, which markets and sells goods or services bought from a manufacturer on a wholesale basis to companies for use or for resale to the end user. For example, The Weather Channel uses the Web distribution company Dalet to manage its digital media systems. Dalet's technology customizes affiliate radio broadcasting stations' weather updates.

- a **Web affiliate**, which forms a partnership with another online business and refers potential customers to the partner's Web site for a commission or fee. For example, suppose you're the editor of an entertainment e-zine. You have a link on your site to Art.com. This art site offers over 100,000 prints, posters, and photographs. By offering the Art.com link on your site, you earn 25 to 30 percent commission on sales from the Web site.

- an **aggregator**, which is a company that distributes a number of products and services from a variety of manufacturers and producers. Many Web aggregators deliver information to customers. Rocketinfo, for example, offers software to research, analyze, and distribute information to subscribers, like a financial services company. Instead of searching the Web for the latest news and information from a variety of sources, the Rocketinfo software tracks the updates and electronically sends the company the updates. NewsGator is a similar plug-in application. You can subscribe to various news syndicates, and each time a syndicate posts something new, NewsGator sends an update to you. Web aggregators can be used to receive updates from news sites and blogs.

- a wholesaler, which receives large shipments of products from many different producers. It breaks the shipment into smaller batches for resale. A company that makes canned peas may sell a truckload of its peas to a wholesaler. The wholesaler, in turn, will sell a few cases of peas to several local supermarkets. DollarDays International is an example of a cybermediary wholesaler. It is a "Web-based virtual warehouse" that offers small businesses deals on thousands of consumer products, from toys to electronics, so they can compete with large companies.

Any business, individual, or organization that sells merchandise, processes information, or conducts transactions on the Internet is a cybermediary. E-tailers, online stock brokers, and online car dealers are all types of cybermediaries. Probably the most familiar type of cybermediary is the search engine.

Section 13-1 Review

RESPOND to what you've read.

1. What is a channel of distribution? How does direct distribution differ from indirect distribution?

2. What does a cybermediary do? Is a wholesaler a cybermediary? Why or why not?

3. What are logistics as they apply to distribution? What are some logistic activities?

4. What is a Web affiliate? How can a Web affiliate be helpful to an online business?

Quick TALK The distributor for an online crafts company went out of business and your holiday shipments are going to arrive late. **With a partner, brainstorm how to word an e-mail to customers explaining why their shipments will arrive late and how you intend to remedy the situation.**

Section 13-2

Physical Distribution

ORDER PROCESSING

Physical distribution involves order processing, transportation, storage, stock handling, and inventory control. Distribution links the company to the customer. Without a well-planned and organized process, an e-commerce company will lose its customers. Your business will have to coordinate with other businesses to make sure everyone's schedule is on the same track.

Purchasing a product initiates the system of physical distribution. Using a successful order fulfillment system, employees can track customer orders, locate products, and resolve any problems that arise as items move from warehouse to shipping company to customer's doorstep. When taking orders online, you have to be vigilant about protecting your customer's account and credit-card information.

Electronic Customer Relationship Management

A critical function of the order-fulfillment system is customer care. **Electronic customer relationship management** (E-CRM) is the cornerstone of good service. Communication is a critical factor to your business's integrity and a way to swiftly resolve conflict as it arises. If a customer is treated well, then repeat business is almost guaranteed, as is word-of-mouth advertising. Customer care is also a place where the company can position itself to gain a competitive advantage in the distribution arena. Good service, in turn, leads to satisfied customers, repeat business, and heftier profits. What exactly does an E-CRM system do? Stand-alone programs, software suites, best-of-breed systems, and hosted E-CRM are a few of the systems that have emerged in the last two decades.

STAND-ALONE PROGRAMS In the early to middle 1990s, most E-CRM systems were **stand-alone programs** used in call centers to track customer contacts and so-called "trouble tickets." A call-center agent would receive a customer phone call or an e-mail message, log it in the E-CRM database, fill in the customer's information, and write a description of the problem. The agent then would attempt to solve the customer's problem and would document those efforts in the program.

If the issue was not resolved on first contact—in response to the first call or an e-mail message—the agent could leave the trouble ticket open and refer it to an agent or supervisor higher up the customer service chain. The second agent would have no need to repeat procedures already tried by the first agent. Instead, the other agent or supervisor could simply read the trouble ticket and pick up where the first agent left off. Vendors of these early E-CRM programs included Remedy and Clarify Inc.

AS YOU READ...

YOU WILL LEARN
- the steps involved in physical distribution.
- how E-CRM works to resolve problems.
- what ERP is and how it works with customer relationship management.

WHY IT'S IMPORTANT

To ensure efficient operations, businesses strive to keep their order processing streamlined. Customer loyalty is closely tied to the satisfaction involved with each transaction and maintaining good service is essential to doing business.

KEY TERMS
- electronic customer relationship management
- stand-alone programs
- enterprise resource planning
- best-of-breed
- hosted E-CRM

PREDICT

Why do you think managing customer relations is vital to business?

CONNECTION
LANGUAGE ARTS

Fix the Attitude

Imagine you've been billed for an item you didn't purchase. The online e-tailer offers two options for complaints, an e-mail address and a phone number. *Pretend you're a customer service representative for an e-tailer and a customer has called in angry about an extra charge. How do you handle this call?*

To illustrate this concept, suppose you order a DVD from an online store. The DVD that arrives at your doorstep is damaged, so you contact customer service to return the product. A customer service agent enters your contact information and a brief description of the problem into an E-CRM database.

SOFTWARE SUITES As time went on, larger vendors began to enter the E-CRM software market and compete with established players like Remedy and Clarify Inc. Many of these larger vendors, such as PeopleSoft and SAP, had already made significant inroads in **enterprise resource planning** (ERP). ERP focused mainly on financial and human resources applications, but the vendors realized their software could be easily adapted to deal with customer relationship-management issues. The E-CRM software suites, or sets of software applications, developed by ERP vendors were relatively complex and expensive compared with their stand-alone predecessors. The ERP vendors had a keyword in their sales pitches to executives: *integration*.

BEST-OF-BREED SYSTEMS Other vendors, such as Siebel Systems, took a different approach. Siebel expanded on the reputation of stand-alone vendors working to perfect E-CRM software and sell it based on a best-of-breed philosophy. **Best-of-breed** is when a company focuses on a single product or service to make it the best in its field. In the case of Siebel and similar vendors, the product is E-CRM. Best-of-breed vendors argued that ERP vendors sold E-CRM systems as an afterthought, rather than as their sole mission, and that those systems were not the best possible.

HOSTED E-CRM Today, best-of-breed and software suite strategies coexist in the marketplace, while stand-alone programs generally have been phased out. A new form of E-CRM, called **hosted E-CRM**, also is gaining customers. With hosted E-CRM, a company does not buy E-CRM software at all. Instead, a company signs a contract with a vendor to "rent" the software on a per-seat basis (usually the number

 SELF HELP Many software producers have developed automated support services to provide their customers with 24-hour troubleshooting systems. When they run into problems with their software, they can log onto the company Web site, run a test to diagnose the problem, and often correct the problem online. *What is the main drawback of automated systems?*

272 Unit 5 Marketing in the Digital World

Figure 13.2

Evolution and Types of E-CRM Systems

E-CRM OPTIONS This table shows the various types of E-CRM systems. *If you were choosing an E-CRM system for your own company, which type do you think you would choose if you were running a small business? A large business? Why?*

Type of System	Vendors	Major Benefits	Major Drawbacks
Stand-alone program	Remedy, Clarify	Relatively easy to install, creates customer history	Relatively limited functionality, usually designed for call centers only
Software suite	SAP, PeopleSoft	Designed to integrate with existing ERP systems from same vendor	More expensive than hosted CRM
Best-of-breed	Siebel Systems	CRM sole focus of system, highly customizable	More expensive than hosted CRM, can be more difficult to integrate than suites
Hosted system	Salesforce.com, PeopleSoft, Siebel	Lower cost because software is rented, feasible for small and mid-sized businesses	Limited customization

of employees). For example, if there are 72 employees in an e-tailer's customer service center, the company could buy 72 licenses. The vendor hosts the software at a central location, and its entire customer companies log in remotely to access it. This approach generally benefits small and mid-sized companies, which often cannot afford to spend large sums of money to purchase in-house software, such as PeopleSoft or SAP. In fact, the hosted approach is becoming so popular that PeopleSoft and Siebel have rolled out their own hosted E-CRM offerings.

Each type of E-CRM has its own advantages and disadvantages. (See **Figure 13.2**.) As you can see, various E-CRM vendors provide different functionality, depending on their target markets, but some of the more common E-CRM functions include:

- logging customer contacts that come in via phone, e-mail message, and live chat;
- tracking shipments from warehouse to customer doorstep;
- tracking and managing the returns process; and
- analyzing trends to identify areas where improvement is needed.

TRANSPORTATION

Transportation involves the actual physical movement of products. It is estimated that up to eight percent of a company's sales are spent on transportation. There are three important questions an e-commerce company must ask itself about what methods of shipping to use:

1. Which common shipping options should the company offer?
2. What are the costs and benefits of each of these options?
3. Should the company offer additional incentives, such as free shipping or same-day shipping?

These three questions should be answered before a company sets up a system of transportation.

Types of Carriers

To answer the first question, a company must familiarize itself with some of the most widely used shipping firms and organizations. At the most basic level is the U.S. Postal Service (USPS). Commercial companies, such as United Parcel Service (UPS) and Federal Express (FedEx) offer additional services beyond the capabilities of the USPS. However, commercial shipping firms also charge more for their services. **Figure 13.3** compares the benefits and drawbacks of these three common shipping methods.

TRANSPORTATION SERVICES AND CARRIERS Offering same-day shipping can give a company a competitive advantage. It can also, however, create the potential for customer complaints if the company is not prepared to follow through on its promise. This dilemma illustrates a time-tested principle of business: underpromise, then overdeliver. Amazon uses this principle well by estimating shipping times of three to

CONNECT

Have you or a family member ever had to ship something to someone? Did you use the post office or did you consider other possible methods? Why does a business need to spend time considering shipping methods?

Figure 13.3

Shipping Methods: Benefits and Drawbacks

COMPARISON SHIPPING This table shows some benefits and drawbacks of commercial shipping companies compared with the U.S. Postal Service. *What tradeoffs does an online business need to consider when choosing which shipping providers to use?*

Shipping Company	Major Benefits	Major Drawbacks
U.S. Postal Service (USPS)	Relatively low cost, signature often not required upon delivery	Limited package tracking with some shipping options
United Parcel Service (UPS)	Detailed package tracking on all shipments	More expensive than USPS, may require signature upon delivery
Federal Express (FedEx)	Detailed package tracking on all shipments	More expensive than USPS, may require signature upon delivery

HAND DELIVERED How does a transportation and delivery service expand its channels of distribution? After UPS bought Mail Boxes Etc. in 2001, FedEx acquired Kinko's in 2004, giving both companies a neighborhood presence and providing customers with easier access to their services. *Why is it important to compare prices when looking for a transportation and delivery service for your e-commerce business?*

seven business days for ground shipments. Shipments rarely take seven days to arrive—and when they arrive sooner, customers are usually pleased. Increasing customers' expectations by promising same-day shipping can lose a company customers if shipments arrive late.

Many shipping companies have begun allowing e-commerce companies to tie their e-business systems into the shipping company's order-fulfillment systems, enabling data to flow easily between the two companies. For example, e-tailers can link their back-end databases—the behind-the-scenes systems that contain information about inventory—to UPS systems and access data about a package's location with ease. This improves customer service and increases the likelihood of repeat business.

This type of integration should appear seamless to the customer, regardless of the complexity of systems behind the scenes. For example, if a customer orders a package and wants to have it shipped via UPS Ground, the customer should receive a confirmation e-mail from the e-commerce company that details the customer's order. When the customer's package is shipped, a second e-mail should be sent by the company listing the UPS tracking number associated with the package. The customer should be able to contact either UPS or the e-commerce company—whichever is more convenient—to find out where the package is. By e-mailing customers at every step of the process, customers feel as though their orders are moving closer toward them, that the company is efficient, and that the company cares about them.

QUESTION
What does it mean to track a shipment?

WAREHOUSE, STOCK HANDLING, AND INVENTORY CONTROL A flexible returns policy, an E-CRM system, and same-day shipping are not enough to give a company a competitive advantage. Online businesses that deal in material goods still need to store and inventory their products. Warehouses can ease the actual movement of products through the distribution channel as products are sold. An e-commerce company's storage facility is usually small but its assembly area is larger. It has to take into consideration how many times workers handle the goods. The more hands that touch them, the more inefficient that part of the business runs.

A warehouse layout should have a strategy built in that focuses on less labor and lower operation costs. The fastest selling items should be stored closest to the loading dock door to speed up the process of moving them in and out. Related items should be stored in the same area so they're easy to locate, as opposed to alphabetizing goods in a warehouse of using an unusual system of sorting them. In general, an e-commerce business needs a warehousing system that makes it easy to take and fill orders quickly.

Receiving, checking, and marketing items for sale are important steps in the physical distribution process. When items are given proper inventory control, this ensures that products are kept in sufficient quantities and available when requested by customers. Buying an e-business system does not guarantee success, either, if a company does not also tie together data from all of its other existing systems. Seamless integration is the key.

Section 13-2 Review

RESPOND to what you've read.

1. How is hosted E-CRM different from E-CRM? What type of company should use hosted E-CRM?

   _____

   _____

   _____

2. Why is it important for businesses to send customers e-mail at various stages of the shipping process?

   _____

   _____

   _____

3. What does it mean to "underpromise and overdeliver?" How can this make a company look good to customers? _____

   _____

   _____

4. What are some ways a warehouse can be organized efficiently? _____

   _____

   _____

Quick TALK

Would you rather purchase a DVD from a company offering low prices but known to have poor customer service, or a company offering higher prices but with excellent customer service? **With a partner, discuss the pros and cons of each and which you would prefer.**

Name _____ Date _____

Worksheet 13-1

Can You Get There from Here?

Delivering the right product to the right consumer quickly and efficiently is extremely important to the success of a business. The plan for how a product will get from one place to another is often referred to as *logistics*. Logistic activities include receiving goods, warehousing, inventory control, vehicle scheduling and control, and distribution of goods.

1. Research careers in logistics. What type of education or training is needed? What skills and knowledge are necessary for a successful career in this field? Would you be interested in a career in logistics? Why or why not?

   _____

   _____

   _____

   _____

2. Imagine you are going to start your own online business. Pick a product to sell and then trace the channel of distribution you will use to acquire the product and then deliver it to your customer.

   _____

   _____

   _____

   _____

3. For the same business, what partnerships could you form with other online businesses to refer business to each other?

   _____

   _____

   _____

   _____

4. Find an example of an online business that sells products through direct distribution and one that uses indirect distribution.

   _____

   _____

   _____

   _____

Name _____ Date _____

Worksheet 13-2

The Customer Wants Delivery Now!

Customer happiness is closely tied to the satisfaction a customer feels when receiving an online order in a timely manner. Quick delivery is often more important than a lower price on the product. Distribution links the company to the customer. Without an efficient delivery system, an e-tailer will lose customers.

1. Using the Internet, research an E-CRM company. What is the name of the company? What software features does the company promote?

   _____
   _____
   _____
   _____

2. How do the features of the E-CRM company you selected improve customer satisfaction?

   _____
   _____
   _____
   _____

3. Using the online business you chose for Worksheet 13-1, describe the transportation services you need to move your products to your customers.

   _____
   _____
   _____
   _____

4. Define integration as it relates to ERP vendors.

   _____
   _____
   _____
   _____

5. Imagine you have been hired as a business consultant for Amazon.com. You have been asked to review its distribution process and make recommendations for improvements. Using only information you find on Amazon's Web site, what would you recommend? Write a one-page summary for Amazon.

Chapter 13 Review and Activities

Chapter Summary

Section 13-1

- In e-commerce, distribution refers to the way the producer or manufacturer delivers products, services, or information to the consumer.

- Some of common methods of distribution used in e-commerce include both direct and indirect distribution. With direct distribution, products are sold directly to customers, or end users. With indirect distribution, products are sold through intermediaries, or third parties.

- Cybermediaries are Internet channels of distribution that help move products from the manufacturer to the consumer or industrial user.

Section 13-2

- The steps involved in physical distribution include order processing, transportation, storage, stock handling, and inventory control.

- E-CRM software is used to resolve problems by logging customer contacts via phone, e-mail, or live chat; by tracking shipments, and tracking and managing returns; and by analyzing trends to see where improvements could be made.

- ERP focuses on financial and human resources applications and integrates with E-CRM software.

Chapter 13 Review and Activities

Review of Key Terms

a. distribution
b. channel of distribution
c. intermediary
d. cybermediary
e. Web distributor
f. Web affiliate
g. aggregator
h. electronic customer relationship management
i. stand-alone programs
j. enterprise resource planning
k. best-of-breed
l. hosted E-CRM

Match each term to its definition.

_____ 1. A business that forms a partnership with another online business and refers customers for a fee.
_____ 2. When a company focuses on a single product or service to make it the best in its field.
_____ 3. A company that combines and offers for distribution a number of products or services from a variety of manufacturers.
_____ 4. A member of the Internet channel of distribution that helps move products from the manufacturer to the consumer or industrial user.
_____ 5. Sells goods or services from a manufacturer on a wholesale basis.
_____ 6. Focused mainly on financial and human resource applications.
_____ 7. Early types of E-CRM systems.
_____ 8. The cornerstone of good service in an online business.
_____ 9. The path a product takes from producer or manufacturer to end user.
_____ 10. A business that acts as a third party or go-between in moving products from the manufacturer to the end user.
_____ 11. When a company "rents" software on a per-seat basis.
_____ 12. The way the producer or manufacturer delivers products, services, or information to the consumer.

Applying Technology to Academics

English Language Arts—Reading
Choose one of the seven stores that were fined by the FTC for not delivering as promised during the holiday shopping period of 1999. Access its Web site and read about the company's current delivery system. Based on what you read, would you be confident that the products you order today would arrive on time?

English Language Arts—Writing
Choose a product you often purchase at a bricks-and-mortar store. Draw a graphic of the channel of distribution that product probably follows from the manufacturer to you. Then, find a Web site that sells the same product online. Is the channel of distribution any different? Write a paragraph with an explanation.

English Language Arts—Speaking
In a team of three or four, choose one of these three shipping firms—U.S. Postal Service, United Parcel Service, or Federal Express. Then, imagine you have been asked to make a presentation that will convince your boss to select that shipper for your company's next big order.

Calculation
It costs a purse manufacturer $20 to make each purse. It sells each purse to a wholesaler at a 40-percent markup. The wholesaler sells each purse to an e-tailer at a 25-percent markup. The e-tailer sells a purse to you at a 50-percent markup. How much does it cost you for the purse?

Chapter 13 Review and Activities

Critical Thinking

1. Do you think the FTC should have fined several stores that were unable to deliver goods as promised during the 1999 holiday season? Explain.
2. When you purchase something online, how important is the distribution process to you? Explain.

COMPETITIVE EVENT

You are to assume the role of assistant manager of a successful music store. The company has a large main store and two branch stores. All three do a brisk business and enjoy a reputation for excellent customer service. The music store's online sales, however, are a different matter. The operation is inefficient and as a result shipping costs are out of control, and there are many customer complaints about late and incomplete deliveries. You have been hired to make the online portion of the business efficient and profitable. You are to make recommendations about improving the service in the online sales division to the store's owner (judge).

EVALUATION

You will be evaluated on how well you meet the following performance indicators:

1. Assess order fulfillment processes.
2. Analyze capabilities of electronic business systems to facilitate order fulfillment.
3. Analyze shipping needs.
4. Select best shipping method.
5. Use an information system for order fulfillment.

INTERNET ACTIVITY

To engage in online activities, visit the *E-Commerce* Web site at **ecommerce.glencoe.com**.

BusinessWeek online

TOPIC: Channels of Distribution

E-commerce has changed the familiar path goods take on their way to the consumer. Once the last link in a long chain of steps from the factory, consumers can now often contact manufacturers directly and buy products from them online.

ACTIVITY Go to the Student Center at **ecommerce.glencoe.com**. Click on *BusinessWeek* Activities and open Chapter 13. There you'll learn more about the channels of distribution, both old and new. You'll also see how businesses are changing the whole flow of merchandise due to e-commerce.

ecommerce.glencoe.com

Chapter 14

Customer Service and Web Site Personalization

Section 14-1
Providing a Customer Interface

Section 14-2
Customizing a Web Site

PREREADING STRATEGIES

Before you read this chapter, finish these statements in your journal:

- Based on this chapter title, I predict...
- The most important thing to remember about e-commerce sales and promotion is...
- Some questions I have about attracting and retaining e-commerce customers are:

BusinessWeek online

Go to *BusinessWeek* online, and check out its customer-service center. Make a list of the services it provides. With your classmates, discuss how these services might be beneficial to an online business.

ecommerce.glencoe.com

A Helping Hand for Customers

There is more to succeeding in business than making sales. A business needs to attract and retain customers. To ensure long-term success, e-commerce companies must make online shopping as pleasant, safe, and efficient as possible.

Drill Down

▸ **1892 Bright Idea**
Dr. Washington Sheffield, a Connecticut dentist, develops the idea of packaging toothpaste in a tube. Four years later, Colgate mass-produces toothpaste.

▸ **1902 Keeping Cool**
Willis Carrier produces the world's first modern air-conditioning unit.

▸ **1906 In an Instant**
While living in Guatemala, English chemist George Constant Washington creates the first instant coffee.

▸ **1947 Transistor Age**
Bell Laboratories physicists John Bardeen, Walter Brattain, and William Shockley invent the first transistor, paving the way for significant advances in computer technology.

▸ **1971 Storage Solution**
Alan Shugart heads a team of IBM engineers to manufacture the first floppy disk. For many years, the invention remains the primary method for storing computer information and files.

Get the Big Picture

The Internet has changed all aspects of business, including how businesses deal with customers. List ways in which shopping online differs from shopping in a traditional bricks-and-mortar store in terms of customer service.

POWER READ

Be an active reader and use these reading strategies:

PREDICT what the section will be about.

CONNECT what you read with your own life.

QUESTION as you read to make sure you understand the content.

RESPOND to what you've read.

Section 14-1

Providing a Customer Interface

AS YOU READ...

YOU WILL LEARN
- why good customer service is vital to e-commerce success.
- about the diverse ways in which customer service can operate.
- how Web sites strive to make online shopping pleasant, safe, and efficient.

WHY IT'S IMPORTANT

The goal of most e-commerce Web sites is to make the online shopping experience as pleasant, safe, and efficient as possible so customers will want to return. Some of the devices Web sites use to build customer loyalty take time and planning to implement and are unknown to the average shopper.

KEY TERMS
- intelligent agents
- semantic Web
- bots
- spiders

PREDICT

What's the most important consideration for online shoppers, and why?

THE IMPORTANCE OF CUSTOMER SERVICE

Customer service is as essential to e-tail success as it is to the profitability of bricks-and-mortar stores. However, the face-to-face contact customers experience in physical environments is impossible to recreate in virtual environments.

Online retailers have worked hard to overcome this disadvantage. In fact, many have succeeded by exploiting some of the Internet's built-in advantages, including convenience, speed, and innovative technologies such as intelligent agents and bots (which are explained later in this chapter). Personalized customer service—dealing with each online shopper as an individual—has helped many e-tailers win over consumers who were resistant to shopping online.

Building Customer Relationships

In 2003 eBay President and CEO Meg Whitman faced a daunting challenge. She needed to find a way to please eBay's customer base—the 85 million people who sold goods on the site. These sellers paid eBay to list their goods. Once an item was sold, they then paid eBay a percentage of the sale price.

On the other side of the auction block were eBay's 42 million shoppers. Since they didn't buy directly from eBay, but from the sellers, these shoppers weren't actually eBay's customers. Even so, Whitman wanted to make sure they were happy because unhappy shoppers wouldn't come back. If they didn't, the sellers would lose their customers, and eBay would lose much of its revenue.

That revenue was significant. In 2003 about $28 billion worth of merchandise was sold on eBay, and the sellers paid the company more than $2 billion. After expenses, eBay had more than $400 million left over—profits it needed to meet Whitman's goal of boosting revenue to $3 billion.

Whitman met the challenge by using some of the company's profits to expand the eBay community. She opened baby eBays in Asia, Europe, and South America and, perhaps more importantly, found new ways to strengthen the bond of trust between buyers and sellers. Today, experts say she made the right move; eBay is the most successful business on the Internet, increasing its revenues year after year.

For most online shoppers, a product's price is a secondary consideration. "Trust is the biggest thing," says Eric Brynjolfsson, director of the Center for eBusiness at the Massachusetts Institute of Technology (MIT).

Shoppers seek out online retailers that can be trusted to keep credit-card information confidential, ship orders promptly, and honor a reasonable return policy.

Brynjolfsson's research supports this observation. When he studied activity on a search engine that allowed shoppers to compare prices from various e-tailers, "less than half of the customers that used [the search engine] actually purchased from the cheapest retailer," he said. The majority bought from the retailers in whom they had the most faith. "Shoppers are reluctant to trust online retailers they've never heard of or dealt with before," he said. Like successful bricks-and-mortar businesses, online retailers strive to create that faith.

On eBay, every transaction between a buyer and seller is transparent—open for all to see. There are no secret deals. Before a sale, buyers and sellers can exchange e-mail and, since 2004, voice mail. After a sale, they can rate each other. Those who don't play by the rules—buyers who don't pay on time, for example, or sellers who take weeks to deliver merchandise—are likely to receive negative feedback, which is posted on eBay's site. **Figure 14.1** shows the type of information eBay posts for its customers.

Whitman also built trust by treating everyone equally. In 2004, manufacturers who sold their products on eBay wanted the right to pay lower commissions than individuals who sold secondhand merchandise. When she refused to meet their request, Whitman emphasized her commitment to her philosophy, explaining, "I am passionate about creating a level playing field."

CONNECT

Are you familiar with eBay? Have you ever considered bidding on an item you saw on the site? Why do you think so many people are intrigued by eBay?

Figure 14.1

Building Customer Confidence

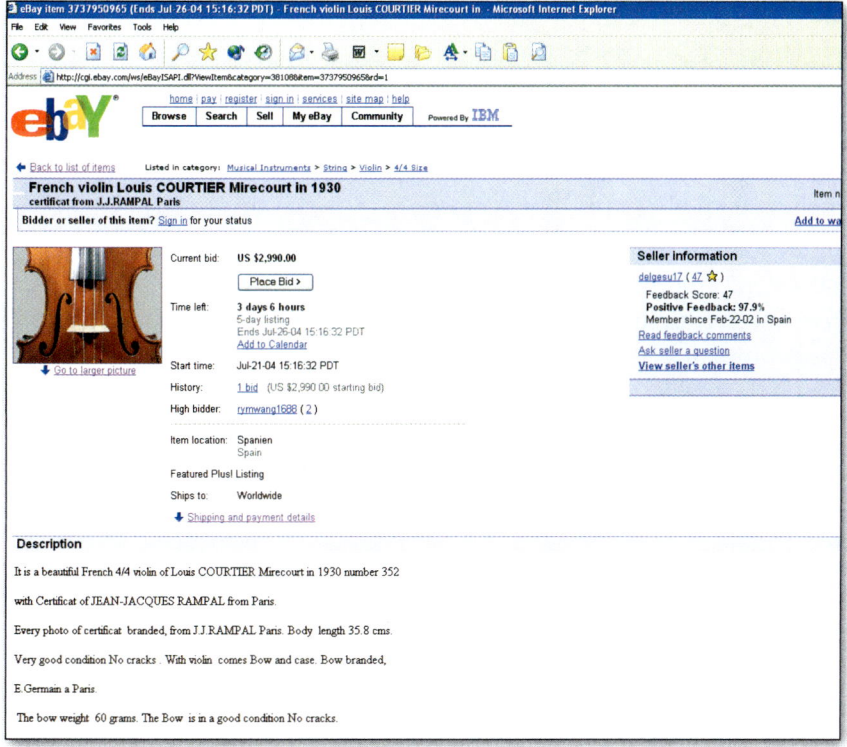

RATINGS GUIDE eBay reinforces customer confidence by keeping transactions between buyers and sellers visible for all to see. Buyers and sellers can exchange e-mail and voice mail, and rate each other when the transaction is complete. *How does posting customer and vendor ratings build confidence in eBay's auction business?*

ADD-ONS Many companies that produce electronic goods such as TVs and cars treat customer service as another product for sale. Customers are required to pay an added fee to receive guaranteed coverage for service and repairs. *What are the benefits for the company in using this approach? What are the drawbacks?*

CUSTOMER SERVICE How many times have you been lost in a grocery store you were visiting for the first time? When you're in unfamiliar territory, nothing seems to be in the right place. On the Web, savvy site designers don't let you get lost. Designing a site that is easy to navigate is essential to creating a satisfying shopping experience and boosting an e-business's bottom line.

Speed is important to shoppers, too. Popular sites, such as Amazon and eBay, enable consumers to get in and out quickly. On average, shoppers who enter these sites find what they're looking for and buy it in less than six minutes. In addition, well-run e-commerce companies deliver merchandise promptly. According to the e-tailing group, inc., an e-commerce consulting company, it usually takes the top companies no more than five days to deliver a purchase to a buyer's door.

Shop.org, an association of Internet retailers, once predicted that revenues from online sales would reach $60 billion in 2006. That's more than three times the amount e-commerce stores brought in five years earlier. This kind of robust growth depends largely on customer service.

CUSTOMER SUPPORT AND ONLINE QUALITY The quality of an e-retailer's site depends, to a large extent, on the organization's support department. Customers often judge a company by the assistance they receive, so support workers must be well trained to handle questions and complaints efficiently, effectively, and pleasantly.

Of the many support-team members, two have the most contact with customers:

- **Customer-Service Representatives.** They take orders over the phone and answer routine service questions. If necessary, they route calls to support technicians.

- **Support Technicians.** They provide solutions to problems customer-service representatives can't solve. Shoppers may be unable to navigate a retailer's site, for example, or get an order form to work properly. Because the same problems crop up again and again, most are easy to solve, either over the phone or in chat rooms. Though rare, customer queries can alert technicians to glitches in their Web site's design, which must be fixed quickly. Online shoppers are in a hurry, and sites that are sluggish or hard to navigate will drive them away.

Customer service can be a drain on a company's resources. To reduce their costs, some online businesses have *outsourced,* or contracted to an outside agency, their customer-support departments to countries like India, where salaries are a fraction of those paid in the United States.

INTELLIGENT AGENTS Have your friends or relatives ever explained how a gadget works, helped you search for a gift, or told you about a sale? Friends like these are not-so-secret agents. They act as your tutors, clerks, advisors, and bankers. These savvy people help you make smart decisions. **Figure 14.2** shows you how.

The Internet has its own agents that can help you in the way your friends do. These **intelligent agents** (also called *Internet agents*) aren't human; they're mini-programs that simplify the work you do online. Intelligent agents can insert your name, address, telephone number, and credit-card information into an order form. With intelligent agents, Barnes&Noble.com can remember you're a mystery buff and suggest new titles you might like to buy. Airline sites can remember you prefer window seats and vegetarian meals.

> **QUESTION**
> Why does the quality of an e-tailer's site depend on its customer-support department?

Figure 14.2

Intelligent Friends as Agents: How They Help You Make Smart Decisions

Consumer Decision-Making Tasks	Roles of Agents	What the Agents Do
Comparing Choices and Identifying Preferences	Tutors	Teach you about the features available in a product category and help you decide which features you prefer.
Searching for Information About Products and Alternatives	Clerks	Help you perform tedious tasks, such as searching for information and screening products.
Evaluating Products	Advisors	Provide expert opinions or advice that fits your particular interests or needs.
Purchasing Products	Bankers	Help you get a good deal on a product or service and may even suggest how to pay for it.

WORD OF MOUTH Friends, relatives, and acquaintances play different roles in influencing your shopping choices. *How are you affected by friends' negative reviews of products or services?*

Source: Adapted from "Agents to the Rescue?" by Patricia M. West, Dan Ariely, et al. *Marketing Letters* 10:3 (1999).

E-commerce business owners like intelligent agents because they provide an important competitive advantage. They create customer loyalty by tailoring the shopping experience to each person's interests and needs. Researchers have found that online shoppers will stick with stores that already know their names, addresses, credit-card information, and preferences.

Intelligent agents can simplify the shopping experience in other ways. For instance, they can screen out products you are unlikely to buy. Suppose you've bought several CDs that feature jazz great Thelonious Monk. An intelligent agent can quickly identify all the jazz artists who play Monk's compositions. You won't have to sift through the store's entire inventory of jazz CDs to find what you want; intelligent agents do tedious tasks like this for you. **Figure 14.3** shows you intelligent agents' capabilities.

In the future, an extension of the current Web may turn intelligent agents into geniuses. Its designers call this extension the **semantic Web**.

Figure 14.3
Capabilities of Electronic Agents

SALES ASSISTANCE An electronic agent is the online equivalent of a sales assistant. *Compare the services provided by electronic agents to the services you receive in a bricks-and-mortar store. Which do you prefer?*

The Goals of Intelligent Agents	How Electronic Agents Meet Each Goal
To help improve the quality of your decisions	• by helping you learn about a product category and decide what you prefer • by reducing the complexity of decision making • by acting as a clearinghouse for good information • by providing a way to evaluate different choices • by adapting to new information about you and feedback from you
To help to make shopping experiences more satisfying	• by remaining flexible and responsive to your requests • by helping you identify products that suit your tastes • by being simple to operate and user-friendly • by eliminating tedious work and allowing you to experience the "fun" of shopping • by setting realistic expectations
To help develop your trust in a seller	• by working to satisfy and protect your best interests • by reassuring you that you are getting the right product at a reasonable price

Source: Adapted from "Agents to the Rescue?" by Patricia M. West, Dan Ariely, et al. *Marketing Letters* 10:3 (1999).

"Semantic" is a fancy way of saying "related to meaning." Programmers trying to create the semantic Web believe that one day they will be able to attach a tag, or a precise meaning, to every bit of information online. If the scheme works, unrelated data could be linked in unexpected ways.

Tim Berners-Lee, the inventor of the semantic Web, gives an example of a teenage girl using it to find and make appointments with two medical specialists. An intelligent agent could tap into several lists of doctors, identify the ones who accept her mother's insurance, and locate the doctors whose offices are no more than 20 miles from her home. The agent would even scan her schedule to find the most convenient time for an appointment.

Many experts say the semantic Web is an impossible dream. Berners-Lee isn't one of them. He is convinced he and thousands of others, working independently, can make this dream come true.

AGENTS AS BOTS Intelligent agents can be sent on missions—to find the lowest price for a product, for example, or to figure out when a friend is online. The intelligent agents already discussed simplify the task of online shopping. **Bots** are more dynamic. Their major use is in mining the Web for information. The word "bot" is short for "robot," which comes from the Czech word "robota," meaning "work." Bots keep working even when your computer is turned off. They scan the Web continuously, looking for computer-program updates, new-product alerts, price changes, and the like. Tell them which information you want, and all by themselves they will scan the Web, find what you want, and report back to you.

Search engines such as Google program bots called **spiders**, or *crawlers,* to sift through billions of Web pages. They crawl through the Web, compiling extensive lists of URLs that you tap into whenever you ask for information.

Bots aren't all the same. *Chatbots* installed in cars and linked to global-positioning satellites tell drivers when to turn right or left. *Knowbots* (or *knowledge robots*) can be programmed to automatically collect news about a particular topic, such as a medical condition, a favorite rock band, or an international bicycle race. *Shopbots* compile databases of products sold at online stores, enabling shoppers to compare prices and features in seconds. A Web site called BotSpot.com lists hundreds of these complex intelligent agents.

COOKIES Intelligent agents depend on files known as *cookies* to remember who you are. (You first learned about cookies as an e-marketing device in Chapter 12.) Cookies keep track of information such as your name, preferences, age, and address. This data is stored in a cookie file in your browser. If your browser doesn't accept cookies, intelligent agents can't do their jobs.

The word *cookie* is an old computer-science term that refers to stored information. Cookies were invented to help make it easier to log onto Web sites. They permit you to enter a site without going through the boring process of identifying yourself every time you log on.

CONNECTION: LANGUAGE ARTS

Worthless Uptime
Due to errors on their sites, e-tailers are suffering from abandoned carts in the virtual aisles. Consumers spend an average of seven minutes coping with a Web site's errors. An e-tailer may think its site works fine because it uploads successfully and may be unaware of problems on the other end. *What kinds of technical errors can occur online that might drive off customers?*

Thanks to cookies, Amazon can remember you like poetry, and Lands' End can remember your dress size. An auction site can remember you're a regular visitor who has bid several times for videos of classic movies from the 1930s and 1940s.

Since cookies are inserted in your browser without your knowledge, your browser may contain some you don't want. Advertising agencies sometimes slip cookies into your browser to discover your interests. If the agency learns you've visited a diet site, for example, you might begin getting a lot of pop-up ads and spam that promote diet plans.

In many ways, a cookie acts much like an intelligent agent. Since it knows who you are, a cookie makes it much easier and faster for you to surf a Web site you've visited before. Unlike an intelligent agent, however, cookies are designed to be more helpful to Web-site developers than to customers.

Section 14-1 Review

RESPOND to what you've read.

1. What does trust mean to online shoppers? What are some examples of this type of trust? _____

2. What are intelligent agents, and how do they help create customer loyalty? _____

3. What is a bot? _____

4. What is a cookie, and what job does it perform? _____

Quick TALK With a partner, discuss whether you would rather search for products by yourself or have an intelligent agent recommend them for you. *Discuss why you feel this way and the advantages and disadvantages you see for each method.*

Section 14-2

Customizing a Web Site

SERVING PEOPLE FIRST

Personalization is the key that unlocks online shoppers' wallets. It is not a new concept. The recommendations of sales clerks have prompted shoppers to buy for centuries. E-commerce has changed the rules, however. Online shoppers can't look sales clerks in the eye. Product recommendations are made automatically by a computer program rather than a human being. What can e-tailers do to acquire customers and make them feel their business is appreciated? This section will answer these questions.

Personalization of Online Customer Service

A happy customer is a repeat customer. There is a good reason for making the shopping experience rewarding for online consumers: Good service sways customers more than price does. **Figure 14.4** on page 292 shows some of the services an e-business should provide its customers.

E-tailers refer to gathering information on customers to identify and target them as **profiling**. The better the customer profile, the better able the retailer is to customize its interactions with that customer.

This intense focus on the consumer is called *customer relationship management* (CRM; see also Chapter 13 for E-CRM). The goal of CRM is to increase customer satisfaction by dealing with each person as an individual. The Internet has given e-businesses an advantage over physical stores because virtual stores have something physical stores don't—powerful software designed to personalize services, provide various customer-support choices, and track customer satisfaction. E-businesses have a number of high-tech strategies to promote customer service:

- **Knowledge Management** for acquiring information about the customer, analyzing it, and keeping tabs on the customer's satisfaction.

- **Database Consolidation** for putting information about a customer in one place, generally a single database. This way, all interactions with a customer can be quickly customized for that customer.

- **Customer Choice** for letting customers choose the way they prefer to receive customer support, whether by e-mail, phone, or online chat. Customers can also decide to subscribe or unsubscribe to electronic newsletters that promote a particular site's products.

The goal of CRM is to trigger sales by satisfying customers. Online businesses can find out how well they are succeeding by counting the number of complaints their customer-service representatives receive.

AS YOU READ...

YOU WILL LEARN
- which methods businesses use to acquire and keep customers.
- about the role marketing plays in e-commerce.
- how e-tailers set prices for the products they sell.

WHY IT'S IMPORTANT

E-commerce businesses use many strategies to promote customer loyalty. Personalized service and quality customer service are expected by customers in traditional stores, but are more difficult to implement online.

KEY TERMS
- profiling
- unique selling proposition
- frequently asked questions
- chat rooms
- message board
- real-time pricing
- bundling

PREDICT

What are some questions customers might ask themselves when visiting a Web site for the first time?

Figure 14.4

Customer Services

E-TAIL BASICS Research by the e-tailing group, inc. suggests that, at a minimum, e-commerce businesses should provide the services described. *Is there anything you would add to this list?*

An E-Commerce Business Should . . .

- enable customers to complete the shopping process on the first try.
- design its site to enable time-starved customers to enter and exit quickly.
- supply a toll-free phone number so that customers can get prompt, personal assistance.
- respond to e-mailed customer questions within 24 hours with a personalized response.
- provide shopping carts or product pages so customers can keep track of their orders as they make them.
- make sure purchases are delivered promptly.
- use a shipping method that allows customers to track or check the status of a shipment.
- send customers e-mailed confirmations of their orders and include order numbers.

A MARKETING STRATEGY Marketing involves selling a product in a marketplace—in this case, the world of online shoppers. Shoppers who are satisfied with a particular online business rarely stray from it. One study found book buyers visited 1.1 online booksellers each month. Obviously, it's important to e-tailers to please their loyal customers. "If managed and maintained," says David Schmittlein, Professor of Marketing at the Wharton School of the University of Pennsylvania, "customer groups cannot be easily copied by the competition."

One thing that can be copied is a company's product line. Amazon sells essentially the same books as Barnes&Noble.com. To compete, companies need to differentiate, or distinguish, themselves from other companies. A major way in which they can do this is through customer service. For example, one e-tailer might offer a refund policy that is faster, easier, or cheaper than another. Some companies will take back any product a customer is dissatisfied with for a fee. Bed Bath & Beyond includes prepaid return shipping labels with every online purchase and accepts refunds and exchanges at any of its stores. (See **Figure 14.5**.) A good marketing strategy to distinguish your company in the e-commerce marketplace should include a good customer-service strategy.

CONSUMER BENEFITS Every business needs a **unique selling proposition** (USP), which is one thing that makes it stand out from the competition. The Web is a crowded place. A USP describes the benefits of buying from one retailer rather than another. Bricks-and-mortar stores have an advantage here. Often, they have to be unique only in their geographical area. E-commerce retailers have to be unique nationally and even globally.

An online merchandiser might offer a wider range of goods than its competitors. The Web site might be more user-friendly and offer customer service in two or more languages. An e-commerce retailer might offer rock-bottom prices. As we have learned in earlier chapters, however, price is not as important as service to online shoppers.

IMPLEMENTING A PERSONALIZATION APPROACH An implement is a tool. To personalize its approach to customers, a virtual business must develop tools to achieve its goals. An obvious tool is the user interface—the way in which the consumer interacts with a Web site. A personalized Web site anticipates the customer's questions: *How do I shop here? How do I look for products? What do I need to know about the products before I buy? How do I know I can trust this merchant?* Smart merchandisers put the answers on their home pages. They also build these answers into their system in a way that leads customers, step-by-step, to pick out a product, put it in their shopping cart, and purchase it.

Another tool that addresses shoppers' needs is an easy way they can get answers to their questions. Online businesses usually recommend users first check their **frequently asked questions** (FAQs) list. If a FAQ doesn't answer their questions, customers are invited to communicate directly to a customer-service representative by e-mail or telephone. The list of options reassures customers they are dealing with a company that cares about them as individuals.

A popular way to personalize the online shopping experience is a greeting to a returning visitor: "Hello, Jan." Behind this greeting is a lot of data about Jan: address, phone number, e-mail address, age, and purchasing

Figure 14.5

Return Policies

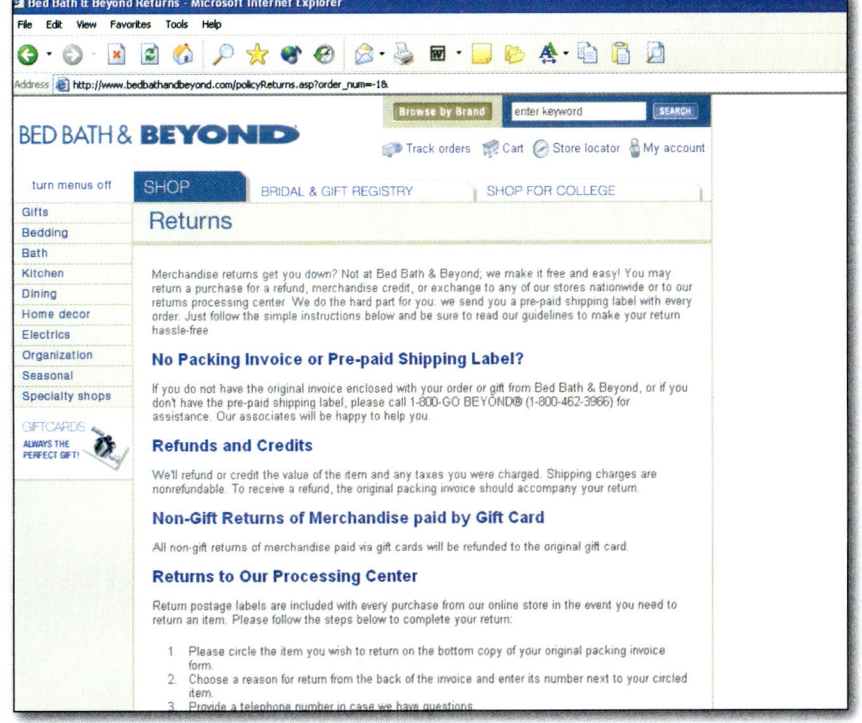

MANY HAPPY RETURNS
Bed Bath & Beyond allows customers to return items free by coming into one of its stores or using prepaid shipping labels. Other businesses require customers to pay a fee to return an item. *Would a return policy of this sort encourage you to shop online?*

> **CONNECT**
> Have you ever used FAQs, chat rooms, or message boards on the Internet to solve a problem or answer a question? Which means do you prefer, and why?

preferences. This allows the merchandiser to recommend products in which Jan might be interested, further personalizing the approach.

These personalized features make online shoppers feel like special guests. That's important, because visitors are more inclined to buy products from a site that "knows" them than one that treats them like strangers.

SOLVING PROBLEMS ONLINE Software programs and advances in technology enable companies to provide customer support through their Web sites. **Chat rooms** are Web pages that allow online shoppers to have real-time, virtual conversations with customer-service representatives. Unlike e-mail requests, which can take up to 24 hours to be answered, "live chats" bring instantaneous assistance. After phone contact with customer-service representatives, chat rooms provide the most personal contact between an online store and its customers.

A **message board** is a page on a Web site where users can share information about common problems. Message boards are virtual bulletin boards. They permit users to scan through questions and responses from other users, add their own questions or answers, and post notes on other subjects. Online businesses like message boards because customers can solve their problems at no cost to the company.

FAQs and message boards are efficient, cost-effective ways to provide customer support. They help reduce the number of questions customer-service representatives have to answer. At the same time, they give customers immediate support for their problems.

Internet Pricing Models

Pricing is a crucial part of a business's strategy. A product's price must cover business expenses and realize a profit. Thanks to the Web, companies can quickly adjust their pricing policies to meet changes in the marketplace—a competitor's lower price, for example, or a drop in demand. Some of the models e-commerce companies use when pricing their products are promotional pricing, real-time pricing, and bundling.

> ▶ **CRAIGSLIST** Craig Newmark created a Web site for sharing information about events in San Francisco. When people around the country started posting job listings and items for sale on the site, Newmark modified his site to satisfy their needs. Today, CraigsList charges fees only for job listings and provides online community services for free. *Visit the CraigsList Web site at CraigsList.org. Other than products for trade and job listings, what other useful customer services are available?*

Web Site Success

SAFETY IN NUMBERS
Shopping.com

Concerns about online shopping have drastically decreased with improvements in customer service, privacy, and return policies. One online axiom remains true: Shoppers don't like to buy from e-tailers that aren't well known. This is where Shopping.com comes in handy. The company was formed in 1999 by combining DealTime, a leading shopping search engine, with Epinions, a popular consumer-review ratings service. Shopping.com lets customers easily find the best possible prices by comparing products, prices, and stores on the Web. A detailed customer-support page answers a variety of questions and also provides a simple contact form if the user cannot find an answer to a particular question.

Thinking Critically
There are potential drawbacks to making an online purchase from an unfamiliar retailer. *What is the worst that can happen?*

PROMOTIONAL PRICING Like bricks-and-mortar stores, e-retailers use a variety of pricing methods to promote products. Both use clearance or closeout sales to sell excess inventory or discontinued products. These sales help to keep a company's inventory fresh and make room in the warehouse for new products. Both types of retailers use newspaper, radio, and television advertising to promote sales.

In addition to conventional methods, online businesses have methods to promote prices unique to e-commerce. They can use eye-catching pop-up ads that link customers to their sites. They can also send personalized e-mail offering promotional prices to their current customers.

REAL-TIME PRICING The price a business should charge for a product can change quickly due to changes in supply or demand, promotional sales by competing companies, or other factors. The ability to change prices instantly to keep up with changes in the marketplace is called **real-time pricing**.

When it comes to real-time pricing, e-tailers have a strong advantage over bricks-and-mortar retailers. Because bricks-and-mortar retailers usually post printed advertisements to promote prices, they have to plan their sales in advance. Through the use of technology, e-tailers can change and post prices instantly at very little cost. Their ability to use real-time pricing gives e-tailers a competitive edge.

BUNDLING A pricing strategy used by traditional retailers that is especially popular with e-tailers is **bundling**. Bundling consists of combining related products or services, which businesses then offer to customers as special package deals. Bundling is typically used for products or services that are in high demand. Online travel agencies regularly

QUESTION
Why is it easier for e-tailers to use real-time pricing than it is for bricks-and-mortar retailers?

Chapter 14 Customer Service and Web Site Personalization **295**

sell bargain packages that include airline tickets, hotel rooms, and rental cars. If you're shopping for a computer system, it is often less expensive to buy a package that includes the CPU, the monitor, and a printer than it is to buy the components individually. Bundled packages save customers money. They also help businesses by getting customers to buy more products.

Section 14-2 Review

RESPOND to what you've read.

1. What is CRM? What are some strategies used by CRM to learn more about customers?

2. What are some ways in which online businesses can provide customer support?

3. What is a unique selling proposition, and why is it essential for an e-commerce business?

4. What is bundling, and how does it help businesses make more money?

Quick TALK With a partner, discuss how each of you would order the following words in relation to the online shopping experience: *trust, service, convenience,* and *price.* **Discuss why each of you chose that order.**

Unit 5 Marketing in the Digital World

Name _____ Date _____

Worksheet 14-1

Secret Agents?

There are agents on the Internet that help you in a variety of ways. These nonhuman agents perform many services, from remembering your interests and previous purchases to locating the lowest prices on a product you want to buy.

1. Using the Internet, conduct research on intelligent agents. Choose three of the following terms, define them, and describe their purpose:

   intelligent agent   infobot
   software agent   chatbot
   softbot   shopbot
   knowbot

2. Create a one-page fact sheet about the three intelligent agents you researched. Use a word-processing program, and include graphics if possible.

3. Locate a shopbot, and use it to compare prices on a product you want to buy. Describe your findings.

   _____
   _____
   _____
   _____
   _____
   _____
   _____
   _____
   _____

4. If you could create a bot, what would it do? What would you call it? Explain in detail.

   _____
   _____
   _____
   _____
   _____
   _____
   _____
   _____
   _____

Name _____ Date _____

Worksheet 14-2

Customer Service at School

Personalized service and quality customer service are expected by consumers, whether they are shopping in a traditional bricks-and-mortar store or online. Web sites that provide services other than retail will also want to do whatever it takes to make sure visitors return again and again.

1. Think about your school Web site. What customers does it serve? (If your school does not have a Web site, choose a college Web site to review.)

   _____
   _____
   _____
   _____

2. Create a list of five FAQs and answers for the school's site. Remember, they must be questions that the site's customers (students, parents, and teachers) would be likely to ask.

   _____
   _____
   _____
   _____
   _____

3. What are some ways in which you could personalize your school's Web site for its users?

   _____
   _____
   _____
   _____

4. Describe how you might add a chat room or message board to your school's Web site to improve service to its customers.

   _____
   _____
   _____
   _____
   _____

Chapter 14 Review and Activities

Chapter Summary

Section 14-1

- Customer service is essential to e-commerce success because there is no opportunity for face-to-face contact with the customer online.

- Online businesses use the convenience, speed, and technologies offered by the Internet to provide personalized customer service.

- Web sites strive to make the online shopping experience pleasant, safe, and efficient by providing superior customer support.

Section 14-2

- Successful e-tailers customize their interactions to suit the needs and desires of individuals because a happy customer will be a repeat customer.

- Marketing plays an important role in e-commerce. Online businesses need to differentiate themselves from their competitors and market those differences.

- E-tailers use promotional pricing, real-time pricing, and bundling to set prices for the products they sell.

Chapter 14 Review and Activities

Review of Key Terms

a. frequently asked questions
b. bots
c. bundling
d. semantic Web
e. real-time pricing
f. chat rooms
g. intelligent agents
h. message board
i. profiling
j. spiders
k. unique selling proposition

Match each term to its definition.

_____ 1. In the future, this may become an extension of the current Web.
_____ 2. Combining related products or services in one package.
_____ 3. Programs that mine the Web for information such as computer-program updates and price changes.
_____ 4. They crawl through the Web compiling lists of URLs.
_____ 5. A page on a Web site that allows users to share information about common problems.
_____ 6. One way in which online shoppers can find answers to questions.
_____ 7. The ability to change prices instantly to keep up with changes in the marketplace.
_____ 8. Internet agents that are not human.
_____ 9. Web pages that allow online shoppers to have real-time conversations with customer-service representatives.
_____ 10. Gathering information on customers to identify and target them.
_____ 11. The one thing that makes a business stand out from the competition.

Applying Technology to Academics

English Language Arts—Reading
Using the Internet, find at least three competitors for each of the following online businesses:
- eBay.com
- Amazon.com
- LandsEnd.com
- Cooking.com

English Language Arts—Writing
Choose a competitor of one of the businesses from the above list. Discuss how the companies distinguish themselves from each other. Describe how the products and services they offer are similar and different.

English Language Arts—Speaking
Using the Internet, research the semantic Web. In a group of two or three, define it, describe who is developing it, and discuss what it will be able to do. Create a short oral presentation and present it to the class.

Graphing Software Program
Make a list of five CDs you want to buy. Go on the Internet and find three online stores that carry those CDs. Look for any promotional sales, package deals, or other incentives to buy. Add up the cost of the CDs plus shipping and handling charges for each of the online stores. Using a graphing software program, create a chart comparing prices at each of the stores to determine which offers the best prices.

Chapter 14 Review and Activities

Critical Thinking

1. To save money on salaries, should online businesses in the U.S. be allowed to outsource jobs to other countries where wages are lower? Explain why or why not.
2. What helps shape your buying decision when making a purchase online?

COMPETITIVE EVENT

Assume the role of Web manager of a luxury chain that sells its own handbags, scarves, and related accessories. Your stores are known for outstanding customer service and quality products. The company is planning to introduce a new line of products and hopes to build demand for it through the company's Web site. Make suggestions to the company's merchandise manager (judge) about how to promote the new merchandise online.

EVALUATION

You will be evaluated on how well you meet the following performance indicators:

1. Describe unique aspects of Internet sales.
2. Develop a plan for selling online.
3. Determine strategies for online customer support.
4. Explain the use of brand names in selling.
5. Develop a plan for online promotional selling.

INTERNET ACTIVITY

To engage in online activities, visit the *E-Commerce* Web site at **ecommerce.glencoe.com**.

BusinessWeek online

TOPIC: E-Commerce in China

Despite its huge population, China currently has a relatively small percentage of Internet users. Nonetheless, with the trend of Internet use growing in China, the country presents great opportunities for e-commerce.

ACTIVITY Go to the Student Center at **ecommerce.glencoe.com**. Click on *BusinessWeek* Activities and open Chapter 14. There you'll learn more about the current state of e-commerce in China and its potential in the future.

Chapter 15

Advertising for E-Commerce

Section 15-1
The Basics of Building an Online Brand

Section 15-2
Advertising Your Web Site

PREREADING STRATEGIES

Before you read this chapter, finish these statements in your journal:

- Based on this chapter title, I predict...
- The most important thing to remember about advertising's role in e-commerce is...
- Some questions I have about advertising and e-commerce are:

BusinessWeek online

Go to *BusinessWeek* online to find examples of branding, sticky content, and dynamic content. In your journal, define these terms, and list at least three examples of each.

Becoming a Brand Name

E-commerce companies seeking to become the brand of choice in a product or service category must work to establish distinct identities within the markets they serve.

Drill Down

1776 Under the Sea
Yale graduate David Bushnell designs and launches the first submarine, the *American Turtle*. The submarine remains a vital tool for scientific exploration and military pursuits around the world.

1791 Start Your Engines
British inventor John Barber patents the first gas turbine, which ultimately allows for fuel engines in automobiles, airplanes, and other vehicles.

1908 Protect and Preserve
Swiss textile engineer Jacques Brandenberger creates cellophane and transforms the world of food preservation and storage.

1922 Diabetes Discovery
Canadian surgeon Frederick Banting and his assistant, Charles Best, discover insulin and ultimately save the lives of millions of people suffering from diabetes.

1955 Remote Control
Zenith Electronics engineer Eugene Polley develops the first wireless remote control for television sets by creating the "Flash-Matic."

1972 At-Home Arcade
Atari engineer Nolan Bushnell creates the video game Pong, which becomes a national phenomenon and begins a new worldwide pastime.

Get the Big Picture
What can you do to create a strong business and distinguish it from the competition?

POWER READ

Be an active reader and use these reading strategies:

PREDICT what the section will be about.
CONNECT what you read with your own life.
QUESTION as you read to make sure you understand the content.
RESPOND to what you've read.

Section 15-1

The Basics of Building an Online Brand

AS YOU READ...

YOU WILL LEARN
- why branding is important to a company and its products.
- how businesses strive to get customers to make purchases on their Web sites.
- why dynamic content is important to a Web site's appeal.
- how and why businesses assess traffic at their sites.

WHY IT'S IMPORTANT
It's not enough for companies to simply create products and make them available to customers. Creating positive feelings about their products and nurturing relationships with customers can ensure brand success and longevity. Businesses use a variety of methods to determine whether or not their brands are selling successfully, and what customers are interested in buying.

KEY TERMS
- brand
- branding
- sticky content
- dynamic content
- unique visitors
- site traffic
- exit page

BRANDING

In the most technical sense, a **brand** is a name, design, symbol, or other feature that distinguishes your products and services from those of other companies. Nike, J.Crew, and Hershey Foods are all examples of successful brands. Brands can encompass entire product lines, single products, various companies, or retail stores. Target is the brand name for a retail store (Target Corporation) that sells numerous other brands. Best Buy is another.

Many brands can also be encompassed under one corporate umbrella. For example, Procter & Gamble produces several popular laundry products, each with its own product line. Tide, Downy, Cheer, and Bounce are all part of the Procter & Gamble family. Though they function as individual brands, they also benefit from the Procter & Gamble affiliation. The Kellogg Company is another example: Apple Jacks, Pop-Tarts, Special K, and Morningstar Farms are all Kellogg brands.

A brand is much more than just a name. It reflects your company's personality, values, and vision, and includes everything customers associate with your products, services, or company—including memories, emotions, packaging, and even Web-site design. Companies want to establish positive personal relationships, and the process of creating these relationships is called **branding**.

As with every element in the marketing mix, branding is all about your target market. Citibank launched a humorous yet memorable advertising campaign featuring frazzled victims of identify theft. According to the ads, these unfortunate victims did not have the protections Citi® Cards users enjoy. The idea was to associate safety and security with the Citi® Cards name in the minds of consumers. Once customers believe that idea, the card has been branded. People are no longer applying for just another credit card; they are applying for security.

Amazon.com emphasizes the conveniences you can't find in a bricks-and-mortar store: the world's largest selection of books, one-click ordering, and personalized recommendations. This strengthens its image as the premier online-shopping destination and gives people a stake in its success. The company wants people to think, What would I do without Amazon? This is not the same as thinking, What would I do without all those books?

APPLE iPOD When Apple's iPod became a big success in 2004, sales for the entire Apple product line increased as well. *Explain how this might illustrate the effects of branding.*

Brands also appeal to social and cultural values. American Apparel is proud of its "Made in the USA" labels, and its socially conscious customers are committed to the brand for that reason. When they purchase t-shirts or yoga pants, they are not only buying the garments, they are buying into an idea.

To create these kinds of relationships, you have to get customers to know you by name. The following sections show you how to capture their attention and keep it.

The Battle for Attention on the Web

Internet users are overwhelmed with choices. You can now find almost anything online—groceries, pet food, clothing, books, electronics, and cell phones. Hundreds of sites compete for your attention during every excursion into cyberspace. As consumers become more tech-savvy, zapping pop-ups and junking spam, businesses must work harder to grab their attention.

Amid this flurry of e-commerce, perhaps no site has succeeded like Amazon, which has built its name on the convenience and selection made possible by online shopping. Today, "Earth's Biggest Selection" and "One-Click Shopping" are synonymous with e-commerce for many consumers. How did this happen? Amazon founder and CEO Jeff Bezos did not simply build his virtual bookstore and sit back, waiting for customers to show up. He used a combination of advertising, promotions, word-of-mouth, Web design, and customer service to build a strong brand.

Amazon started out as a bookstore, so Bezos needed to attract readers to his site. If he could get them to visit just once and experience the site's vast selection and convenience, he knew they would come back for more. He understood it was not the books he was selling; it was the Amazon brand.

CREATING A STICKY SITE Once customers click onto your site, you must entice them to stick around long enough to make a purchase. Remember, it is not very hard for customers to leave; they can click over to the competition in a matter of seconds to compare prices, selection,

PREDICT

Why might an e-commerce company spend extra money to maintain free interactive games and features on its Web site?

CONNECTION
MATH

A One-Click Wonder

Typically, you only have to click once in an e-mail to get the information the sender intended for you to read. DoubleClick analyzed retailers and catalog companies' e-mail marketing campaigns. For every 1,000 pieces of e-mail sent, only 2.65 purchases resulted. In 2002 the average order size was $102, and in 2003 the average order size was $98. *What is the average percentage of decline over the two years? Why would it decline?*

Figure 15.1

Interactive Features

STICKY FEATURES Interactive games are effective at keeping viewers from leaving a site. They also provide an opportunity to present viewers with product advertisements. *What sorts of features might convince you to "stick around" a Web site?*

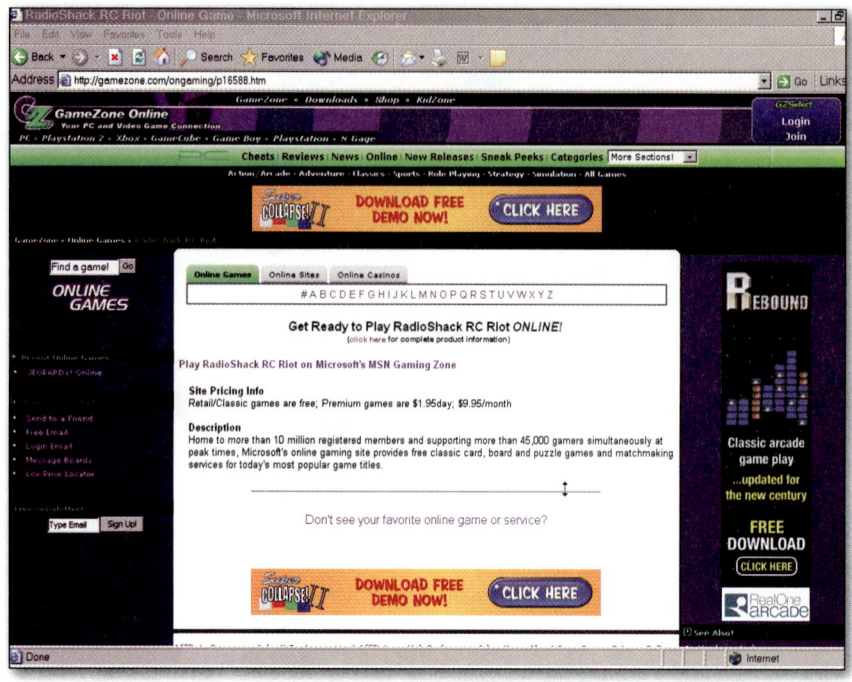

CONNECT

What is your favorite brand of sneakers? Would you consider buying them over the Internet?

and shipping rates. To keep customers' attention and to keep them from clicking away, many e-commerce companies have enhanced their Web sites with sticky content.

Sticky content is information or features that entice visitors to stay for long periods of time and return frequently. There are several ways to make your site sticky: chat rooms, localized weather reports, personalized news, games, user reviews, 3-D content, and frequently updated products or promotions. The concept is simple: If people enjoy the experiences a site provides, they will spend longer periods of time there. Also, once they have taken time to register or learn games, they are much more likely to return.

RadioShack.com is an excellent example of a sticky Web site. (See **Figure 15.1**.) RadioShack Corporation teamed up with MSN to create RC Riot, an interactive online game in which players race cars in several virtual environments—a backyard, a playground, and a RadioShack store. At the end of the game, players can click to the RadioShack site. According to MSN, the site receives up to 70,000 unique visitors each day during promotional periods.

Another way e-commerce companies create sticky sites is with dynamic content. **Dynamic content** is information that frequently changes or is updated. On ESPN.com, users can find up-to-the-second scores from their favorite sports, read articles, and find breaking news about players. They can also enter contests and shop for t-shirts, hats, and tickets for upcoming games. The content keeps visitors glued to the site and entices them to return after they leave. Meanwhile, the online store scores a slam dunk.

Your site must constantly adapt to changing market conditions to stay sticky. Imagine a gym that never invests in new machines. The rickety cables and tears in the seats will drive customers away, as will the outdated dance moves in the aerobic classes. The same is true online. Customers want sites that are easy to navigate, easy to find, and meet all of their needs. Since those needs change over time, they also want a site that can change with them. Fortunately, the very server that hosts your site can produce information to help you do this.

EVALUATING TRAFFIC SOURCES In the digital world, every action leaves a footprint, and every user leaves a trail. When visitors connect to a Web site, the host server immediately records several pieces of information: unique IP (Internet Protocol) address, computer platform, browser type and version, time of connection, referring URL, page views, and cookies. These records are called the *log files*. You can use them to improve design and functionality, as well as determine the best places for advertisements, as illustrated in **Figure 15.2**.

Figure 15.2

Log File Users

SITE FEEDBACK *Different people in a company who work on a Web site are able to use different information from log files to improve the Web site's performance.* *What are some specific examples of information available in log files?*

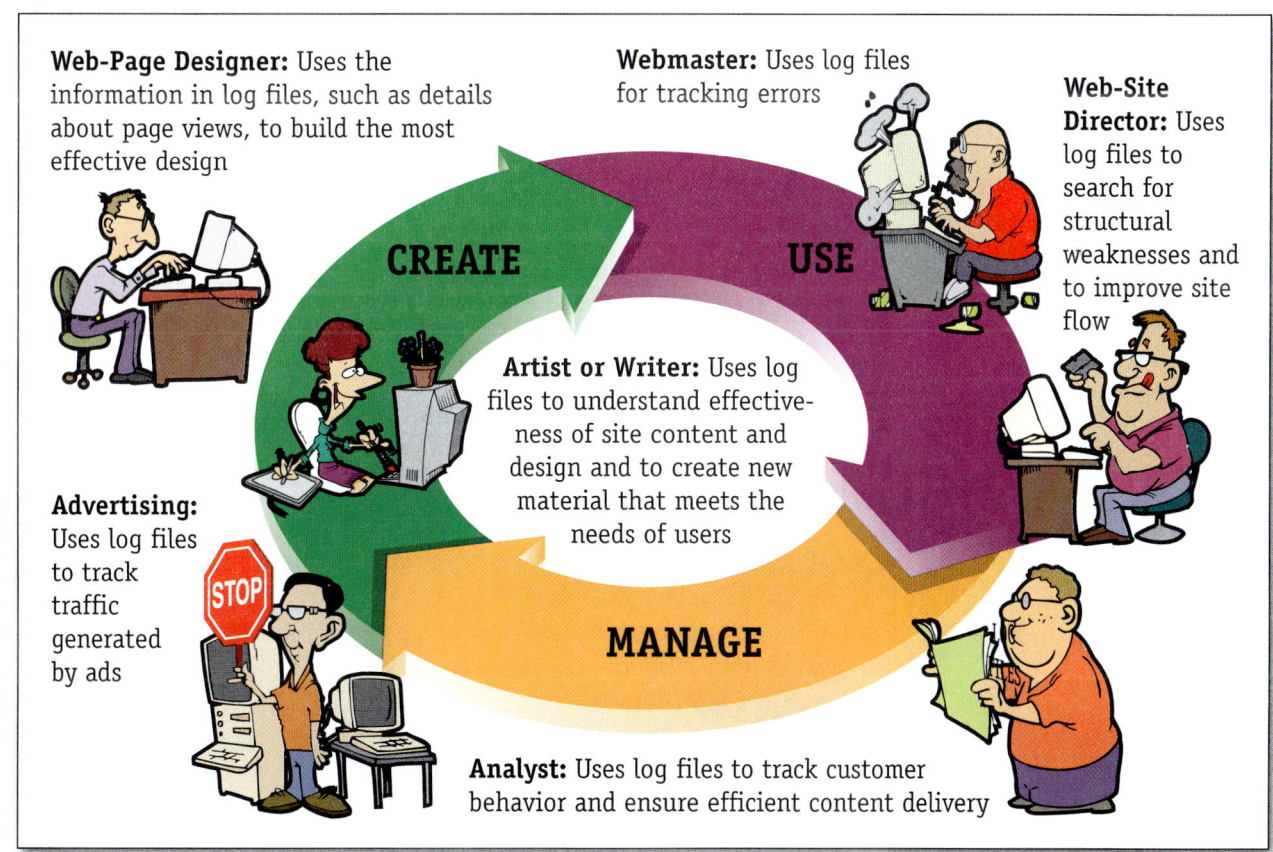

Web-Page Designer: Uses the information in log files, such as details about page views, to build the most effective design

Webmaster: Uses log files for tracking errors

Web-Site Director: Uses log files to search for structural weaknesses and to improve site flow

CREATE

USE

Artist or Writer: Uses log files to understand effectiveness of site content and design and to create new material that meets the needs of users

Advertising: Uses log files to track traffic generated by ads

MANAGE

Analyst: Uses log files to track customer behavior and ensure efficient content delivery

To use log files, you must understand the information they contain. Whenever a browser connects to a site, one hit is recorded in the server log. A *hit* is any single request to a server, including requests for images, text files, or HTML pages. Since many pages contain several images or files, loading just one page can produce many hits. Consequently, the number of hits does not tell you much about how many people actually visit your site.

Much more important is the number of **unique visitors**. This is the number of individual IPs connecting to your site. It can give you a much clearer picture of how many people actually visit. It is important to remember, however, that even this number is not exact because most Internet users are assigned a random IP every time they connect. What may appear to be several unique visitors may actually be only one person. There are some users with static IPs, but you cannot assume most users have this. You can use the number of unique visitors to track spikes and dips in **site traffic**, which is the total activity on a site. Did your traffic increase on the day of your big sale? If so, then you know your advertising campaign—and your sale—was a success.

Log files also contain referring URLs and search terms, which tell you exactly where customers were surfing when they found your site. If your online store sells juniors' apparel and a high percentage of visitors arrive by searching for cargo pants, your next promotional campaign might be related to that product. If those searches were made on MSN.com, you might also consider placing an ad on the MSN shopping portal.

Some of your advertisements will be more successful than others, and this will be reflected in the log files. If more people are clicking the banner on the *Seventeen* magazine Web site, you should invest more advertising dollars there and in similar places. You might even consider placing a print ad on Seventeen.com to reach more readers.

Log files reveal more than just raw data. They can also reveal patterns, such as how customers actually move through your site. Every time someone loads a page, the log files record a one- page view. The sequence and timing of page views, otherwise known as the clickstream, will tell you which pages customers view first, which links they click, and the amount of time that elapses in between. Are they clicking on pictures? Which categories are most popular? Which pages seem to hold their attention?

You can even find the most common **exit page**, or the last page viewed as part of a visit. Is there something specific about this page that drives users off the site? Perhaps customers leave at the exact moment you ask for their credit-card information. Could it be that your privacy policy is unclear? Perhaps your shopping cart does not seem secure, or shipping is too expensive or slow.

Strategic use of the log files will help you respond to customer needs, making it possible to offer the right design and content to attract visitors and keep them there. This, in turn, will help you build a strong brand.

QUESTION

Why is information about customers' exit pages important to an online business?

Section 15-1 Review

RESPOND to what you've read.

1. What is branding? Why is it so important to businesses?

2. Why must Web sites adapt quickly to changes in market conditions?

3. Why are Web sites concerned about how many unique visitors they have?

4. What is "dynamic content"? What is an example of this type of material on a Web site?

Quick TALK With a partner, pick the two Web sites each of you visit most often. *List the specific reasons you keep coming back to your favorite sites and which dynamic content you look forward to accessing.*

Section 15-2

Advertising Your Web Site

AS YOU READ...

YOU WILL LEARN
- why advertising is important for a successful Web site.
- which strategies and techniques companies use to advertise their products and services on the Internet.
- why some types of online advertising can turn potential customers off.

WHY IT'S IMPORTANT

Businesses face unique obstacles when advertising on the Internet. While direct online advertising is vital to a business's success, there are other methods companies employ when attempting to reach customers.

KEY TERMS
- spoof e-mail
- banner ad
- skyscraper banner
- opt-in
- opt-out
- online coupons
- interstitials

PREDICT

Why do businesses still need to advertise on the radio or television?

WEB SITE PROMOTION

If people don't know your Web site exists, how can you expect it to be successful? Promoting your e-commerce site is perhaps the most important factor in determining the success or failure of your business. There are many ways online companies can market their products or services, from advertising in various media to strategically positioning their Web sites with search engines. (See **Figure 15.3**.) When determining how to promote your online business, you can choose among traditional media such as newspapers, television, and radio, or from the new markets available online. A major success factor is selecting the vehicles that will best promote your business to potential customers.

Types of Online Advertising

Online advertising presents many unique challenges. First, your message can only reach those people who have access to a personal computer and Internet connection. While the number of households with personal computers has increased, it still trails radio and television by a significant margin. NationMaster.com reports that U.S. households own more than 500 million radios and 220 million televisions—and only 150 million personal computers.

Therefore, it is critical for online companies to include media outlets such as radio, television, and print in their advertising and promotional campaigns. Vehix.com creates television commercials that emphasize the speed and ease of searching for cars online. Since these commercials reach at least some non-Internet users, they also emphasize the added value of the online medium. Just as with Amazon, the idea is to build positive associations with the brand experience—not just with the products or services.

Among the problems with online advertising are the perceptions that it violates consumer privacy, and that e-mail advertising can carry computer viruses. Another problem is **spoof e-mail**—e-mail in which the sender's address is forged to appear as if it came from a legitimate source. The message usually contains an urgent warning for the customer to update credit-card or other personal information, with a link provided to a "trusted" company site. When customers click the link, they are taken to a fake page where identity thieves wait to steal their private information. Since the spoof site looks identical to a trusted business's, many customers fall into the trap. Amazon.com, PayPal.com, eBay.com, and BestBuy.com have all been victims of these fraudulent communications. These actions have reduced the number of potential

Figure 15.3

Advertising Online

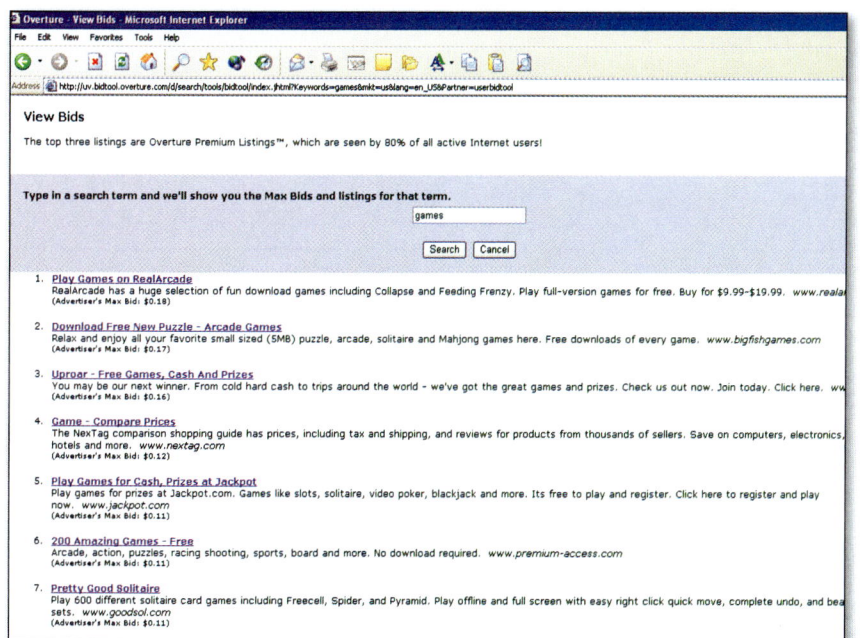

Reproduced with permission of Yahoo! Inc. © 2004 by Yahoo! Inc. YAHOO! and the YAHOO! logo are trademarks of Yahoo! Inc.

PLACEMENT FEES For a fee, Overture.com promises to improve your company's positioning at several top search engines. This figure shows Overture's paid listing results for the search term "games." *What steps does an e-commerce site have to take before moving to dramatically increase its traffic?*

customers who will read unsolicited marketing messages or click on e-mail links.

These problems aside, online advertising has many advantages over other media. First, it is quicker. Building a brand used to be a painstaking process that evolved slowly to meet changing market conditions. Now, when a brand needs to change its tactics or techniques, it can move quickly on a Web campaign. Second, the virtual world provides a level of interactivity never before seen. Since branding relies on customer perceptions and feelings, getting customers actively involved in the process can actually strengthen and deepen your brand identity. Third, you can respond to criticism much more quickly online. If a competitor makes claims against your brand, you can respond instantly. Don't forget, in the online world, you have more power to reach customers where they already like to hang out.

There are many reputable and trusted ways to reach your customers. The following sections explain the options, as well as how to use them to build your brand.

BANNER ADS A **banner ad** is a small graphic designed to promote a product, service, or company. When you click the banner, you are taken to the company's Web site. These ads derived their name from the banners you see at events such as fundraisers, races, and rallies. **Figure 15.4** on page 312 shows a typical banner ad.

Traditional banner ads appear horizontally near the top of a browser window, usually measuring 468 pixels long and 60 pixels high. *Pixel* is an abbreviation of the term *picture element* and refers to the smallest element of a digital image. Think of pixels like atoms: Alone, each one

CONNECT

When have you seen banner ads while surfing the Internet? Have you clicked on them? Why or why not? How are you likely to respond to this type of advertising?

Figure 15.4

Banner Ad Options

TYPES OF BANNERS Like ads in magazines or newspapers, banner ads come in all sizes, including horizontal (standard), vertical (skyscraper), and reduced size (mini). *What makes a banner ad effective as a sales tool?*

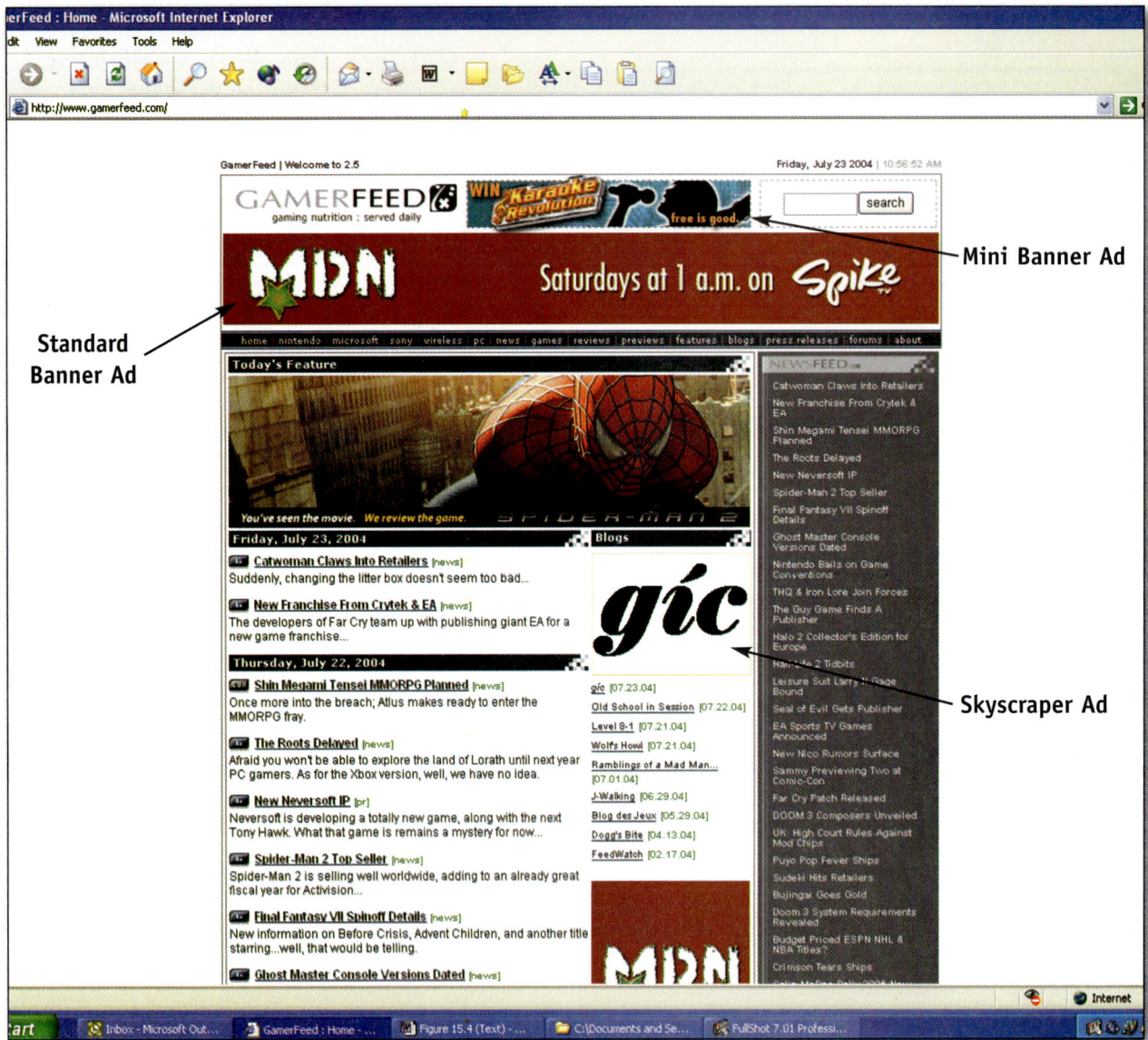

is extremely small, but together, they constitute a whole image. Another way to think of them is like grains of sand on a beach. The number of pixels in the length and width is the *resolution*. The higher the resolution, the larger your image will be.

Banner ads are one of the most effective types of online advertising. Media Metrix and Microsoft both report that, on average, banner ads increase traffic by nearly 40 percent. They must be used strategically, however. For example, teen movies are advertised on sites such as Seventeen.com to reach teenage girls. IBM purchases skyscraper banners on *BusinessWeek*'s Web site to reach its target market of business professionals.

A **skyscraper banner** stands tall like a big-city tower, aligned vertically on the page. The Interactive Advertising Bureau (IAB) has set standard sizes for these ads at 160 pixels wide by 600 pixels tall, making them very easy to notice. According to the IAB, skyscraper banners are 25 percent more effective than traditional banner-size advertisements.

An even more personal way to reach customers is through e-mail. In the next section, you will learn *industry best practices* for permission marketing.

E-MAIL E-mail marketing is one of the most cost-effective ways to reach your customers, but because of the tidal wave of spam drowning consumer inboxes, this option is also loaded with risk. Some users report spam to their ISPs, many of whom offer anti-spam software. E-commerce companies, however, can still use e-mail in positive ways to advertise their products and services. Amazon has built its business in part on personalized e-mail recommendations. How can you achieve similar success? The best way to avoid offending customers is to follow e-mail industry best practices. This means you must gain permission, keep permission, and always give customers a way out.

First, you must provide an opportunity to **opt-in** to e-mail lists, newsletters, and promotions. This means that customers must actually check a box granting you permission to send marketing messages. This is more ethical and straightforward than **opt-out**, in which customers receive marketing messages unless they specifically request not to receive them. In opt-out, customers may accidentally grant permission without even knowing it.

SPAM ALERT Businesses that advertise via e-mail need to be aware that many consumers view any unsolicited e-mail as spam and react unfavorably toward it, no matter how good the offer. *What steps can a business take to avoid negative responses to e-mail ads?*

Web Site Success

ONLINE BRAND OVERTAKES STOREFRONT SALES
REI.com

Try to track down a retail store or restaurant that doesn't have a Web site. It's getting more difficult to do. Such sites are used as resources for store location, hours, sales fliers, menus, and more. A growing number of retail stores have started selling their wares on the Web, creating an online brand they hope will benefit from their traditional one. Can a long-standing company's Web site prove more profitable than its storefront? Recreational Equipment, Inc. (REI), founded in 1938 with 70 retail stores throughout the United States, has proven it can. The popular retailer now enjoys more sales through its Web site than at all of its bricks-and-mortar stores combined.

Thinking Critically
REI.com has been successful in growing its online brand by including in-store kiosks that allow consumers to purchase online from directly inside the store. *What other approaches might a retailer use to successfully combine traditional and online sales?*

QUESTION
For an e-commerce company with a limited advertising budget, what is the most effective way to spread the word about a new product or service?

Confirmed opt-in is a higher standard for permission marketing. With this method, customers receive one confirmation e-mail after checking a box. Inside the message, they find a link that allows them to immediately unsubscribe. This prevents others from secretly signing them up to the list and gives them a chance to change their minds.

The highest standard is *double opt-in,* in which customers receive a confirmation message requesting an immediate response. If they fail to respond, they are *never* subscribed to the list. Because users invest more time in subscribing, this is considered the most targeted, effective approach. People who complete this process really want to see your marketing messages.

Even when people initially grant permission to receive e-mail, they may change their minds later. Therefore, it is equally important to give customers a way out. Every message should contain information about how to unsubscribe. If you follow these practices, your permission-marketing campaigns will have a much greater chance of success.

ONLINE COUPONS **Online coupons** are just like traditional coupons, except they are digital. These paperless discount offers usually take the form of short codes that customers enter at checkout. An online coupon might read "TAKETEN" or "DC-30," much like the bar codes on paper grocery coupons. In the same way a clerk scans a bar code, an e-commerce site records the discount information.

Buy.com, a leading online electronics retailer, often e-mails coupon codes to its customers. These coupons entitle users to free shipping, or $10 off one purchase of $100 or more. The e-mail contains information about products customers have previously purchased or viewed on the site, which can help create customer loyalty.

E-mail is not the only way to deliver coupons. You can also use them to reach new customers through site partners such as CouponMountain.com and CouponCabin.com. This is a less targeted approach, however, because customers who search these sites are more interested in good deals than particular brands. However, if deal-seekers enjoy your site, they are likely to return for future purchases.

There are problems with using online coupons. Some online stores require customers to enter a coupon code to get a discount. This often causes customers who don't know the code to abandon their online shopping carts. One solution to avoid this is to create special links from e-mail or from coupon sites that automatically generate a discount. This way, customers who lack the coupon code will not be left out.

INTERSTITIALS **Interstitials** are online advertisements that automatically open a new browser window. These ads came to be called pop-ups because of the way they literally burst onto the screen. According to *BusinessWeek,* pop-ups entice customers to click at four times the rate of banner ads. Even so, these ads are the mosquitoes and flies of the Internet. Nobody likes them. As soon as you swat one away, another one pops up in your face. In response to consumer complaints, many ISPs and browsers, including EarthLink, AOL, MSN, and Mozilla, now provide pop-up blocking technology.

Still, some businesses use them effectively. Orbitz.com, a popular travel site, created a series of pop-ups with games and puzzles, one of which convinced 13 percent of viewers to visit the site. The key is to make pop-ups inviting instead of flashy and annoying. Remember, you want to create a positive association with your brand.

SWEEPSTAKES AND GIVEAWAYS Sweepstakes and giveaways can be used to tap into wider markets, attract new customers, and create excitement about a new product or service. They can also create immediate positive feelings toward your brand, since customers associate your name with prizes and having fun.

Apple Computer, Inc. teamed up with PepsiCo, Inc. to give away 100 million free songs at the iTunes Music Store. Customers merely had to purchase a bottle of Pepsi and twist off the cap to see if they'd won. The idea was, as consumers reached into the refrigerated soft-drink case, they would remember the iTunes Music Store promotion and grab a Pepsi first, hoping for a free song download.

This kind of giveaway also builds both brands. The iTunes/Pepsi promotion tapped into widespread consumer disappointment over the music industry's prosecution of illegal music downloading. In the advertisement, one young woman declares that she can still download free music, while the Green Day version of "I Fought the Law" plays in the background. Music fans feel as though Pepsi and Apple are on their side.

Contests can also make your site sticky, especially when they require registration. Registered users at Zappos.com are entered into a weekly drawing for free shoes. This tempts customers to register, and once they register, they are much more likely to return.

Section 15-2 Review

RESPOND to what you've read.

1. What advantages does online advertising have over other media? _____

2. Why is a double opt-in message considered to be the highest standard in marketing? How does this type of marketing message work? _____

3. How can online contests and sweepstakes help a business's sales? _____

Quick TALK Present the following five marketing techniques in proper order according to the likelihood you would pay attention to each: *banner ads, unsolicited marketing e-mail, online coupons, interstitials, sweepstakes, and giveaways.* **Compare your list with that of a partner, and discuss whether your views are similar or different and why you think this is so.**

Name _____ Date _____

Worksheet 15-1

Careers in Digital Advertising

Advertising and marketing, whether for e-commerce or traditional business, require art, design, and creative talent. Some of the job titles in the online advertising world are Web advertising specialists, service specialists, designers, writers, editors, video producers, animators, content creators, communications consultants, researchers, and promotions specialists.

1. List five specific skills you think someone would need to be successful in the online advertising world.

   _____
   _____
   _____
   _____
   _____

2. Choose two of the above job titles, and research them on the Internet. Answer these questions:

   a. What education is necessary for each?

   _____
   _____
   _____

   b. Are there any schools in your state that offer training and/or a degree in these fields? Where?

   _____
   _____
   _____

   c. What are the specific duties/job responsibilities for these jobs?

   _____
   _____
   _____

   d. What do you think you would like or dislike about these jobs?

   _____
   _____
   _____

3. Prepare a chart with the information about the two jobs you researched.

4. Imagine your school offers training in one of the job titles you researched. Prepare a marketing flyer advertising the training program to students.

Name _____ Date _____

Worksheet 15-2

Brand: New

Marketing is perhaps the most important component of running a business. If people don't know you exist, then nothing else matters. As a successful business owner, you know advertising is key and the creation and development of your marketing strategy will help you be competitive in the online world.

1. As the owner of an online business, you have decided to add a new product line to your business. Select the product line you plan to add, and explain why you think it will sell.

   _____

2. Design the advertising campaign you will use to introduce your new product line. Include timelines, strategies, and tools (such as banner ads, e-mail, and pop-ups) you will use.

3. Using a computer software program, create a banner ad, an e-mail marketing strategy, and an online coupon.

4. Create a PowerPoint presentation of the entire campaign and present it to your class.

Chapter 15 Review and Activities

Chapter Summary

Section 15-1

- A company's brand reflects its personality, values, and vision. When a company creates positive, personal relationships with its customers, it is called branding.

- Sticky content—such as chat rooms, localized weather reports, personalized news, games, user reviews, 3-D content, and frequently updated promotions—entices visitors to spend time and return to a Web site.

- Dynamic content, or content that changes frequently, helps keep customers glued to the site and entices them to return.

- Log files, which record hits to a site, the number of unique visitors, and site traffic, help businesses improve the design and functionality of their sites.

Section 15-2

- Since not everyone owns a computer, it is important for online companies to advertise in traditional media such as radio, television, and print.

- Banner ads, e-mail, online coupons, interstitials, sweepstakes, and giveaways are some of the strategies companies use to advertise their products and services on the Internet.

- E-mail marketing is one of the most cost-effective ways to reach customers. Permission to send customers e-mail marketing messages should be gained by allowing them to opt-in to e-mail lists.

- Due to the increased use of spam and pop-ups, some types of online advertising can turn potential customers off.

Chapter 15 Review and Activities

Review of Key Terms

a. sticky content
b. site traffic
c. banner ad
d. online coupons
e. opt-out
f. spoof e-mail
g. brand
h. opt-in
i. exit page
j. dynamic content
k. skyscraper banner
l. unique visitors
m. branding
n. interstitials

Match each term to its definition.

_____ 1. Online advertisements that automatically open a new browser window.
_____ 2. Total activity on a site.
_____ 3. A small graphic designed to promote a product, service, or company.
_____ 4. Customers receive marketing messages unless they specifically request not to.
_____ 5. The last page viewed during a visit to a Web site.
_____ 6. Information or features that entice Web-site visitors to stay longer and return frequently.
_____ 7. An e-mail in which the sender's address is forged.
_____ 8. The process of creating a positive personal relationship with your customers.
_____ 9. Customers check a box to grant a business permission to send marketing messages.
_____ 10. Digital discount offers.
_____ 11. The number of individual IPs connecting to a site.
_____ 12. Information on a Web site that changes frequently.
_____ 13. A name, design, symbol, or other feature that distinguishes a company or product.
_____ 14. An advertisement aligned vertically on a page.

Applying Technology to Academics

ELA **English Language Arts—Reading**
Research the Web site of one major corporation from which you or your family has purchased a product, such Procter & Gamble, The Kellogg Company, or Nike, Inc. Find out which other products or services are marketed by the corporation. Describe how the corporate brand is included on each product.

ELA **English Language Arts—Writing**
Select a Web site, and analyze its sticky content. Write a paragraph describing what you like and dislike about it. Then, describe the information or features you would add to make it "stickier."

ELA **English Language Arts—Speaking**
Imagine you own an online company and have decided to advertise it on the radio. Write a 30-second commercial advertising your online business, and then present it to your class. Be creative. Incorporate music and a jingle.

MATH **Calculation**
You are responsible for monitoring the traffic on the Web site of a large bricks-and-mortar business. Last week your log files recorded 18,368 hits. The week before, you had 15,467. By what percent did your hits increase from last week to this week?

Chapter 15 Review and Activities

Critical Thinking

1. Some e-mail ads are located at the top of the Web site, and others are located at the bottom of the site. Which do you think are most effective? Why?
2. Most people don't like pop-ups. Some browsers even offer pop-up blocking. Why do you think pop-ups are still being used? Are there any that you think are being used effectively?

COMPETITIVE EVENT

Assume the role of assistant manager for a local family-owned clothing store. The store's owner (judge) is considering expanding the company's operation to include a Web site and would like to do some advertising via the Internet. Make recommendations to the owner (judge) about how to advertise online.

EVALUATION

You will be evaluated on how well you meet the following performance indicators:
1. Explain the nature of online advertisements.
2. Select strategies for online advertising.
3. Explain the nature of e-mail marketing.
4. Select techniques for promoting the Web site.
5. Make oral presentations.

INTERNET ACTIVITY

To engage in online activities, visit the *E-Commerce* Web site at **ecommerce.glencoe.com**.

BusinessWeek online

TOPIC: Careers in Internet Advertising

As the most recent innovation in the widespread exchange of information, the Internet is fertile ground for advertisers. With its relatively low cost and quick turnaround, Internet advertising presents options for advertisers that other media can't match.

ACTIVITY Go to the Student Center at *ecommerce.glencoe.com*. Click on *BusinessWeek* Activities and open Chapter 15. There you'll learn more about Internet advertising and its potential as a career versus more mainstream media.

SELF-ASSESSMENT

A Day in the Life of the Internet

On an average day in 2003 and 2004, roughly 72 million adult Americans used the Web. Below is a list of 25 activities they did online at least once, represented as a percentage of Americans with Internet access.

To Do: (1) Study the table. Estimate the number of times you perform each activity on any given week. Write that number (even if it is zero) in the middle column. (2) Find the percentage of your classmates who did each activity at least once in an average day. Write the answers (even if it is zero) in the third column.

Source: Pew Internet & American Life Project Tracking Surveys
[1] Figures are from 2003 and 2004. [2] If you rarely use a computer, estimate the number of times you performed these activities in a month.

25 Activities American Adults with Internet Access Do Online During an Average Day[1]	Percentage of all "wired" American adults who do each activity at least once in an average day	Number of times you did each activity at least once during an average day[2]	Percentage of your class who did each activity at least once during an average day
Go online	56%		
Send e-mail	48%		
Use a search engine to find information	31%		
Get news	28%		
Surf the Web for fun	23%		
Look for information on a hobby or interest	21%		
Check the weather	20%		
Research a product or service before buying it	16%		
Look for information about movies, books, or other leisure activities	13%		
Check sports scores and info	11%		
Watch a video clip or listen to an audio clip	11%		
Send an instant message	10%		
Do research for school or training	10%		
Look for information from a government Web site	9%		
Play a game	9%		
Bank online	9%		
Look up phone numbers or addresses	7%		
Look for health or medical information	6%		
Download files such as games, videos, or pictures	6%		
Share files from own computer with others	4%		
Chat in a chat room or in an online discussion	4%		
Listen to music online at a Web site	4%		
Participate in an online auction	3%		
Buy a product	3%		
Download music files	1%		

A Year in the Life of the Internet

In 2004, 63 percent of American adults—about 128 million people—logged onto the Internet at least once. Here are 25 activities they did, shown as a percentage of the total number of Americans with Internet access. This figure is much higher than 128 million.

To Do: (1) Study the table below. Then estimate the number of times you performed each activity over the past year. Write that number (even if it is zero) in the middle column. (2) Working with your classmates, find the percentage of them who did each activity at least once during the year. Write the answers (even if zero) in the third column. (3) How do your class's percentages compare with those of all teens?

25 Activities American Adults with Internet Access Do Online During the Year[1]	Percentage of all "wired" American adults who do each activity at least once during the year	Number of times you did each activity at least once during the past year	Percentage of your class that did each activity at least once during the past year	Percentage of all "wired" Americans age 12–17 who did selected activities at least once[2]
Send e-mail	91%			92%
Use a search engine to find information	88%			
Search for a map or driving directions	88%			
Research a product or service before buying it	78%			66%
Look for information on a hobby or interest	76%			69%
Check the weather	75%			
Look for information about movies, books, or other leisure activities	73%			83%
Get news	71%			68%
Surf the Web for fun	67%			84%
Look for health or medical information	66%			26%
Look for info from a government Web site	66%			
Buy a product	66%			31%
Look up phone numbers or addresses	54%			
Do research for school or training	53%			
Watch a video clip or listen to an audio clip	52%			
Check sports scores and info	43%			47%
Download files such as games, videos, or pictures	42%			
Send an instant message	38%			74%
Play a game	39%			66%
Listen to music online at a Web site	37%			59%
Bank online	34%			
Chat in a chat room or in an online discussion	25%			55%
Participate in an online auction	24%			31%[3]
Share files from own computer with others	20%			
Download music files to your computer	14%			53%

[1] Figures are from 2002 to 2004.
[2] By the year 2000, the last time such a survey was taken. The teen survey did not contain as many options as the adult survey.
[3] Visited, not necessarily participated.

BusinessWeek It's Your Project

SPRINGBOARD: POP QUIZ

According to a recent survey by the National Endowment for the Arts, what percentage of Americans engaged in any kind of creative writing activity during the past year?
- **a.** 7%
- **b.** 27%
- **c.** 47%

This lab will require some research. Go to BusinessWeek online for useful articles and resources.

What's Your World?

Everyone lives in a number of worlds. Some worlds, such as the worlds of our country and family for example, are chosen for us. Others we choose to be in. These include the world of work, a circle of friends, a team, club, or other world. How do these worlds shape your life? This project will help you find out.

Pick two worlds that interest you. Take them from the list or from any other source. Avoid two that are obviously similar, such as two types of school groups. Learn all you can about each world, letting the seven steps serve as your guide.

Here are a few worlds for you to consider. You may be:

- a citizen of a nation
- a citizen of a state
- a citizen of the world
- a club or team member
- a follower of a particular religion
- a member of a family
- a member of the workforce
- a member of a peer group
- a resident of a city or town
- a resident of a neighborhood
- a student in a school
- an artist
- an athlete
- one of millions of drivers
- one of millions of English speakers
- part of the U.S. economy
- members of an online chat group

Step 1 Start with *BusinessWeek*

Orient your project by conducting research in the *BusinessWeek* online archives. Use key words that relate to your topic to find articles. The articles may offer some insight as you progress through this project.

Step 2 Investigate and Engage

- How would you describe each world in writing?
- What do you already know about these worlds?
- What do you want to learn from this project?

Step 3 Identify the Obstacles

- In what ways are these worlds alike and different?
- What roles do you play in these worlds?
- How do the worlds shape your life?
- How do you shape those worlds?

Step 4 Conduct Research and Seek Solutions

- Where will you look for answers to the questions in Step 2?
- How might books, computers, and other technologies help?
- After researching the issues, what problems have you been unable to solve?

Step 5 Select Information and Analyze Data

- How can you determine which information is accurate and relevant to this issue?
- Turn your research results into statements that cast light on the issue.

Step 6 Connect to the Real World

Create a way to present your project—an audio-visual presentation, a talk, a Web page, an essay, or something else. It's your choice. Be sure to address these questions:

- How do your findings help you to better understand these worlds and your role in them?
- How do the skills and knowledge you acquired from this project prepare you for the future?

Step 7 Self-Evaluation

Assess the quality of your work. Use an evaluation tool such as a checklist, a small-group discussion, a log, or something else. Be honest. That way you'll learn where you need to work harder the next time.

Glossary

ad-hoc research research such as mail, telephone, and online surveys that focuses on specific questions during a single interview

advertising research consumers evaluate everything from fonts and image quality to the feelings they evoke to measure consumer response to both the message and the medium

affiliate other online businesses with which an e-commerce company shares a business relationship

affiliate program a partnership through which an online business delivers customers to other online businesses

aggregator a company that offers for distribution a number of products and services from a variety of manufacturers and producers

analog modem communicates over standard telephone lines by converting computer (digital) data into sine waves; at the receiving end, the data is then converted back into digital

antivirus software software designed to monitor and eliminate viruses

applets a mini program that can carry out a specific function on a Web page

application layer how an application interacts with network operating systems; this is an advanced security system with a more complex set of rules, allowing traffic to pass through only after a rigorous hurdle of tests

application-level firewall evaluates network packets for valid data at the application layer before allowing a connection

assets all of the things owned by a business

attributes formatting codes that describe an element

asymmetrical encryption provides file encryption and also issues a public and private key, or code; the public key can be used by anyone to encrypt anything, but only the private key can decrypt it

B2B see **business-to-business**

B2C see **business-to-consumer**

B2G see **business-to-government**

back-end management behind-the-scenes operations of a business; the infrastructure of your online company—the part that helps people find, buy, and receive what they're seeking

balance sheet shows how much a business is worth at a specific time; it lists *assets* and *liabilities* to show the *net worth*, or actual value of the business

bandwidth determines the amount of traffic that the server will allow on your site

banner ad a small graphic designed to promote a product, service, or company; when

Glossary

users click the banner, they are taken to the company's Web site

basic partnership a business that is formed and equally shared by two or more individuals

bastion host a heavily fortified computer server that handles all incoming requests over the Internet

best-of-breed when a company focuses on a single product or service to make it the best in its field

beta testing sending out an early version of software to selected users to catch coding errors or system crashes prior to public release

big-box retailers companies that build single-story structures of approximately 10,000 square feet or more; many of these stores have their own warehouse and distribution systems that lower their operating costs and result in lower prices

blog a public online journal kept by a writer, or blogger; short for Weblog

blogger a writer who maintains a public online journal

board of directors a group of people elected by stockholders to keep an eye on the management team

bot programs that scan the Web continuously, looking for computer-program updates, new-product alerts, price changes, etc.; short for "robot"

brainstorming activity that involves a group of people generating and sharing ideas

brand a name, design, symbol, or other feature that distinguishes one company's products and services from those of other companies

branding establishing a positive personal relationship between the customer and the company providing the product or service

brand loyalty refers to a customer's preference for a particular product

bricks-and-mortar business a business with physical storefronts

broadband enables large amounts of electronic data to be transmitted quickly

broken link a link to a Web page or file that is no longer available or cannot be found

bundling combining related products or services, which businesses then offer to customers as special package deals

business method patent patents issued to software and Internet companies that have devised novel ways of doing business; these patents usually combine software with business methodology; also known as an Internet patent

business model a system of policies, operations, resources, and technologies used to generate revenue

Glossary

business plan a detailed description of a business's objectives, products and services, operations, potential customers, and financial resources

business-to-business (B2B) business model that applies when a business transacts information, goods, or services with another business

business-to-consumer (B2C) business model that applies to any business or organization that uses the Internet to sell its products or services to consumers

business-to-government (B2G) business model that connects the private sector to the government marketplace

button graphic feature, usually a *GIF* file, that helps visitors navigate quickly and easily from one area to another

C2B *see* **consumer-to-business**

C2C *see* **consumer-to-consumer**

C corporation *see* **corporation**

cable modem service requires a special cable modem and possibly a network interface card to connect to the Internet; provides a faster Internet connection than dial-up

cascade describes the order in which styles are applied on a Web page

Cascading Style Sheets (CSS) enables Web authors to define colors, fonts, link colors, layout, and other aspects of Web design; a single style sheet applies a cohesive design to every Web page, allowing authors to change the look of a site by editing only one document; works in conjunction with HTML

cash flow the difference between the amounts of money a company generates and spends each month

category killers large stores that specialize in a particular type of product, such as toys, hardware, books, or sporting goods

channel of distribution the path a product takes from producer or manufacturer to consumer

chatbots programs installed in cars and linked to global positioning satellites to assist drivers

chat rooms Web pages that allow online shoppers to have real-time, virtual conversations with customer-service representatives

Children's Online Privacy Protection Act legislation designed and enforced by the Federal Trade Commission to prevent the collection of personally identifiable information from children without their parents' consent

circuitry switching system by which early telephone communications functioned

clickstream data a Web site visitor's electronic footprints, recorded in most server logs, including their unique IPs,

Glossary

page views, downloaded images, and URLs

collateral something valuable, such as a car or house, used to secure a loan and that the creditor can take if the loan is not repaid

commission fee earned for selling a product; usually a percentage of the amount of the sale

comparison shopping comparing products and prices before making a purchase to ensure the best deal

compression method of saving graphics in a compacted file type to save space

conceptualizing sketching samples of the look and feel of a Web site based on goals and ideas generated during the brainstorming process

confirmed opt-in a high standard for permission marketing in which customers receive one confirmation e-mail after checking a box; inside the message, they find a link that allows them to immediately unsubscribe

consumer the end user when a product is purchased for personal use

consumer goods physical items that people buy for personal or household use

consumer panels groups of consumers who are paid to answer questions or provide feedback during product development

consumer-to-business (C2B) business model in which the customer initiates the transaction, posting an intent or desire to buy a certain product at a certain price

consumer-to-consumer (C2C) business model that allows consumers to interact with one another online to transact goods and services

continuous research research in which consumer panels are asked questions over a period of time; in response to the panels' feedback, companies change products or services in time for the next panel meeting

contrast the difference in brightness between two related graphic elements, such as text and background

converting the act of changing the format of a file

cookies small text files written directly onto visitors' hard drives, containing information such as names, shopping-cart contents, user preferences, or unique identification codes; using cookies, businesses can collect data and analyze the behavioral patterns of their customers to market personally to them

copyright a form of protection provided by U.S. law that grants exclusive rights for the original works of an author

copyright notice information sometimes attached directly to intellectual property that includes the word "Copyright" or the symbol ©, the year the

Glossary

work was copyrighted, and the name of the copyright owner

corporation a business that is considered an entity separate from its owners; also known as a C corporation

crawler-based search engines search engines that use automated computer programs to scan Internet databases in search of new or revised Web pages

creditors people who lend money

CRM *see* **customer relationship management**

cross-sell hyperlink hyperlink that takes a viewer to an item associated with the item they're currently viewing online

cross selling generating another sale by offering related items

CSS *see* **Cascading Style Sheets**

culture a way of life that includes behaviors, beliefs, values, and generally accepted symbols for a group of people; culture can describe a religion, race, gender, or a geographic location

customer choice letting customers choose the way they prefer to receive customer support, whether by e-mail, phone, or online chat

customer relationship management (CRM) intense focus on the consumer, the goal of which is to increase customer satisfaction by dealing with each person as an individual

customer service representatives company employees responsible for taking orders over the phone, answering routine service questions, and routing calls to support technicians

cyber law an evolving legal framework that governs Internet activities; covers topics ranging from copyright infringement to e-mail privacy, identity theft, and interstate e-commerce

cybermediary an Internet channel of distribution that helps move products from the manufacturer to the consumer or industrial user

cyber squatter person who registers domain names of well-known companies or individuals with the sole purpose of selling the names back to the rightful owners to make a profit

database consolidation a strategy for putting information about a customer in one place, generally a single database, to quickly customize all interactions with a customer

database-driven Web site search engine that organizes search information, pulls together items from a database that fit a keyword description, and generates a Web page displaying all the relevant products

demographics information about the characteristics of customers a business serves, such as their age, average income,

Glossary

level of education, and where they live

department stores stores that offer a variety of products and choices within each product line and a floor plan that provides specialized departments such as men's apparel, housewares, and appliances

DHTML *see* **Dynamic HTML**

digital certificate computer file used to verify to customers that a company is what it claims to be

digital portfolio an electronic collection of your goals, achievements, honors, and reflections that demonstrate career growth over a period of time

digital subscriber line (DSL) Internet connection that typically provides ten times faster download speeds than a standard dial-up modem; also referred to as a T line

direct distribution occurs when goods and services are sold from the producer directly to the customer

discount stores stores that offer reduced prices

distribution the way the producer or manufacturer delivers products, services, or information to the consumer

domain name an addressing scheme employing words and phrases to identify and locate computers on the Internet

double opt-in the highest standard for permission marketing, in which customers receive a confirmation message requesting an immediate response; if they fail to respond, they are not subscribed to the list

double taxation when income is taxed twice

DSL *see* **digital subscriber line**

dynamic content Web site information that frequently changes or is updated

Dynamic HTML (DHTML) a hybrid technology that combines HTML, CSS, and JavaScript to add dramatic effects and animation to Web pages

e-business (electronic business) any process a business conducts over a computer network

e-cash a legal form of computer-based currency that allows for the purchase of items by credit card, check, or money order

e-checks electronic checks

e-commerce (electronic commerce) the conducting of business and communication transactions by electronic means

economic indicators statistics reported on a quarterly basis by the Census Bureau of the Department of Commerce tracking key statistics used to analyze business conditions and make forecasts; for example, retail e-commerce sales, the unemployment rate, the

Glossary

inflation rate, consumer spending, and the balance of trade

E-CRM *see* **electronic customer relationship management**

EFT *see* **Electronic Funds Transfer**

elastic demand when pricing changes create a change in the amount of goods or services consumers are willing to buy at a certain price

Electronic Communications Privacy Act of 2000 legislation that allows companies to monitor an employee's e-mail if either the sender or the recipient has given consent or the organization can demonstrate a reason why the company and/or its employees could be harmed by the electronic communication

electronic customer relationship management (E-CRM) software designed to manage and improve communication between a business and its customers electronically

Electronic Frontier Foundation (EFF) a nonprofit organization of lawyers, volunteers, and contributors who are working to protect your digital rights

electronic funds transfer (EFT) provides electronic payments and collections for online sales

electronic résumé an employment document that uses key words to provide an employer with information regarding a job candidate's professional experience, education, and job qualifications

e-mail (electronic mail) a system of worldwide electronic communication through which a user can compose a message at one computer and send it to a recipient at another computer

e-mail marketing inexpensive advertising method with high response rate that allows a business to establish personal relationships with customers, send customized messages, tell customers about new products and sales, or remind them to order gifts for holidays or special events in their profiles

e-marketplace the online shopping location

embedded style sheet a style sheet that is placed in the page itself

empty element an element that contains an instruction or an image rather than text

encryption the scrambling of data from plain text into code once it is sent from a computer

enterprise resource planning (ERP) software originally developed to focus mainly on financial and human resources applications, but adapted to deal with customer relationship management issues, thus competing with E-CRM

e-résumé *see* **electronic résumé**

ergonomics the study of the physical, environmental, and emotional areas of work; it examines how workers perform tasks by studying their

Glossary

capabilities and limitations, and the tools, equipment, and furniture they use

ERP *see* **enterprise resource planning**

e-store Internet business that provides back-end management tools for other Web businesses

e-tailing the buying and selling of retail goods on the Internet

eWallet a software application that stores a customer's data for easy retrieval during online purchases

exchange rate price at which one currency can buy another currency

executive summary the part of a business plan that contains a detailed description of a company's mission statement that defines the purpose of a business and highlights key facts about the company, including its goals and objectives, products and services, target market, competitive position, marketing strategies, management, and financial plans

exit page the last page visited by a customer before leaving a Web site

export product sold in a country other than the one in which it was produced

external style sheet a separate document that can be linked to a page or to an entire Web site

e-zine electronic magazine

FAQs *see* **frequently asked questions**

feasibility assessment takes ideas from the brainstorming process and examines how practical they would be to implement

file links links that enable visitors to download a music file or a product brochure created in a word-processing program

file transfer protocol (FTP) system by which users can obtain files and data over the Internet; works well when e-mail attachments are too large for a server

firewall a combination of hardware and software used to block potential hackers from gaining access to a computer system

font a typography term that categorizes a typeface or family of typefaces

forward-looking statements the parts of a business plan that describe the future of a company in both the short and the long term

frames allow for the display of more than one Web page in a single browser window

free trade trade among countries without barriers such as tariffs or quotas

frequently asked questions (FAQs) a list of questions found on some Web sites that addresses common user concerns in regards to a product or service

Glossary

G2C see **government-to-consumer**

general partner in a limited partnership, the partner who assumes the management responsibilities for the business

GIF see **Graphic Interchange Format**

globalization enhancing connectivity and interdependence among the world's markets and businesses

goals general targets a person wants or expects to reach sometime in the future

government-to-consumer (G2C) business model that allows consumers to easily access relevant information from government agencies

graphic interchange format (GIF) format that allows images to be compressed with no loss of quality; commonly used for animations, cartoons, and logos

growth plan describes how a business will expand over time

hacker a person who uses technical expertise to break into computer systems for malicious purposes

hit any single request to a server, including requests for images, text files, or HTML pages

home page a site's main page; usually the first page that appears when you log on to a site

horizontal hub an *intermediary hub* focused on providing the same types of products or services across various industries

hosted E-CRM arrangement where a company signs a contract with a vendor to "rent" E-CRM software on a per-seat basis (usually based on the number of employees)

hot spots positions on a graphic that serve as links to other locations online

HTML see **hypertext markup language**

HTTP or *http://* see **hypertext transfer protocol**

human translation the use of a multilingual person to change one language into another

hyperlink also called *hypertext links* or simply *links*; connects the current Internet document with another location in the same document, another document on the same Web site, or another document somewhere else on the Web; a blue, underlined font usually identifies hypertext links

hypertext link see **hyperlink**

hypertext markup language (HTML) the standard language for the Web browser; an easy-to-learn standard that uses tags to structure text and display visual elements

hypertext transfer protocol (HTTP or *http://*) the language that moves hypertext

Glossary

files across the Internet and defines the rules for transferring those files, which may include text, graphic images, sound, video, and other multimedia

icons easy-to-spot text links on a Web page that usually features a graphical representation of what the link leads to

identity theft the practice of running up bills or committing crimes in someone else's name

image map a graphic that has several different *hot spots*; the hot spots serve as links to more than one location

imports goods that are manufactured in one country, then shipped into another country for sale

income statement summarizes a company's revenue and expenses for a period of time and shows profits or losses

indirect distribution when a business gets the goods or services it sells from other producers and then sells them to consumers; indirect distribution involves one or more intermediaries

industrial user the end user when a product is purchased for business use

inflation measures the rise in the cost of consumer goods

initial public offering (IPO) first sale of a company's shares to the public

inline style sheet used for applying style to a Web page by adding style rules on a line-by-line basis

integrated services digital network (ISDN) the oldest form of broadband; operates over standard telephone lines and fiber-optic circuits

intelligent agents programs that simplify the work you do online by automatically performing tasks; also called Internet agents

intermediary *see* **intermediary hub**

intermediary hub a business that acts as a third party or go-between in moving products from the manufacturer to the end user; *see also* **horizontal hub** *and* **vertical hub**

Internet a global network of computers, communication tools, and information resources; also known as cyberspace

Internet directory a comprehensive listing of Web sites

Internet filter tool used by organizations to secure computer systems; a software program that limits access to Web sites on the Internet

Internet patent *see* **business method patent**

Internet service provider (ISP) a company that provides other companies or individuals with access to or a presence on the Internet

Glossary

internship typically unpaid job opportunity to work over a period of time sampling the career tasks in a field by completing hands-on projects

interstitials online advertisements that automatically open a new browser window; also called "pop-ups" because of the way they burst onto the screen

IPO *see* **initial public offering**

ISDN *see* **Integrated services digital network**

ISP *see* **Internet service provider**

Java a programming language; distinct from JavaScript

JavaScript short bits of code that add functionality to a Web page; a language which must run in connection with a Web page; distinct from Java

job work that a person does for pay

job shadowing learning about a job by following a competent worker for a day to witness firsthand the work environment, the skills needed, and the tasks performed in a career area of interest

Joint Photographic Experts Group (JPEG) format that provides designers with high-quality images that can be saved in very small files; best suited for photographs

JPEG (or JPG) *see* **Joint Photographic Experts Group**

knowbots (or knowledge robots) bots that can be programmed to automatically collect news about a particular topic

knowledge management high-tech strategy for acquiring information about the customer, analyzing it, and monitoring the customer's satisfaction

layout refers to the arrangement of the elements on the page, such as text, graphics, and headlines

liabilities debts or lawsuits

licensing the granting of permission to use intellectual property, such as music, photos, software programs, and inventions

licensing agreement an agreement that describes the cost and conditions of use for the right to market and profit from another company's licensed product

limited liability company (LLC) business ownership model that combines the features of a sole proprietorship, a partnership, and a corporation; like a corporation, an LLC is considered a legal entity separate from its owners

limited partner a partner who is not involved in the management but puts up the money to finance the company and shares in its profits and losses

Glossary

limited partnership a business structure in which partners do not have equal rights; a limited partnership consists of at least two people: at least one general partner and at least one limited partner

link *see* **hyperlink**

LLC *see* **limited liability company**

log files information recorded by the host server when visitors connect to a Web site, such as unique IP (Internet protocol) address, computer platform, browser type and version, time of connection, referring URL, page views, and cookies

logic bomb programming code added to a software program that lies dormant until a predetermined period of time passes or an event occurs, triggering the typically malicious code into action; also called slag code

logistics the plan for how a product will get from one place to another; logistic activities include receiving goods, warehousing, inventory control, vehicle scheduling and control, and distribution of goods

machine translation translating English text into another language using a software application

mailto: hyperlink that launches computer users' e-mail programs and allows them to send e-mail directly to an addressee

management plan the part of a business plan that provides a detailed description of the business, its philosophy, and its goals, as well as information about a company's form of ownership, legal structure, and a brief history

market economy where individual buyers and sellers interact with one another to exchange goods, services, and money; also called a capitalist economy

marketing the process of planning and carrying out the production, distribution, promotion, and pricing of products and services

marketing mix consists of four critical elements, collectively known as the four Ps: product, price, place, and promotion

marketing plan the part of a business plan that explains in practical terms how a company will launch and develop a business

marketing research focuses specifically on consumer behaviors and the reasons behind buying habits; it collects information about consumers, especially a company's target market, and analyzes demographics, buying habits, product preferences, and concerns about technology; distinct from *market research*

market research encompasses all of a business's efforts to collect data regarding the economic climate, competitor prices or strategies, governmental regulations and policies, consumer demographics, and

Glossary

emerging technologies; distinct from *marketing research*

market segments distinct groups of people who share certain characteristics

mass customization the production of goods that offer specialized choices to mainstream buyers

mentorship a paid work experience that allows employees to receive professional guidance and training from an experienced worker in a selected career field

merchandising activities such as acquiring products for sale, setting prices, displaying products, and making them available for purchase

message board a page on a Web site where users can share information about common problems

meta data information about a document contained within a *meta tag*

meta tag provides information about a Web page without altering the display

metrics patterns found in clickstream and other data

mission statement a short, general description that explains the purpose of a Web site and what its creators hope to accomplish with it, for example, the purpose and goals of a business

MP3 high quality audio format that works on the Web

multichannel retailer a retailer that sells its products via traditional channels (catalog, bricks-and-mortar, and telephone) as well as via an online channel

multimedia a visual presentation that combines text, graphics, motion, and sound to convey a message; usually a computer software program is used to create the multimedia presentation

navigate find a way around a site, including getting from page to page and getting to and back from other sites

navigational bars elements on a Web page that enable navigation through a Web site, such as a horizontal or vertical column, buttons, photos, or color-coded tabs

navigation scheme shows how different Web pages relate to one another and link to other sites

net worth the actual value of a business

non-store retailers businesses that use means other than traditional storefronts to sell their products, such as infomercials, catalogs, door-to-door solicitation, trade shows, and vending machines

objectives specific targets to be reached by a certain time, the progress toward which can be measured

Glossary

Occupational Outlook Handbook (OOH) a career reference written and updated by the U.S. Department of Labor, Bureau of Labor Statistics

Occupational Safety and Health Administration (OSHA) the department of the U.S. government responsible for ensuring the safety and health of the work environment

ODR *see* **online dispute resolution**

online coupon a digital coupon

online dispute resolution (ODR) the process of resolving cross-border disputes in the electronic-business environment

OOH *see* **Occupational Outlook Handbook**

operations plan the part of a business plan that includes information that describes the business processes: how products are marketed, how customers place orders, how orders are processed, how products are delivered, and how inventories and customer service are managed

opt-in when customers check a box granting a company permission to send marketing messages

opt-out when customers must specifically request not to receive marketing messages from a company

organizational chart describes a company's culture and management hierarchy

OSHA *see* **Occupational Safety and Health Administration**

outsourced when work is contracted to an outside agency, sometimes in another country

packet-level firewall a software firewall used mostly in home settings to analyze network traffic at the transport protocol layer to see if it matches the rules that define acceptable data flows

packet switching network describes the network design of the early Internet, wherein one machine delivered packets of data communications to numerous machines at one time

pass-through tax treatment when a corporation pays no taxes, but its investors pay taxes on their share of the profits and can use their share of the losses to offset any other personal income they've had for the year

patent a property right granted to the inventor of a product or process by the U.S. Patent and Trademark Office

pay-for-performance sponsored advertising that appears in connection with specified search keywords

pay-per-click revenue model wherein companies only pay for qualifying clicks to the destination site based on a pre-arranged rate

PDA *see* **personal digital assistant**

Glossary

PDF *see* **portable document format**

permission marketing when a potential customer has *consented* to receive marketing messages and will only receive advertisements about products and services they have requested

personal digital assistant (PDA) handheld computer providing wireless Internet connectivity

personalization refers to establishing one-to-one relationships with your customers, for example greeting customers by name or offering book and movie recommendations based on previous visits

PGP *see* **pretty good privacy**

PHP a scripting language that can be embedded in HTML

picture element *see* **pixel**

pixel refers to the smallest element of a digital image; an abbreviation of the term *picture element*

point size a measure used for calculating the size of type

portable document format (PDF) a file format created by Adobe that allows formatted documents to be transferred and viewed over the Internet without the need for the program originally used to create them

position in e-commerce, refers to the placement of a URL in search-engine results

pretty good privacy (PGP) the most popular program for encrypting and decrypting e-mail messages; it requires both the sender and recipient to have an encryption key, which is a code used to translate the message

price the amount of money a business receives in exchange for a product or service

primary research data gathered directly through one-to-one interviews, focus groups, and surveys

Privacy Act of 1974 designed to regulate the collection, maintenance, use, and dissemination of personal information by federal executive-branch agencies

privacy policy a written, legally binding statement informing users about how their personal information will be managed and maintained

product any item manufactured for sale

product line a group of related products that complement one another

product mix the full set of products offered for sale, including all product lines and categories

product quality describes the durability, reliability, and aesthetic features of the construction materials, as well as the service customers receive from a business, including warranties, guarantees, and secure shipping

product research research that focuses on the best way to produce and distribute products or

Glossary

services in the current technological or environmental context

profiling gathering information on customers to better customize a business's interactions with that customer

profits money made once a business's operating costs are paid

promotion hyperlink hyperlink that refers you to a product or service the site is currently offering

promotions advertisements intended to inform potential customers about products or services

protectionists those who favor government protection for domestic producers

purchase orders written documents that serve as an offer to buy certain items from a supplier at a specified price

pure-play retailers retailers who sell primarily through the Internet

quotas prescribed quantities or maximum amounts of imports allowed into a country

real-time pricing the ability to change prices instantly to keep up with changes in the marketplace

recommendation hyperlink hyperlink that takes Web users to a product promotion that might interest them based on products they have purchased before

registered trademark (®) trademark that has been registered with the federal and/or state government and can be secured by the U.S. Patent and Trademark Office for a fee

resolution the number of pixels per inch in an image

retailers establishments that sell goods and services to the general public

return on investment (ROI) how much money the company will make over a specified period of time in relation to how much money was invested

revenue model how a company generates income

robots automated programs that follow links to visit Web sites on behalf of search engines or directories

S corporation *see* **Subchapter S corporation**

sales research research that studies sales and market data to determine how well a product or service should do in the marketplace

satellite broadband service transmits high-speed data via satellite to a dish antenna at a home or business

search engines programs that retrieve information on the Web, using key words to access lists of documents containing those key words

Glossary

secondary research gathered from literature, publications, broadcast media, and other nonhuman sources

secured creditor a creditor whose loan is backed by collateral

Secure Sockets Layer (SSL) helps encrypt and protect the information that customers enter into Web pages when making a purchase; this protocol is built into most browsers and is supported by most Web servers

semantic Web a theorized extension of the current Web that may allow programmers to attach a tag, or a precise meaning, to every bit of information online

server a computer system that provides access to the Internet; also called a Web host

service mark similar to a trademark, it is used on services to identify their uniqueness and is also protected by trademark law

services work done in exchange for payment, such as printing brochures, tailoring clothing, and editing video

services retailer a business that provides services

shares the units of ownership held by each stockholder

shopbots bots that compile databases of products sold at online stores, enabling shoppers to compare prices and features in seconds

site map a diagram of a Web site's overall structure

site traffic the total activity on a site

skyscraper banner an advertisement that is aligned vertically on a Web page

smart card credit card with an embedded microchip, which is loaded with data that can be programmed for various applications

sole proprietorship a business owned and operated by one person

spam unsolicited "junk" e-mail sent by companies to potential customers

specialty stores stores that specialize in specific kinds of products or product lines and offer a wide assortment within their given categories

spiders (or crawlers) bots that search the Web, compiling extensive lists of URLs to keep search engine databases up to date

spoof e-mail e-mail in which the sender's address is forged to appear as if it came from a legitimate source

SQL (Pronounced "sequel") see **Structured Query Language**

SSL see **Secure Sockets Layer**

stand-alone programs E-CRM systems popular in the early to middle 1990s; used in call centers to track customer contacts

Glossary

sticky content information or features that entice visitors to stay for long periods of time and return frequently

stockholder a person who owns part of a corporation

storyboards pencil sketches or computer-generated images that show what individual Web pages will look like, illustrating everything that will appear on the Web site, including text, graphics, color schemes, links, and menus

streaming audio format that enables audio files to play in real time rather than requiring a complete download

Structured Query Language (SQL) a standardized language for requesting information from a database

Subchapter S corporation a business with 75 or fewer stockholders that is set up like a corporation but is taxed like a partnership

supply and demand the relationship between the amount of a good or service that is available and how much people are willing to pay for it

support technicians employees who provide solutions to problems customer-service representatives can't solve

supporting infrastructure how a company lays out the foundation for its e-commerce activities; includes the physical tools needed to get the job done, such as equipment, support services, and people

T-line *see* **digital subscriber line**

tables originally developed to display lists of information that have multiple rows and columns of data, tables are now commonly used to create Web-page layouts

tags formatting bits of code that define Web-page elements

target market the specific group of customers a business wants to attract

tariffs taxes or fees that various governments place on selected imported products

TCP/IP *see* **Transmission control protocol/Internet protocol**

terms of sale describes the methods of payment a business will accept as well as a company's refund policies, and warranties or guarantees

title element an HTML element for the lead section of the page code

trademark the use of a word, phrase, symbol, product shape, or logo by a manufacturer or business to identify its goods and to distinguish them from others on the market

trademark infringement when someone uses a name or phrase trademarked by another company or person to identify his or her own product

Glossary

Transmission control protocol/Internet protocol (TCP/IP) the common underlying language or protocol through which systems communicate on the Internet

Trojan virus refers to a program that appears safe but actually contains something harmful

uniform resource locator (URL) indicates the address of a Web site; it consists of two primary parts the hypertext transfer protocol and the domain (or server)

unique selling proposition (USP) the one thing that makes a business stand out from the competition; a USP describes the benefits of buying from one retailer rather than another

unique visitors the number of individual IPs connecting to a site

unsecured creditor someone who lends money without collateral

upsell hyperlink hyperlink that refers you to a location that presents a similar but more expensive item

URL *see* **uniform resource locator**

USP *see* **unique selling proposition**

value chain the sequence of design, production, and marketing efforts a business conducts to deliver its products at the right price and time

values formatting codes which describe an attribute

variable pricing e-commerce revenue model that allows customers the option of naming their own price for a product or service

vertical hub an *intermediary hub* that matches buyers and sellers within a particular industry

virtual auction Web site where buyers are able to name their own price as well as compete against other bidders; the site charges sellers a nominal listing fee and takes a small percentage of the sale

virtual marketing a cost-effective strategy that uses customers to help promote a business

virus a program written to inflict harm on a computer system and to interfere with its normal operation

Web affiliate a business that forms a partnership with another online business and refers potential customers to the partner's Web site for a commission or fee

Web browser a program used to view, download, surf, or access Web documents

Glossary

Web distributor a business that markets and sells goods or services bought from a manufacturer on a wholesale basis to companies for use or for resale to the end user

Web globalists consultants who advise companies on how to design their Web sites and market and sell their products effectively to international audiences

Web host provider a business that allows customers to house their Web site documents on the company's servers

Weblog *see* **blog**

white space blank space on a page

wholesalers businesses that sell products to distributors or retailers and not usually to the end-user or consumer

wireless Internet system that works much in the same way as a cordless telephone, adding short-range radios to stationary computers, laptops, and hand-held *personal digital assistants* (*PDAs*)

wireless Internet service providers (WISPs) companies that offer wireless connection services to the public

WISPs *see* **wireless Internet service providers**

World Wide Web (WWW) a global collection of graphical and hypertext Internet pages that can be read, viewed, and interacted with via computer

worms programs that replicate themselves exponentially, causing malicious actions against the computer's resources and files

WWW *see* **World Wide Web**

Index

Note: The italicized *f* following page numbers refers to figures.

Abate, Joseph T., 149
Ace Hardware, 49
AceHTML, 204, 210
achievement and recognition, 94
ad-hoc research, 255
Adobe Acrobat Reader, 28, 195
Adobe PageMaker, 204
ad sales
    advertising revenue, attracting, 160–161
    building up customers, 161
    getting started, 160–161
    online advertising, 160, 160*f*
    prominent placement on Yahoo!, 160
    trading advertisements, 161
Advanced Research Projects Agency (ARPA)
    ARPANET, creation of, 7
    defined, 7
advertisements, 30, 31
advertising research, 256
advertising a Web site, 310–315
Affiliated Foods Midwest, 46
affiliate programs, 251
    affiliates, 161
    commission, 161
    defined, 161
agents as bots
    bots, description of, 289
    chatbots, 289
    crawlers, 289
    knowbots, 289
    shopbots, 289
    spiders, 289
aggregator, 269
Allen, Paul, 93
AltaVista, 14
Amazon.com, 36, 36*f*, 150, 158, 159, 161, 304
American Apparel, 305
AmericanEagle.com, 34
analog modem
    data-transmission rates, 13
    defined, 13
    modem's speed, 13
    second phone line, 13

antivirus software
    antivirus-protection actions, 127*f*, 128
    types of viruses, 127, 128
    virus, 126
    virus attacks, 127
AOL, 12
Apollo Hosting, 12
Appert, François, 245
Apple Computer, 163, 315
Apple's QuickTime, 218
applets, 212
Armstrong, Neil, 5
ARPANET
    beginnings of, 7, 7*f*
    expansion of, 7
    first international connection, 7
    increased commercial activity, 7
    Internet today, 7
    packet switching network, 7
articles of incorporation, 143
Aspdin, Joseph, 185
assembly-line workers, 19
assets, 170
asymmetrical encryption, 57
Atlantic Monthly, 19
atmospheric steam engine, 45
AT&T, 6
attention on the Web
    Amazon's strategies for success, 305
    sticky site, creating, 305–307
    traffic sources, 307–308
attributes, 206
AT&T WorldNet, 12
auction-style models, 37
audience, knowing
    demographics, 189
    Matrix, The, 189, 190
    movie studio Web sites, 189, 190
    Nielsen/NetRatings survey, 189, 191*f*
    Seabiscuit, 190
    sites with similar target markets, 189, 191*f*
    statistics on Internet users, 189, 190*f*
    target market, 189
    type of customers, 189

audience, reaching
    accessing your site, 194, 195
    customers, communicating with, 195
audio on web
    audio cues, 217
    converting, 217–218
    MP3 files, 217, 218
    streaming audio, 218
    traditional sources, 217
Automated Clearing House (ACH) Network, 54
Autoweb, 161

BabelFish.com, 69
BabyBunz.com, 216
back-end management
    contents of, 230
    databases, 232
    defined, 230, 231*f*
    e-stores, 232, 233, 233*f*, 234
    servers, 230, 231
    shopping-cart software, 232, 233, 233*f*, 234
background color
    balanced color, 214
    contrast, 214, 215*f*
Bain, Alexander, 203
balance sheet, 170
bandwidth, 231
bank tellers, 19
banner ads
    description of, 311
    effectiveness of, 312
    options, 311, 312*f*
    picture element, 311
    pixel, 311, 312
    resolution, 312
    skyscraper banner, 313
Banting, Frederick, 303
Barber, John, 303
Bardeen, John, 283
Barnes&Noble.com, 150
Barnes & Noble, 47
basic partnerships
    definition of, 141
    equal authority and responsibility, 141
    investments in company, 141
    law firms as example, 141
    leaving business, 141
Bechtel, 143

Index

Bed Bath & Beyond, 292, 293f
Behaim, Martin, 27
Bell, Alexander Graham, 5
Ben and Jerry's, 142, 159, 186
Berners-Lee, Tim, 8, 45, 289
Best, Charles, 303
Best Buy, 47, 49, 304
BestBuy.com, 310
best-of-breed systems
    best in field, 272
    defined, 272
Best of the Web (Forbes), 150
beta testing, 256
Bezos, Jeff, 250, 305
big-box retailers, 47, 48f, 49
Binney, Edward, 245
Birch, John, 265
Birdseye, Clarence, 45
Birkenstock.com, 250
Biro, George, 225
Biro, Laszlo, 225
Black Entertainment Television (BET), 256
blog, 149, 161
blogger, 161
Blogger (company), 149
blogging, 149
BlueNile.com, 32
board of directors, 143
Bohlin, Nils, 225
Boingo, 14
Bookfinder.com, 147
books, 29
Borders, 49
bots, 289
BotSpot.com, 289
brainstorming
    collaborating, 188
    defined, 188
    importance of, 188
Bramah, Joseph, 139
Brandenberger, Jacques, 303
branding
    brand, definition of, 304
    brands under one corporate umbrella, 304
    description of, 304
    social and cultural values, 305
    successful brands, 304
brand loyalty, 31
Brattain, Walter, 283
Brearley, Harry, 225
Breyers ice cream, 142
bricks-and-mortar businesses, 19, 50

Brin, Sergey, 169
broadband connections, 137
broadband services
    broadband options, 13–14
    cable modem, 13
    defined, 13
    digital subscriber line (DSL), 13
    Integrated services digital network (ISDN), 13
    satellite broadband service, 14
    T line, 13
broken links, 208
Brown, Dan, 147
browser search tools, 14
Brynjolfsson, Eric, 284–285
bundling, 295
Bushnell, David, 303
Bushnell, Nolan, 303
business method patents (Internet patents), 117
business model, 34
business ownership, types
    aspects of business, understanding, 145
    corporation, 142–143, 144
    incorporation in America, 144–145
    partnership, 141, 142
    registering business, 144–145
    sole proprietorship, 140–141
business plan
    allocates resources, 164
    appraisal of obstacles, 164
    defined, 164, 165f
    informative, 164
    putting plan into action, 170–172
business plan, elements of
    e-business plan, 170
    elements of, 164–165, 165f
    executive summary, 166
    financial plan, 169–170
    growth and contingency plan, 170
    management plan, 166, 168f
    marketing plan, 166, 167
    operations plan, 168, 169
    Small Business Administration (SBA) recommendations, 164
    web and retail competition, 168
    web-site demographics, 167, 168

business plan, putting into action
    comprehensive outline, 171f
    conferences and seminars, 171
    draft, feedback, 170
    loans or investments, 172
    Service Corps of Retired Executives (SCORE), 170
    Small Business Development Center (SBDC), 170
business-to-business (B2B) model, 34
    description of, 35
    energy chain, 35
    how it works, understanding, 35, 36
    value chain, 35, 35f
business-to-consumer (B2C) model, 34
    business loyalty, 36
    customer satisfaction, 36
    defined, 36
    examples of, 36, 36f
    keeping customers busy, 36
    predicted growth, 36
business-to-government (B2G) model, 34, 37
BusinessWeek, 29, 37, 44, 243, 312, 315
buttons, 208
    button, defined, 216
    purpose of, 216
    types of, 216
Buy.com, 314
buyer protection, 123, 124f
buying and selling online, 34–38
buying trends, 19

cable modem service, 13
career opportunities
    content-producing career, 104
    e-commerce marketing, 103
    Internet and e-business, types of jobs, 102, 102f–103f
careers in e-commerce
    e-résumé and portfolio, creating, 105, 105f, 106
    finding, 104
    friends and family, 104
    newspaper classifieds, how to search, 104
    Web information, 104

Index

careers in e-commerce, training and education
    art or business, 97
    communication skills, 96–97
    courses, appropriate, 96
    elective courses, 97
    math skills, 97
    multimedia presentations, 97
career strategy, selecting
    internship, 96
    job shadowing, 96
    mentorship, 96
careers *versus* jobs, 94–95
Carlson, Chester F., 185
Carpal Tunnel Syndrome, 98
Carrier, Willis, 283
carriers, types of
    transportation services and carriers, 274, 275
    warehouse, stock handling, and inventory control, 275
Cascading Style Sheets (CSS)
    advantages and disadvantages of, 209 210
    cascade, 209
    cascading, 209
    defined, 209
    drawbacks, 210
    embedded style sheet, 209, 211*f*
    external style sheet, 209
    inline style sheet, 209
    printer-friendly pages, 210
cash flow, 170
category killers, 47, 48, 49
C corporation, 142
CDNow.com, 267
Census Bureau of the Department of Commerce, 149, 150
CERN, 8
certificate of stock ownership, 142
channel members
    aggregator, 269
    cybermediary, 269
    Web affiliate, 269
    Web distributor, 269
channels of distribution
    chain links, 268*f*
    channel members, 269
    defined, 267
    direct and indirect, 268*f*
chatbots, 289
chat rooms, 19, 31, 294

Cheap Tickets, 228
Children's Online Privacy Protection Act of 1998 (COPA), 120
chips, ID, radio-powered, 3
chronological résumé, 105
Circuit City, 49
circuitry switching, 7
Citi® Cards, 304
CitySearch, 18
clickstream data, 255
CNN, 31
Coca-Cola, 31
CoffeeCup Button Maker, 212
CoffeeCup HTML Editor, 204, 210
CoffeeCup Image Mapper, 209
Cohen, Ben, 142, 159
collateral, 141
color, background
    balanced color, 214
    contrast, 214, 215*f*
color consistency
    background color, 214, 215*f*
    browser-safe palette, 214
    colors, using, 214
    VisiBone.com, 214
combination résumé, 105
commission, 161
communications systems, 70
comparison shopping, 147, 147*f*
competitive landscape
    ability to compete, improving, 74
    pros and cons, 74
Computer Economics, 126
Computer Industry Almanac, 66
computer revolution
    introductory stage, 18
    permeation stage, 18
computers
    first freely programmable electromechanical, 27
    replacing humans in workplace, 19
Computing Technology Industry Association, 102
conceptualizing
    defined, 189
    personality of site, 189
    sketching samples of site, 189
confirmed opt-in, 314
connections. *See* Internet connections
Constitution's Commerce Clause, 147

consumer, 266
consumer benefits, 292–293
consumer goods, 158
consumer panels, 256
ConsumerReports.org, 33
consumer-to-business (C2B) model, 34
    description of, 37
    reverse-auction scenario, 37
consumer-to-consumer (C2C) model, 34
    auction-style models, 37
    consumer interaction, 36
    description of, 36
    online auctions, 36, 37
consumer-to-government (B2G) model, 34, 37
contingency plan, 170
continuous research, 255, 256
cookies, 251
    advantages of, 290
    description of, 289
    inserted in browser, 290
    invention of, 289
copyright
    copyright notice, 114–115
    defined, 114
    information, 115, 115*f*
    lawsuits, 114
    permission to copy, 114
    protection, 114
    registration of work, 115
    works created after 1978, 114
copyright notice, 114–115
corporations
    bankruptcy, 144
    board of directors, 143
    C corporation, 142
    disadvantages of, 144
    double taxation, 144
    formation of, 143
    general corporation, 142
    initial public offering (IPO), 143
    liability, 142
    owners of, 142
    private, 143
    public, 142, 143
    sales, volume in, 142, 143*f*
    shares, 142, 143
    stock, 143
    stockholder, 142, 143
    structure of, 143, 144
    types of, 142–143

Index

CouponCabin.com, 315
CouponMountain.com, 315
coupons, 30, 31
Covisint, 38
CraigsList.org, 294
crawler-based search engines, 226
   current listings, 15
   defined, 15
   disadvantages, 15
   updates, 15
crawlers, 289
credit-card protection (Gap online), 123, 124*f*
credit cards
   Automated Clearing House (ACH) Network, 54
   merchant account, 54
   total transaction, 54
creditors, 141
cross-continent electronic communication, 17
cross-sell hyperlink, 52
cross selling, 248
CSS, 209
culture
   cultural differences, 67
   defined, 67
   foreign language sites, 67, 68*f*
   Web site for global business, issues, 67
currency
   converted, 71
   exchange rates, 71
   fluctuations, 71
customer choice, 291
customer demographics, 19
customer needs, meeting
   made-to-stock business model, 36
   mass customization, 32
   value chain, 32
customer relationship management (CRM), 291
customer relationships
   agents as bots, 289
   cookies, 289–290
   customer service, 286
   customer support, 286–287
   intelligent agents, 287–289
   online quality, 286–287
   research, 284–285
   service warranties, 286
   trust, importance of, 284–285

customers, communicating with
   Contact Us link, 195
   customer-service representatives, 195
   fax numbers, 195
   Guest List, 195
   phone numbers, 195
   toll-free numbers, 195
customer service
   customer relationships, building, 284–290
   easy site navigation, 286
   face-to-face contact *versus* virtual environment, 284
   prompt delivery, 286
   speed, 286
customer support and online quality
   customer-service representatives, 286
   outsourcing, 287
   support technicians, 286–287
customization of Web site
   Internet pricing models, 294–295
   personalization of online customer service, 291–294
customize, 32
CuteHTML, 204
CuteMap, 209
cyberlaw
   copyright, 114–115, 115*f*
   defined, 114
   domain name disputes, 116
   knock-offs, 116
   patent, 117, 117*f*
   trademark, 116, 117
cyberlocations, 66
cybermediary
   defined, 269
   forms of, 269
cyberspace, 6
cybersquatters, 122
cyber stores, 17

D

database consolidation, 291
databases
   database-driven Web sites, 232
   PHP, 232
   Structured Query Language (SQL), 232

data-transmission rates, 13
Da Vinci Code, The (Brown), 147
DealTime, 295
debit cards, 54
Dell, 3
Dell, Michael S., 3
Dell.com, 36
demographics, 167
   defined, 189
   of internet users, 189, 190*f*
department stores, 49
descriptive file types. *See* file types, descriptive
design and interactivity, adding
   Cascading Style Sheets (CSS), 209–210
   DHTML (Dynamic HTML), 211–212
   Java, 212
   JavaScript, 210, 211
developing countries, 76–77
DHTML (Dynamic HTML), 209
   defined, 211–212
   disadvantages, 212
   drop-down menus, 212
dial-up connections
   analog modem, 12–13
   broadband connection, 13
   data transmission rates, 13
   modem's speed, 13
   second phone line, 13
Dickson, Earle, 245
dictionary, first English language, 27
Diesel, Rudolf, 65
digital automated voice mechanisms, 19
digital certificates
   asymmetrical encryption, 57
   description of, 56–57
   encryption, 57
digital portfolios, 106
digital protection, 122–128
digital subscriber line (DSL), 13
Diller, Barry, 18
direct and indirect channels of distribution, 268
Direct Marketing Association, 103
discount stores
   big-box retailers, 47, 48*f*, 49
   category killers, 47, 48*f*, 49
   defined, 49

Index

dispute resolution
    online dispute resolution (ODR), 77
    third-party negotiators, 78
    Web-based negotiation, 78
distribution
    channel members, 269
    channels of, 267–269, 268f
    consumer, 266
    defined, 266
    direct and indirect channels of distribution, 267
    distribution process, 266–267
    e-marketplace, 266
    industrial, 266
    intermediary, 268
    logistics, 266
    missed delivery deadlines, 267
    process, 276f
    Project Toolate.com, 267
Division of Corporations, 144
domain name
    companies, use of, 10
    defined, 9
    Domain Name System (DNS), 10
    extensions, commonly used, 10, 10f
    uniform resource locator (URL), 10
domain name disputes, 116
DoubleClick, 305
double opt-in, 314
double taxation, 144
Dove ice cream, 142
DreamWeaver, 210
Drucker, Peter F., 19
Drudge, Matt, 161
Drudge Report, 161
duty-free shops, 71
dynamic content, 306

E

Earthlink, 12
Eastman, George, 93
eBay, 37, 50, 78, 146, 310
    boosting revenue, 284
    customer base, pleasing, 284
    equal treatment, 285
    ratings guide, 285, 285f
e-book experiment, 162

e-business plan
    launching business, investors and, 170
    sample of Web site, 170
    Web server or Web host, 170
e-business *versus* e-commerce
    differences between, 29f
    e-business, 28
    electronic transactions, 28
e-cash, 55
e-checks (electronic checks)
    defined, 55
    third-party account server, 55
e-commerce, benefits and characteristics
    advertisements, 30, 31
    analysis, 30, 31
    brand loyalty, 31
    chat rooms, 31
    coupons, 30, 31
    inventory, managing, 31, 32
    limitless buyers, 30
    market, 30, 31
    meeting customer needs, 32
    online surveys, 31
    open 24/7, 30, 31
    pricing competitively, 33
    purchasing habits, tracking, 31
    starting out, 33
    statistics, tracking, 31
    Web technologies, evolution of, 29
    why buy online, 30, 30f
e-commerce, characteristics
    advertisements, 19
    bricks-and-mortar business, 19
    chat rooms, 19
    distribution, 19
    e-commerce sector, up-to-date numbers on, 19
    geographic boundaries, 19
    growth rate, 19
    leveling playing field, 19
    multichannel retailer, 19
    online surveys, 19
    pure-play retailers, 19
    revenue increases, 19
    selling, 19
    specials, 19
    traditional businesses, 19
e-commerce, future of
    careers *versus* jobs, 94–95
    lifestyle, influences on, 94, 95f
    workplace, 95

e-commerce, infrastructures, and developing countries
    benefits of, 76
    developing countries, 76–77
    information and communications technology (ICT)-based economy, 76
    infrastructure, 76
    regulating activities online, 76
    world economy, 76
E-Commerce and Development Report 2003, 76
e-commerce business models
    business model, defined, 34
    business-to-business (B2B) model, 34, 35, 35f, 36
    business-to-consumer (B2C) model, 34, 36, 36f
    business-to-government (B2G) model, 34, 37
    consumer-to-business (C2B) model, 34, 37
    consumer-to-consumer (C2C) model, 34, 36, 37
    consumer-to-government (B2G) model, 34, 37
    government-to-consumer (G2C) model, 34, 37, 38f
    intermediary hub, 34, 38
    transacting business online, 34
    on the Web, 34
e-commerce (electronic commerce)
    benefits and characteristics of, 29–33
    characteristics of, 19
    communication with customers, order status, 20
    computer revolution, stages of, 18
    cyber stores, 17
    defined, 17
    electronic communication, 17–19
    payment processing, 20
    purpose of, 20
    transaction processing, 20
    transactions, examples of, 17
    transmitting orders, 20
    virtual stores, 20
e-commerce environment
    e-business compared with, 28–29
    electronic economy, 28

Index

microchip, 28
supporting infrastructure of company, 28
e-commerce workplace environment
   approaching, 95
   equal opportunity employer, 98
   flexibility, 97, 98
   traditional business office, 98
   working at home, 97
economic indicators, 149, 150
economics, impact of e-commerce
   economic impact, 149
   economic indicators, 149, 150
   inflation, 149
   local and state economies, 147–148
   market economy, 146, 147, 147f
   new ways to compete, 150
   online savings, 149
   savings, differences in, 149
Edison, Thomas, 45
editing site, 196
education for e-commerce career, 96–97
elastic demand, 33
electronic commerce. *See* e-commerce
electronic communication
   business and personal use of, 18
   commercial use restrictions, 18
   commercial Web sites, 18
   computer revolution, classifying, 18
   computer size, changes in, 18
   computers replacing humans, 19
   cross-continent electronic communication, 17
   digital automated voice mechanisms, 19
   electronic telegraph system, 17–18
   evolution of, 17–19
   global interdependence, 18
   Internet, 18
   Morse Code, 17
   planned Internet usage, survey, 18
   prevalence of, 18
   self-service environment, 18
   telegraphy, 17
   World Wide Web, 18
Electronic Communications Privacy Act of 2000, 120

electronic customer relationship management (E-CRM)
   best-of-breed systems, 272
   communication, 271
   customer care, 271
   good service, 271
   hosted E-CRM, 272, 273, 273f
   software suites, 272
   stand-alone programs, 271–272
   types, 273f
electronic economy, 28
Electronic Frontier Foundation (EFF)
   activities, 120
   defined, 120
electronic funds transfer (EFT), 54
electronic résumé, 105, 105f, 106
electronic telegraph system, 17–18
electronic transactions, 28
Elizabeth II (queen), 27
e-mail
   defined, 10
   file-management issues, 10
e-mail marketing, 251, 252
   confirmed opt-in, 314
   customers, avoid offending, 313
   double opt-in, 314
   opt-in, 313
   opt-out, 313
   unsolicited e-mail, 313
   unsubscribing, 314
e-marketplace, 266
employee's e-mail, monitoring, 120
encrypted digital signature, 126
encryption, 57
   credit-card data, 126
   defined, 125, 126
   encrypted connection, checking for, 126
   key, 126
   padlock, 126
   pretty good privacy (PGP), 126
   security enforcement service, 126
energy chain, links, 35, 35f
Englebart, Douglas, 157
enterprise resource planning (ERP), 272
Epinions, 295
e-résumé and portfolio, creating
   "buzzwords," 106
   chronological résumé, 105
   combination résumé, 105
   contents of, 106
   digital portfolio, 106

electronic résumé, 105, 105f, 106
   functional résumé, 105
   keyword summary, 106
   résumé, defined, 105
e-RFx tools, 54
ergonomics
   defined, 98
   Occupational Safety and Health Administration (OSHA), 98
   studies, recommendations, 98–99
   working long hours, discomforts, 98
e-stores
   AACart, 234
   annual fee, 234
   features of, 234
   Quikstore, 234
   SalesCart's FrontPage Ecommerce, 234
   software, 234
   Yahoo! Stores, 234
e-tailing, 47, 48, 50
   advantages and disadvantages, 57, 57f, 58
   online purchasing process, setting up, 53–54
   order fulfillment and customer service, 56
   payment options, 54–55
   product merchandising, 52–53
   security issues and concerns, 56–57
ethical considerations on internet, 117
Evans, Oliver, 203
eWallets, 54, 55, 55f
exchange rates, 71
Excite, 14
executive summary
   defined, 166
   overview of total presentation, 166
exit page, 308
Expedia.com, 18, 36, 228
export, 75–76
e-zine, 161

F

Fahrenheit, Gabriel, 27
Fair Credit Billing Act, 123f
Farnsworth, Philo, 265

Index

Fast Company, 252
feasibility assessment
    conceptualizing, 189
    defined, 188
    feasible ideas, 188
    Onstar, 188
    rejected ideas, 188–189
    subscription-based satellite communication network, 188
federalism, 147
Federal Trade Commission Act, 119
Federal Trade Commission (FTC), 119, 267
FedEx, 18
file links, 208
file transfer protocol (FTP), 11
file types, descriptive
    file size, 215
    GIF, 215
    JPEG (or JPG files), 215
    resolution, 215
financial plan
    assets, 169
    balance sheet, 169
    capital, 169
    cash flow, 169
    income statement, 169
    liabilities, 169
    return on investment (ROI), 170
    wages and salaries, 170
firewalls
    application layer, 123
    application-level firewall, 123
    bastion host, 123, 124
    defined, 123
    packet-level firewall, 123
Flash, 204, 209
Fleming, Alexander, 157
fluctuations, 71
Foley's, 49
fonts
    categories, 216
    font, 216
    font's point size, 216
Forbes magazine, 150
Ford, Henry, 5, 185
foreign language sites, structure, 67, 68*f*
Forrester Research, 36, 66, 73, 167
forward-looking statements, 167
Fox Broadcasting Company, 18
frames
    advantages, 206
    defined, 206
    disadvantages, 206
free trade, 76
frequently asked questions (FAQs), 293
Frick, Eugen, 225
FrontPage, 210
FTP, 11
functional résumé, 105

G

Galvin, Paul, 139
gap.com, 48, 124*f*
Gates, Bill, 93, 122
general partner
    number of, 141
    responsibilities, 141, 142
    terms of partnership agreement, 142
geographical barriers
    challenges and advantages, 73
    removal of, 73
GIF
    defined, 215
    images, 215
    resolution, 215
    uses, 215
Gillette, King Camp, 65
Glaser, Rob, 183
global e-commerce
    business before Internet, 66
    currency, 71
    cyberlocations, 66
    globalization, 66
    language, 69, 69*f*, 70, 70*f*
    pricing, 71
    worldwide online trade, 66
global e-marketplace
    movie studios, 73
    regulation and taxation, 77
    revenue concerns, 73
    trade policies, 75–76
globalization
    communications system, 70
    culture, 67, 68*f*
    currency, 71
    cyberlocations, 66
    defined, 66
    in e-business world, 66
    going global, 66
    language, 69, 69*f*, 70, 70*f*
    pricing, 71
    worldwide online trade, 66
GlobalReach.com, 69
goals, 186
Google.com, 15, 169, 233
government-to-consumer (G2C) model, 34, 37, 38*f*
Grainger, W.W., 19
Graphic Interchange Format (GIF), 215
graphics
    buttons, 216
    defined, 215
    descriptive file types, 215
    fonts, 216
    GIF, 215
    JPEG, 215
    resolution, 215
    using, 215
Greenfield, Jerry, 142, 159
growth and contingency plan
    contingency plan, 170
    growth plan, 170
Gutenberg, Johann, 5

H

hackers, 122
Harris, Ethan S., 149
Harrod, Henry Charles, 157
health information, 36
Helix, 183
Hershey Foods, 304
hit, 308
Hoff, Ted, 203
Holmes, Edwin T., 113
holographic tags, 31
Home Depot, 19
home page
    defined, 192
    elements of, 192
    layout, 192
Home Shopping Network, 50
honor system, 162
horizontal hubs, 37
hosted E-CRM
    advantages and disadvantages, 273, 273*f*
    defined, 272
    functions, 273
    renting software per seat basis, 272, 273
Hotels.com, 18
hot spots, 208, 209

Index

HTML
    attributes, 206
    brackets, 9
    code, 205
    defined, 8–9, 204
    document, 205, 205f
    dynamic HTML, 211
    elements, 205
    empty elements, 205
    sample HTML document, 9f, 205, 205f
    tag name, 205
    tags, 9, 205
    values, 206
    Web browsers, 9
    XHTML, 9
    XML, 9
HTML editor, 204
HTML Kit, 204
Hubert, Conrad, 245
human translation, 69
    translation services, 70
    web globalists, 70
hyperlinks
    broken links, 208
    buttons, 208
    defined, 52, 207
    e-commerce Web site, 208
    file links, 208
    icons, 208
    interconnectedness of, 207
    mailto: link, 208
    text links, 208
hypertext links, 207
hypertext markup language. *See* HTML
hypertext transfer protocol (HTTP or *http://*), 8
HyperTracker, 31

I

IBM, 65, 312
icons, 208
ID chips, radio-powered, 3
identity theft, 123
illegal business practices, 120
image map
    creating, 209
    defined, 208
    hot spots, 208, 209
imports, 75
income, 94

income statement, 170
Incorporation Commission, 145
indirect distribution, 268
industrial user, 266
Infinity Host, 12
inflation, 149
infrastructure, 76
infringement, 116, 117
Ingram Book Group, 160
initial public offering (IPO), 143
Integrated services digital network (ISDN), 13
intelligent agents
    capabilities of, 288, 288f
    description of, 287, 287f
    e-commerce businesses, advantages for, 288
    semantic Web, 288, 289
Interactive Advertising Bureau (IAB), 313
InterActiveCorp (IAC), 18
intermediary, 268
intermediary hub
    horizontal hubs, 37
    vertical hubs, 37
International Data Corporation (IDC), 18
international e-commerce
    competitive landscape, 74
    developing countries, 76–77
    e-commerce, 76
    geographic barriers, removal of, 73
    global e-marketplace, 73
    infrastructures, 76
    trade policies, 75–76
    worldwide online trade growth, estimates, 73, 74f
Internet
    birth of, 7
    companies and, 6
    defined, 6
    demographic trends, 190f
    governing groups, 7
    government agencies/groups and, 6
    high-speed networks, 6
    history and, 5
    World Wide Web *versus*, 8
Internet addresses
    ARIN, 9
    domain name, 9–10
    Internet Protocol (IP address), 9

    regional Internet registry (RIR), 9
    uniform resource locator (URL), 10
Internet advertising, 160
Internet agents, 287
Internet anatomy
    e-mail, 10
    FTP, 11
    Internet address, 9, 10
    transmission control protocol/Internet protocol (TCP/IP), 8
    World Wide Web, 8, 9
Internet connections
    broadband services, 13–14
    connecting, 12
    dial-up connections, 12–13
    Internet service provider (ISP), 12
    web hot providers, 12
    wireless internet, 14
Internet directories, 15
Internet Engineering Task Force, 7–8
Internet Explorer, 8
Internet filters
    defined, 125
    employee Internet usage, monitoring, 125, 125f
    Internet Policy Consulting, 125, 125f
    updates, 125
    using, reasons for, 125
    virus infection, 125
Internet patents, 117
Internet Policy Consulting, 125, 125f
    employee usage, 125f
Internet pricing models
    bundling, 295
    pricing, importance of, 294
    promotional pricing, 294, 295
    real-time pricing, 295
Internet Protocol
    IP address, 9
    regional Internet registry (RIR), 9
Internet service provider (ISP)
    e-mail accounts, 12
    points of Internet access, 12
    prevalence, 12
    Web site development, 12
internships, 96
interstitials, 315

Index

inventory, management of
    merchandise protection, 31, 32
    online retailer, advantages, 31
iTunes Music Store, 163, 315

J

Java, 209
    applets, 212
    defined, 212
    dual navigation system, 212
    dynamic navigation buttons, 212
Java virtual machine, 212
JavaScript, 209
    advantages of, 210
    description of, 210
    functionality to Web page, 210
    interactivity, 210
    processing order forms, 210
    script library, 211
    totals items in shopping cart, 210
    using, 210, 211
JavaScript Source, 211
J.C. Penney, 47, 49
J.Crew, 304
Jenner, Edward, 65
jewelry, 29, 30, 32
job shadowing, 96
jobs in e-commerce
    career opportunities, 102, 103, 104
    Technical Careers Chart, 102, 102f, 103f, 103, 104
    trends, 100, 101f
Johnson, Samuel, 27
Joint Photographic Experts Group (JPEG), 215
journalism, timeliness of, 31
JPEG (or JPG files), 215
junk mail, 120

K

Kellogg Company, 304
King, Stephen, 162
Kmart, 47
knowbots, 289
knowledge management, 291
knowledge robots, 289

L

Land's End, 48
language
    human translation, 69, 70
    machine translation, 69
    online population by language, 69, 69f
    translation services, 70
    Web content by language, 70, 70f
    Web globalists, 70
Larson, John, 65
law and ethics
    cyberlaw, 114–117
    ethical considerations on Internet, 117–118
layout, 192
Levi Strauss, 32
liabilities, 170
licensing
    agreement, 162
    benefits, 162
    defined, 162
    Internet stealing, 162
lifestyle influences
    achievement and recognition, 94
    on income, 94
    social and leisure time activities, 94
limited liability company (LLC)
    company debts, 144
    defined, 144
    number of people, 144
    pass-through taxation, 144
    profits and losses, 144
    state tax, 144
limited partner
    number of, 141
    obligations of, 141, 142
limited partnerships
    definition of, 141
    general partner, 141, 142
    limited partner, 141
limitless buyers, 30
links, 207
local and federal economies, e-commerce impact
    bricks-and-mortar retail stores, 148
    federalism, 147
    free trade among every state, 147–148
    Internet growth, 148
    taxation and Internet transactions, 148
log files
    advertisement success, 308
    exit page, 308
    hit, 308
    patterns, 308
    site traffic, 308
    unique visitors, 308
    URLs and search terms, 308
    using, 307, 307f, 308
logic bomb
    defined, 128
    effects of, 128
logistics, 266

M

machine translation, 69
Macintosh, Charles, 157
Macromedia Dreamweaver, 204
Macromedia Flash, 204
Macy's, 49
Macys.com, 267
Magee, Carl C., 139
management plan
    day-to-day business operations, 167
    defined, 167
    experience, 167
    forward-looking statements, 167
MapQuest, 27
market economy
    comparison shopping, 147, 147f
    defined, 146
    e-commerce, economic impact of, 149
    economic indicators, 149, 150
    how it works, 146
    local and state economies, 147–148
    new ways of competing, 150
    supply and demand, 146
marketing, 246–253, 247f
marketing mix
    defined, 246
    elements of, 246, 247f
    personalization, 257, 258, 259
    place, 255, 256
    price, 255
    product, 246, 247, 248, 249
    promotion, 256–257

Index

marketing plan
    competitive environment, 167
    defined, 167
    selling, 167
    terms of sale, 167
marketing research
    advertising campaign, testing, 256
    clickstream data, 255
    conducting, 255, 255*f*, 256
    defined, 255
    metrics, 255
    product research, 256
    sales research, 256
marketing strategy
    customer-service, 292, 293*f*
    return policies, 292, 293*f*
    standing out from other companies, 292
market opportunities, 254–258
market research
    defined, 254
    online customers, 257–258
    primary research, 254
    secondary research, 254
market segments, 247, 248*f*
mass, 32
mass customization
    defined, 32
    delivery cycle, 32
    production, 32
    value chain, 32
Match.com, 18
Matrix, The, 189, 190
McAfee, 127
McDonald's, 47
MCI, 6
McNamara, Frank, 265
Media Metrix, 312
meeting customer needs, 32
mentorships, 96
merchandising activities, 52
merchandising cues, 52, 53*f*
mercury thermometer, 27
message board, 294
meta data
    crawler-based search engines, 228
    defined, 227
    keywords, 228
    meta tag, 227, 229*f*
    robots, 227
    title elements, 227–228
meta tag, 14, 227, 229*f*

metrics, 255
microchip, 28
Microsoft, 312
Microsoft FrontPage, 204
Microsoft's MSN, 15
Microsoft's Windows MediaEncoder, 218
mission statement, 166, 186
Mondera.com, 32
Monk, Thelonious, 288
Monster.com, 104
Montgomery Ward, 46
Morse, Samuel F.B., 17, 93
Morse Code, 17
MP3, 217, 218
MSNBC, 31
MSN Maps, 27
multichannel retailers, 19, 50
multimedia
    audio cues, 217
    converting, 217, 218
    MP3, 218
    streaming audio, 218
    using audio on web, 217–218
    using video on web, 218
multimedia presentations, 97
musculoskeletal disorders (MSDs), 98
music, 29
music licensing rights
    music compositions, 162
    public performance and Internet transmission, 162
    radio and television, 162–163

N

NASDAQ, 143
National Aeronautics and Space Administration (NASA), 6
National Maintenance Services, 35
National Science Foundation, 18
NationMaster.com, 310
navigating, 192
navigational bar, 194
navigation schemes
    defined, 193
    how Web pages relate, 194
    navigational bar, 194
Nestlé, 68*f*
NetBank.com, 36
NetCom, 6, 12
NetGrocer.com, 19
Netscape Navigator, 8

network engineers, 19
Newcomen, Thomas, 45
Newmark, Craig, 294
news, timeliness of, 31
New York Stock Exchange (NYSE), 143
New York Times, 93, 161
New Zealand Treasury department, 37, 38*f*
Nickerson, William, 65
Nielsen/NetRatings survey, 189, 191*f*
Nike, 304
Nolo.com, 145
non-store retailers
    bricks-and-mortar businesses, 50
    description of, 50, 51*f*
    e-tailing, 50
    growth, factors for, 50
    multichannel retailers, 50
    pure-play retailers, 50
Norton, 127

O

objectives, setting
    attainable objectives, 187
    prototype of product, 187
    site's objectives, 187
    specific objectives, 187
Occupational Outlook Handbook (OOH), 100
Occupational Safety and Health Administration (OSHA), 98, 100
One-Click Shopping, 305
online advertising, types of
    advantages, 311
    banner ads, 311, 312–313, 312*f*
    e-mail, 313, 314
    intersitials, 314–315
    media outlets, 310
    online coupons, 314
    placement fees, 311*f*
    problems with, 310, 311
    spoof e-mail, 310, 311
    sweepstakes and giveaways, 315
online auctions, 36, 37
online banking, 36
online bill payment, 29
online collaboration, 53
online coupons, 314
online dispute resolution (ODR), 77

Index

online producers, 19
online savings, 149
online schmoozing, 91
online search market, 233
online surveys, 19, 31
online trade growth, 74*f*
Onstar, 188
OOH, 100
open 24/7, 30, 31
operations plan
    defined, 169
    elements of, 168*f*, 169
opt-in, 313
opt-out, 313
"or," 15
Orbitz.com, 228, 315
order fulfillment and customer service, 56
order processing
    best-of-breed systems, 272
    electronic customer relationship management (E-CRM), 271–273
    hosted E-CRM, 272, 273, 273*f*
    software suites, 272
    stand-along programs, 271–272
OSHA, 98, 100
Oshman's, 47
Otis, Charles, 157
Otis, Norton, 157
Overstock.com, 147

P

packet switching, 8
packet switching network, 7
Pacscene, 251
Page, Larry, 169
Paramount Pictures Corporation, 18, 73
parental control, online information, 120
Parks, Alexander, 139
partnerships
    basic, 141
    limited, 141, 142
pass-through taxation, 144
pass-through tax treatment
    defined, 144
    disadvantages of, 144
    double-taxation, 144
patent
    business method patents (Internet patents), 117

1-Click system, Amazon.com, 117, 117*f*
    defined, 117
    patented business process, 117, 117*f*
Paychex, 35
pay-for-performance
    cheap tickets, 228
    how they work, 228
    reaching customers, 229
    search results, 228
    "Sponsor Results," 228
PayLess ShoeSource©, 47
Paymentech, 71
payment options
    checkout process, 54
    credit cards, 54
    debit cards, 54
    e-cash, 55
    e-checks, 55
    electronic funds transfer (EFT), 54
    smart cards and eWallets, 54, 55, 55*f*
PayPal.com, 55, 310
pay-per-click
    advertising, 251, 252*f*
    affiliate search engines, partners with, 228
    defined, 228
payroll service, 35
PepsiCo, Inc, 31, 315
permission marketing, 253
personal data, government protection of
    e-mail monitoring, 119–120
    Federal Trade Commission (FTC) Act, 119
    Privacy Act of 1974, 119
personal digital assistants (PDAs), 14
personalization
    cookies, 251
    defined, 251
    e-mail marketing, 251, 252
    permission marketing, 253
personalization approach, implementing
    frequently asked questions (FAQs), 293
    greeting to returning visitors, 293, 294
    user interface, 293

personalization of online customer service
    consumer benefits, 292–293
    customer choice, 291
    customer relationship management (CRM), 291
    database consolidation, 291
    implementing, 293, 294
    knowledge management, 291
    marketing strategy, 292, 293*f*
    personalization, importance of, 291
    profiling, 291, 292*f*
    solving problems online, 294
PHP, 232
physical distribution
    defined, 271
    order processing, 271–273
    transportation, 273–276
picture element, 311
pixel, 311, 312
Pizza Hut, 18
place
    defined, 250
    impact of, 250
Plant, The (King), 162
Pollard, William, 93
Polley, Eugene, 303
pop-ups, 315
portable document format (PDF) files, 195
Poulsen, Valdermar, 185
Powell's City of Books, 150
Powells.com, 150, 267
pretty good privacy (PGP), 126
price
    defined, 249
    dynamic pricing, 249
    image, 249
    pricing structure, 249
    variable pricing, 249
    virtual auctions, 249
Priceline.com, 37, 142–143
pricing
    in international arena, 71
    real-time transaction processing, 71
pricing competitively
    competitor's Web site, monitoring, 33
    elastic demand, 33
    online intelligence for pricing, 33
primary research, 254

Index

Privacy Act of 1974, 119
privacy matters, 91
privacy online
    Children's Online Privacy Protection Act of 1998 (COPA), 120
    crime, 119–120
    e-mail monitoring, 119–120
    privacy policy, 119–121
    protection groups, 120
    rights, protecting, 119–120
    selling personal data, 119
    spam, 120–121
    terrorism, 119–120
privacy-protection groups
    defined, 120
    Electronic Frontier Foundation (EFF), 120
problems online, solving
    chat rooms, 294
    message board, 294
Procter & Gamble, 304
product
    cross selling, 249
    defined, 246
    demographically, 247
    geographically, 247
    line, 247, 248
    market segments, 247, 248f
    mix, 249
    product benefits, 247
    psychographically, 247
    quality, 246, 247
    services, 246
product merchandising
    categorizing products, 52
    hyperlinks, 52
    merchandising activities, 52
    merchandising cues, 52, 53f
    presentation of products, enhancing, 52
product preferences, 19
product research, 256
product sales
    products you buy, 159
    products you make, 159
    wholesalers, 159, 160
profiling, 291, 292f
promotion
    affiliate programs, 251
    defined, 52, 250
    hyperlink, 52
    newsletters or e-zines, 251

pay-per-click advertising, 251, 252f
tell-your-friends, 250–251
user reviews, 251
virtual marketing, 250
word-of-mouth, 250
promotional pricing, 294, 295
Pronto, 14
protectionists, 75
public key, 57
purchase orders
    contents of, 53
    defined, 53
purchasing process
    MarketMax Inc., 53
    online collaboration, 53
    requests for information, 54
purchasing process online
    online collaboration, retailers and suppliers, 53–54
    purchase orders, 53
pure-play retailers, 19
Pyra Labs, 149

quotas, 75

radio-powered ID chips, 3
RadioShack, 306
Ralphs, 46
real estate, 29, 36
RealNetworks, 183
real-time pricing, 295
real-time transaction processing, 71
recommendations hyperlink, 52
Reebok, 18
regional Internet registry (RIR), 9
registered trademark (®), 116
regulation and taxation, 77
REI, 49
REI.com, 314
REMAX.com, 36
Reno, Jesse W., 93
requests for information, 54
resolution, 215, 312
résumé, defined, 105
retailers, 46
retail experience, 48
retailing, history of
    anchor tenants, 47
    big-box retailers, 47

category killers, 47
department and chain stores, 46
e-tailing, 47
first U.S. retailers, 46
general store, 46
shopping malls, 47
strip malls, 46–47
retailing before e-commerce, 46–51
retailing today
    department stores, 49
    discount stores, 49
    non-store retailers, 50, 51f
    retail experience, 48
    services retailers, 49
    shopping methods, 47–48
    speciality stores, 49
    types of retailers, 48–50, 48f, 51f
return on investment (ROI), 170
revenue models
    ad sales, 160–161
    affiliate program, 161
    book and subscription services, 158, 159
    consumer goods, 158
    content, 158
    definition of, 158
    Intuit, 158
    licensing, 162
    music licensing rights, 162
    product sales, 159, 160
    subscriptions, 158, 161
    types of, 158–163, 159f
reverse-auction scenario, 37
RichFX, 52
Ritty, James, 265
Roberts, Larry, 5
robots, 227
Roentgen, William, 185
Rollerblade.com, 9, 10
Rowling, J.K., 116

SalesCart's FrontPage Ecommerce, 233
sales research, 256
Salon.com, 158, 159
San Francisco Museum of Modern Art Web site, 115f
satellite broadband service, 14
Schmittlein, David, 292
S corporation, 144
script library, 211
Seabiscuit, 190

Index

search engines
    browser search tools, 14
    description of, 14
    meta tags, 14
    tips, 15, 15f
SearchEngineWatch, 15
search tools
    benefits of, 16
    crawler-based search engines, 15
    differences between, 15
    Internet directories, 15
    symbols, 15
    wildcards, 15
Sears, Roebuck, and Co., 46, 47, 49, 50
secondary research, 254
secured creditor, 141
Secure Sockets Layer (SSL), 56, 123f
    asymmetrical encryption, 57
    description of, 56
    encryption, 57
Securities and Exchange Commission (SEC), 168
security enforcement service, 126
security guards, 31
security issues and concerns
    asymmetrical encryption, 57
    digital certificates, 56–57
    encryption, 57
    public key, 57
    Secure Sockets Layer (SSL), 56
security of data
    antivirus software, 126–128, 127f
    company data, loss of, 122
    credit card protection (Gap online), 123, 123f
    cybersquatters, 122
    encryption, 125, 126
    firewalls, 124, 125
    hacker, 122, 124
    identity theft, 124
    Internet filters, 125
Seidenberg, Ivan G., 137
servers
    bandwidth, 231
    choosing, 230, 231
    defined, 230
    managing stock and sales, 231
    reliability, 231
    storage space for Web site, 231

Service Corps of Retired Executives (SCORE), 170
services, 246
services retailers, 49
Seventeen.com, 312
Seventeen magazine, 308
shares, 142, 143
Sheffield, Washington, 283
Shockley, William, 283
shopbots, 289
shoplifting, 31
Shop.org, 286
shopping-cart software
    credit cards, 233
    PayPal, 233, 233f
    taking payments, 232, 233
Shopping.com, 243, 295
short term goals, 186–187
Shugart, Alan, 283
Silver, Bernard, 265
Simjian, Luther George, 265
SimpleText for the Macintosh, 204
site maps and storyboards
    categories, 192–193
    site map, 193, 193f
    storyboards, 193
site traffic, 308
skyscraper banner, 313
slag code, 128
slander, 120
Small Business Administration
    considerations before business plan, 164
    defined, 164
    publications and services, 164
Small Business Development Center (SBDC), 170
smart cards
    defined, 54
    stored-value cards, 55
    uses, 54, 55
Smith, Harold, 245
social and leisure time activities, 94
social-networking sites, 91
software, 29
software, antivirus
    antivirus-protection actions, 127f, 128
    types of viruses, 127, 128
    virus, 126
    virus attacks, 127
software suites
    description of, 272

    enterprise resource planning (ERP), 272
    integration, 272
sole proprietorship
    bank loans, 141
    business insurance, 140–141
    collateral, 141
    credit-card debt, 141
    creditors, 141
    debts, 141
    defined, 140
    IRS views on, 140
    liabilities, 140–141
    loans, 141
    secured creditor, 141
    taxes, 140
    unsecured creditor, 141
Sony.net, 75
spam
    blocking, 121
    combatting, 121
    defined, 120–121
    junk mail, 120
    reporting, 121
    "spammers," 121
specialty stores, 49
    defined, 49
    sales strategies, 49
    staff, 49
Spencer, Percy L., 65
spiders, 289
spoof e-mail, 310, 311
Sprint, 6
SquareTrade Inc., 78
stand-alone programs, 271–272
starting out
    business model, 33
    domain name, 33
    e-commerce site, building, 33
    virtual storefront, 33
statistics, tracking, 31
steam-powered locomotion engine, first, 27
sticky site, creating
    changing market conditions, 307
    dynamic content, 306
    sticky content, 306, 306f
stockholders, 142, 143
stored-value cards, 55
storyboards, 192–193
Strauss, Levi, 245
streaming audio, 218

Index

Structured Query Language (SQL), 232
subchapter S corporation (S corporation)
    pass-through tax treatment, 144
    when to choose, 144
subscription-based satellite communication network, 188
subscriptions
    blogs, 161
    e-zine, 161
    print and online, 161
    weblog, 161
Supercuts, 46
supply and demand, 146
supporting infrastructure of company, 28
surveillance cameras, 31
SVG (Scalable Vector Graphics), 201
sweepstakes and giveaways, 315
Symantec.com, 124
symmetrical encryption, 57
systems analysts, 19

T

tables
    description of, 206
    layouts, 207
    popularity of, 207
    table-based formatting, 207, 207f
tags, 205
    brackers, 9
    Web browser, 9
Target, 46, 47, 48, 304
target market, 189
tariffs, 75
taxes and internet, 118
Taylor Nelson Sofres (TNS), 250
TCP/IP (Transmission control protocol/Internet protocol)
    defined, 8
    Internet protocol, 8
    transmission control protocol, 8
Technical Careers, 102, 102f–103f, 103, 104
telegraphy, 17–18
telephone operators, 19
tell-your-friends, 250–251
terms of sale, 167
The Java Boutique, 212
third-party negotiation, 78
Thurman, J.S., 45
Ticketmaster, 18
title element, 227
T line, 13
T-1 line, 13
T-3 line, 13
Tolia, Nirav, 243
Toys "R" Us, 47, 49
Toysrus.com, 267
trademark
    defined, 116
    infringement, 116, 117
    registered trademark (®), 116
    service mark, 116
trade policies
    export, 75–76
    free trade, 76
    imports, 75
    protectionists, 75
    quotas, 75
    tariffs, 75
    World Trade Organization (WTO), 76
traffic sources, 307–308
training for e-commerce career
    art courses, 97
    business courses, 97
    communication skills, 96–97
    courses, appropriate, 96, 97
    elective courses, 97
    math skills, 97
    multimedia presentations, 97
traits of e-commerce workers
    decisive, 99
    problem-solvers, 99
    risk-takers, 99
    self-starters, 99
transportation
    carriers, types, 274, 274f, 275–276
    defined, 274
    issues before setting up system, 274
    services and carriers, 274, 275
travel, 29
travel services, 36
trends in e-commerce
    data management skills, 100
    expected growth, 100
    job growth opportunities, 100, 101f
    retailing jobs, 100
    Web site, development and maintainence, 100
Trevithick, Richard, 27
Trojan viruses, 127
Tupper, Earl, 203
Turbotax.com, 158

U

uniform resource locator (URL), 10
Unilever, 142
unique selling proposition (USP), 292–293
unique visitors, 308
Universal Studios, 73
unsecured creditor, 141
upsell hyperlink, 52
URL
    domain (or server), 10
    hypertext transfer protocol, 10
    suffix, 10, 10f
URL placement
    crawler-based search engines, 226
    getting site closer to top of results page, 226
    meta data, 227–228
    pay-for-performance, 228, 229
    pay-per-click, 228
    position, defined, 226
    title element, 227
U.S. Census Bureau, 19, 28
U.S. Department of Commerce, 150
U.S. Department of Defense, 7
U.S. Department of Labor, 100
U.S. National Science Foundation, 6
U.S. Treasury Department, 77

V

Vaaler, Johann, 139, 203
value chain, 32, 35, 35f
value-chain model, 35–36
values, 206
Vehix.com, 310
Verizon Communications Inc., 137
vertical hubs, 37

Index

video on web
    Apple's QuickTime, 218
    guidelines, 218
    Microsoft's Windows MediaEncoder, 218
    as powerful tool, 218
    RealMedia's Helix Producer, 218
    software, 218
    using, 218
virtual marketing, 250
virtual storefront, 33
virtual stores, 17, 32
virus, 126
    effects on computer, 127f, 128
    self-defense measures, 127f
virus attacks, 127
virus detection, 127
virus infections, 126
VisiBone.com, 214
visual balance
    defined, 214
    white space, 214
Volta, Alessandro, 5

Wall Street Journal, The, 161
Wal-Mart, 35, 46, 47, 49
warehouse, stock handling, and inventory control, 275
Washington, George Constant, 283
Web affiliate, 269
Web and retail competition
    competitive analysis, 168
    financial press, 168
    industry consultants, 168
    rivals, details of, 168
    Securities and Exchange Commission (SEC), 168
    stock and industry analysts, 168
Web-based negotiation, 78
Web browser
    hypertext markup language (HTML), 8–9
    hypertext transfer protocol (HTTP), 8
Web content, 183
Web design, specifics of
    frames, 206
    HTML, 204–206
    HTML editor, 204
    hyperlinks and image maps, 207–208, 209
    Macromedia Flash, 204
    simple text editor, 204
    tables, 206, 207, 207f
    writing Web pages, 204
    writing your own code, 204
    WYSISYG software, 204
Web design, visual elements in
    background color, 214, 215f
    color consistency, 214
    graphics, 215–217
    multimedia, 217–218
    visual balance, 214
Web-design basics
    specifics of, 204–209
    tools, advanced, 209–212
Web design tools, advanced
    adding style and interactivity, 209
    cascading style sheets (CSS), 209–210, 211f
    design and interactivity, technologies for adding, 209–212
    DHTML (Dynamic HTML), 211–212
    Java, 212
    javascript, 210, 211
Web distributor, 269
Web globalists, 70
Web host providers, 12
weblog, 161
WebMD.com, 36
Web site, accessing
    HTML file with search words, 194, 195
    major browsers, 194
    portable document format (PDF) files, 195
    readable site, 194
    search words, 194, 195
    video streaming, fancy graphics, and sound effects, 195
Web site, editing
    changes, creating and editing, 196
    testing Web site, 196
Web site, establishing goals
    goals, 186
    mission statement, 186
    objectives, 186
    short term and long term, 186
    Web sites with similar goals, 187
Web site, know your audience
    demographics, 189, 190f
    different sites appeal to different markets, 189, 190
    movie studio Web sites, 189, 190
    sites with similar target markets, 189, 191f
    target market, 189
Web site, organizing
    editing site, 196
    home page, 192
    layout, 192
    navigating, 192
    navigation schemes, 193, 194
    reaching audience, 194, 195
    site maps and storyboards, 192–193
Web site, setting objectives
    attainable objective, 187
    brainstorming, 188
    conceptualizing, 189
    feasibility assessment, 188–189
    kind of Web site, 187
    writing objective of, 187
Web site demographics
    demographics, 167
    identifying customers, 167
    site's special features, 167
Web site for global business
    content, 67
    design, 67
    navigation, 67
Web site hosting service, 12
Web site in search engine, maximizing
    placement of URL, 226, 227–229
    registering with search engine, 226, 227f
Web technologies, evolution of, 29
Wetzel, Don, 265
Whitman, Meg, 284, 285
wholesalers, 46, 159, 269
wildcards, 15
Windows' Notepad, 204
Windows operating system, 128
wireless Internet
    personal digital assistants (PDAs), 14
    wireless Internet service providers (WISPs), 14
Woodland, Joseph, 265
word-of-mouth, 250

360 Index

Index

working in field, traits for, 99
workplace, 95
WorldPay, 71
World Trade Organization (WTO), 76
World Wide Web (WWW)
    creator of, 8
    development of, 45
    hypertext markup language (HTML), 8, 9, 9*f*
    hypertext transfer protocol (HTTP or *http://*), 8
    Internet *versus*, 8

Web browser, 8
XHTML, 9
XML, 9
worms
    defined, 127
    effects on computers, 127
    hackers, 127
    patches and updates, 128

XHTML, 9
XML, 9

Yahoo!, 15, 233
Yahoo!Maps, 27

Zappos.com, 315
ZDNet.com, 33
Zuse, Konrad, 27

Credits

Cover Photography © Corbis; © Paul Barton/Corbis; © Peter Sterling/Corbis; **2** Image Source Limited/Index Stock Imagery; **3** Courtesy of Dell Computer Corporation; **4** PhotoDisc; **6** ThinkStock LLC/Index Stock Imagery; **14** Bob Daemmrich/PhotoEdit; **26** PhotoAlto/eStock Photo; **32** Ed Bock/Corbis; **32** David Young-Wolff/PhotoEdit; **47** Time Life Pictures/Getty Images; **56** Francisco Cruz/SuperStock; **64** Hartmut Schwarzbach/Peter Arnold, Inc.; **71** Altrendo Images/Getty Images; **77** ImageState; **89** Robert E Daemmrich/Stone/Getty Images; **90** Jules Frazier/PhotoDisc; **91** Noel Hendrickson/Masterfile; **92** Digital Vision; **97** Thinkstock LLC/Index Stock Imagery; **98** Robert E. Daemmrich/Stone/Getty Images; **112** Corbis; **116** Alexander Nemenov-STF/AFP/Getty Images; **126** Imagemore/SuperStock; **135** Brendan Smialowski/Getty Images Editorial; **136** PhotoDisc; **137** Courtesy of Verizon; **138** Jon Feingersh/Masterfile; **142** Lon C. Diehl/PhotoEdit; **148** Courtesy of Feldman & Associates; **156** Mattei Michele/Corbis/Sygma; **162** AP/Wide World Photos; **166** A. Ramey/PhotoEdit; **181** Billy Hustace/Stone/Getty Images; **182** Michael Newman/PhotoEdit; **183** Michael Newman/PhotoEdit; **184** David Young-Wolff/PhotoEdit; **187** Andreas Buck/Peter Arnold, Inc.; **195** Bonnie Kamin/PhotoEdit; **202** David Raymer/Corbis; **208** Corbis; **217** Grantpix/Index Stock Imagery; **224** J. Feingersh/Masterfile; **228** Laurence Dutton/Image Bank/Getty Images; **232** Andrew Brown/Ecoscene/Corbis; **241** Mark Gibson/Index Stock Imagery; **242** Tim Boyle/Getty Images; **243** Felicia Martinez/PhotoEdit; **245** Peter Byron/PhotoEdit; **250** Eric Fowke/PhotoEdit; **xii, 257** Stephen Frink/Corbis; **265** Ed Kashi/Corbis; **272** Noel Hendrickson/Masterfile; **275** Michael J. Doolittle/The Image Works, Inc.; **282** Andersen Ross/PhotoDisc; **286** James Leynse/Corbis; **294** John Chapple/Hulton Archive/Getty Images; **302** James Leynse/Corbis; **305** James Leynse/Corbis; **313** Michael Cogliantry/Image Bank/Getty Images; **325** Lori Adamski Peek/Stone/Getty Images